The future of mankind depends on ever-increasing agricultural production to provide food, fibre, fuel and other essential commodities. This can only be achieved through a sound knowledge of the plants which feature prominently in agriculture. The purpose of this book is to describe these plants in detail, together with the products that are obtained from them. Economically important field crops and pasture plants of temperate and subtropical regions are included, but for a treatment of fruit crops, flowers and trees the reader is requested to look elsewhere. There is an opening chapter on world population and food supply, followed by a general introduction to plant structure, and the book concludes with a discussion of physiological principles of crop growth and yield.

AGRICULTURAL PLANTS

Agricultural Plants

SECOND EDITION

R.H.M. LANGER & G. D. HILL
PLANT SCIENCE DEPARTMENT, LINCOLN UNIVERSITY, NEW ZEALAND

ILLUSTRATIONS BY KAREN MASON

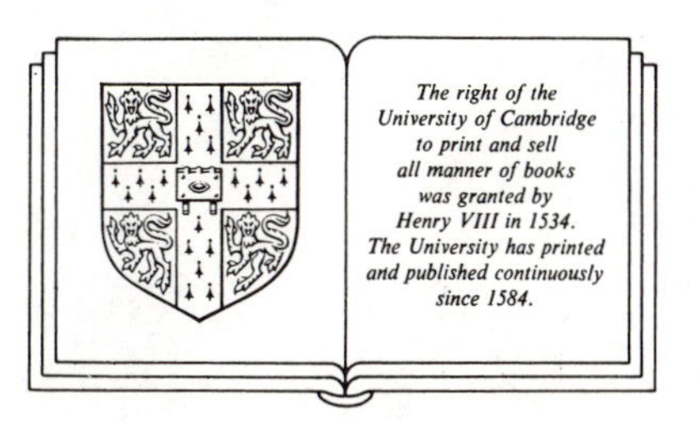
The right of the
University of Cambridge
to print and sell
all manner of books
was granted by
Henry VIII in 1534.
The University has printed
and published continuously
since 1584.

CAMBRIDGE UNIVERSITY PRESS
CAMBRIDGE
NEW YORK PORT CHESTER MELBOURNE SYDNEY

Published by the Press Syndicate of the University of Cambridge
The Pitt Building, Trumpington Street, Cambridge CB2 1RP
40 West 20th Street, New York, NY 10011-4211, USA
10 Stamford Road, Oakleigh, Victoria 3166, Australia

First published 1981
Second edition 1991

Printed in Great Britain at the University Press, Cambridge

British Library cataloguing in publication data

Langer, R.H.M.
Agricultural plants. – 2nd ed.
1. Crops
I. Title II. Hill, G.D.
630

Library of Congress cataloguing in publication data

Langer, R.H.M.
Agricultural plants / R.H.M. Langer, G.D. Hill; illustrations by Karen Mason. – 2nd ed.
p. cm.
Includes index.
ISBN 0-521-40545-9. – ISBN 0-521-40563-7 (pbk.)
1. Crops. I. Langer, R.H.M. (Reinhart Hugo Michael), 1921–
II. Langer, R.H.M. (Reinhart Hugo Michael), 1921– Agricultural plants. III. Title.
SB91.H55 1991
631–dc20 91-2246 CIP

ISBN 0 521 40545 9 hardback
ISBN 0 521 40563 7 paperback

VN

Contents

Preface to the first edition

This book is intended to serve as an introduction to agricultural plants, their structure, botanical characteristics, their place in agriculture, their cultivation and uses. It is not a treatise on economic botany, nor does it cover details of farming practice. What we have attempted to do is to provide a description of important crop and pasture plants, illustrated wherever possible, as a basis for a foundation course in agricultural botany or plant science.

Although it was our plan originally to restrict ourselves to plants of the temperate zone, we soon realised that plant introduction and adaptation have helped to break down rigid geographic boundaries in many climatic regions. The emphasis on temperate species has been retained, but at the same time we have attempted to cover most of the important world crops at least in subtropical areas. On the other hand, we have concentrated on field crops to the exclusion of tree species. Another problem of definition facing us was the somewhat artificial division between agriculture and horticulture. Where the distinction is mainly one of scale of production, such as in many vegetable crops, we have included the species concerned, but for a treatment of fruit crops, flowers and herbs the student should refer to appropriate horticultural texts.

Taxonomic arguments have been avoided as being out of place in an introductory book of this kind. In arranging the families that we have included, we have adopted the simplistic approach of alphabetical order, except that we have placed the two representatives of the Monocotyledons at the beginning, to be followed by the Dicotyledons. Similarly, within each family we have refrained from going into details of taxonomy, and we hope that the more discerning readers will forgive us. The description of the economically important species within these families is preceded by an opening chapter which puts agricultural plant production within the context of world population and food supply, followed for those who need it by a summary of plant anatomy and morphology. The final chapter discusses some physiological principles of crop growth and yield. Throughout, the coverage includes both crop and pasture species.

We hope that, despite its undoubted limitations, this book will be of value both to students of agricultural botany and their teachers in a range of countries and that in some small way it will contribute to food production on this overcrowded planet.

R.H.M. Langer
G.D. Hill

Preface to the second edition

This new edition of *Agricultural Plants* incorporates significant changes which have occurred during the last few years. New information has been added, including description of some crop plants which first have come into prominence in recent times, while material which is no longer relevant has been deleted. Changes in the botanical name of some plant families have forced a rearrangement of chapters, because the alphabetical order of presentation has been retained, and one further family has been included. The introductory chapter on world population and crop production is presented in a modified form as is the final chapter on the physiological basis of yield. However, the main purpose of the book remains the same, and we hope that it will continue to play a useful part as a teaching aid and guide for students of all ages.

R.H.M. LANGER
G.D. HILL

1 World population and crop production

World population statistics make terrifying reading, so rapid has the rise in numbers become. It was not always like this, because for many centuries war and famine, disease and natural disaster kept increases at a reasonable level. Not only that, but any local pressure on land and other resources could always be countered by migration and the colonisation of new areas. But, in the last few decades, the picture has changed dramatically, and we are now acutely aware of the dire consequences of overcrowding, both regionally and internationally. The population explosion, which Malthus so gravely predicted and described, has occurred and, whether we are prophets of doom or confident technocrats, we have to face the future with realism.

As recently as the middle of this century, well within the life span of many people living today, the total world population was about 2400 million. Forty years later it had reached 5300 million, and the curve continues to rise, climbing inexorably into the future (Fig. 1.1). The present increase per day is about 250 000, the size of a medium city, and at this rate 90 million new people are added every year, more than enough to populate an entirely new country.

United Nations' estimates predict that there may be 10 000 million people on Earth by the year 2050, but this rate of increase could well be exceeded. Although more optimistic forecasts suggest that world population may begin to level off at this total, there remains the enormous problem that human fertility rates may have declined in relatively prosperous countries, which are capable of sustaining larger numbers, whereas poor, disaster-prone regions of the world suffer from over-population and famine. Such is the age structure of the human population that further very considerable increases are inevitable, no matter how

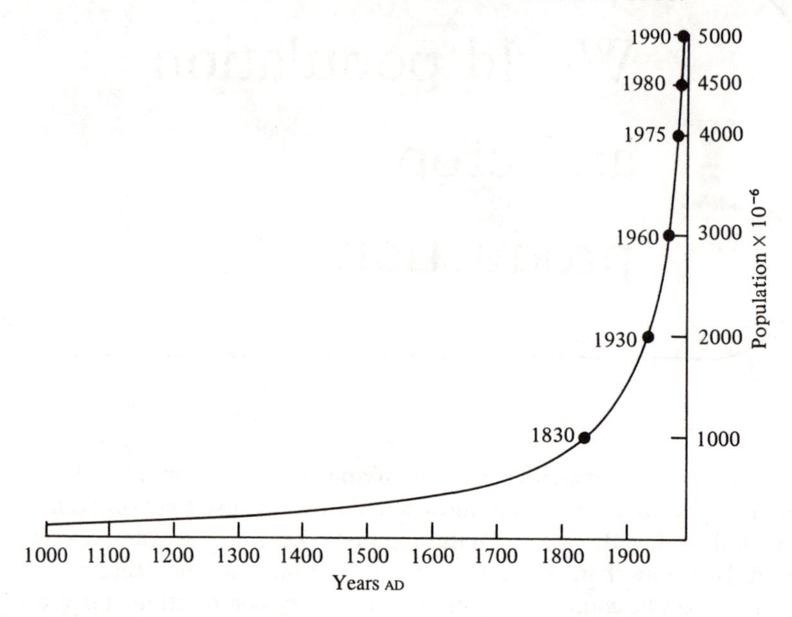

Fig. 1.1 Growth of the world population from AD 1000 to the present and projection into the future.

effective birth control measures prove to be. This is the problem that faces mankind.

Let us for the moment ignore future projections and consider the requirements of the present world population of about 5300 million. This number of people creates enormous pressures and demands for food and raw materials for industry, and it is to crop plants, one of our few renewable resources, that we must look to meet most of these needs. In highly developed countries we may not always be aware of our dependence on plants, because much of our diet is of animal origin and, as Table 1.1 shows, only about one-half to a third of our protein intake comes directly from plant sources. However, since plants are essential for the maintenance of herbivorous animals, we are in the long run concerned with plant production just the same. In any case, on a world scale about 88% of the calorie requirements and 90% of the proteins come directly from plant sources.

To meet these demands, man has developed a range of crop plants but, considering the long history of domestication, it is very noticeable that the number of species involved is strictly limited. It has been estimated that on a dry-weight basis, some 80% of edible plant material is derived from only 11 species, of which two-thirds are cereals. We must therefore get to know

Table 1.1. *Proportion of diet coming directly from plant sources*

Land mass	Calories (%)	Protein (%)
Asia	95	78
Africa	93	79
Europe	81	53
North America	68	30
Oceania	65	30

these plants thoroughly in order to safeguard and preferably improve their productivity. But, even if numerically only a few species predominate, there are many others which play an important part in the maintenance of mankind. Leguminous grain crops provide nearly one-quarter of the world's dietary protein requirements at present and, although they have been described condescendingly as poor man's meat, the ever increasing costs and unfavourable energy balance of animal production may well force us to place greater reliance on these plants in the future. Potatoes do not occupy a large proportion of the total cultivated area of the world, yet their productivity per unit area is very high. Many other crops do not serve as food directly but find their way into our diet after processing. Sugar is obtained from sugar cane or beet, and oil or fats from such plants as ground-nut, soya bean or sunflower. Still other species enrich and improve our diet, even if statistically they do not feature very prominently, while others again, like barley, are used for the production of alcoholic beverages. There are thus many species which have an important place as food plants and, even if they are not listed as major crops in Table 1.2, they fully deserve our attention.

The importance of plants as primary sources of food for man should not obscure their essential role in maintaining his domestic animals. As we have seen in Table 1.1, a high proportion of the diet in developed countries is of animal origin, and we only need to remind ourselves of such commodities as eggs, butter, milk and cheese to appreciate how important they are. However, by far the highest input of animal products into our diet comes from meat. On a world scale, total annual consumption is over 110 million tonnes, of which over 75% consists of beef and pork, while poultry and sheepmeat contribute another 23%. The vast population of animals required to produce these goods rely on plants for their subsistence. Some, like pigs and poultry, are fed predominantly on grain, but all the others are grazing animals and depend on natural or sown pastures

Table 1.2. *Major food crops (from* FAO Production Yearbook*)*

Crop	World production ($t \times 10^{-6}$)	Contribution to total world food production (%)
Wheat	508	16.3
Rice	485	15.6
Maize	405	13.0
Potatoes	266	8.5
Barley	170	5.5
Cassava	137	4.4
Sweet potatoes	111	3.6
Soya beans	92	3.0
Tomatoes	63	2.0
Sorghum	61	2.0
Leguminous grains	55	1.8
Oats	39	1.3
Millet	31	1.0
Rye	29	0.9
Total food crops	3116	100.0

or on forages. To meet these requirements, the Poaceae, which also include the cereals, come into prominence as the family containing the herbage grasses, but equally important are the pasture legumes which not only provide nutritious feed but also enrich the ecosystem with symbiotically fixed nitrogen. We shall need to be acquainted with the more important of these species.

Apart from their function of providing food for man, plants also serve a great variety of other purposes. Many of them have been in use since the early days of civilisation, others have been recognised in fairly recent times, and there may be others again which still await development. In an energy-hungry world, crops capable of producing oil are assuming greater prominence, not only in the tropics where the oil palm flourishes, but also in more temperate regions in which soya beans, cotton, sunflower, rape, or linseed provide valuable raw materials for industry. The oil is used in the manufacture of very many products, such as paints, varnishes, plasticisers for plastics, nylons, lubricants or soaps, quite apart from the commodities made by the food industry. Some of the species concerned are also grown to produce valuable plant fibres, and both cotton and linen flax deserve mention in this connection. World cotton production alone amounts to over 12 million tonnes annually. In addition to these crops cultivated as sources of industrial oil and fibre, there are many other plants that contain substances useful to man. Some of these, like the hop, provide flavours, while others like tobacco or coffee are used as stimulants. Certain species have been used since antiquity in herbal medicine, and literally thousands of biochemical components have now been isolated which are used widely in chemo-therapy, as anti-fertility agents, insecticides and a great many other purposes. New compounds are being added every year, and many of them are of sufficient interest to the pharmaceutical industry to deserve further development.

This account of plants in the service of man could not be complete without mentioning important developments in the field of energy. Dwindling resources of petroleum and the necessity by many nations to import all their requirements have stimulated research into the use of crop plants for the production of liquid fuel. In some countries great strides have already been made in setting up new industries concerned with the extraction and fermentation of sugars from such plants as sugar cane or beets. Starchy plant products from potatoes or cereals are also under investigation, although this would involve an additional step in alcohol production. Although initially ethanol is intended as an additive to petrol, there are no great technical problems involved in dispensing with conventional fuel altogether, and some countries like Brazil are well ahead with such developments.

It should not be necessary to enumerate further uses of plant products, nor to construct an impressive list of essential crop species. Enough has been said to demonstrate the great dependence of man on plants and his ever increasing demands for plant products. If we now cast our eye into the future and take into account the inevitable and rapid rise in world population, we are coming face to face with the outstanding problems of our time. How to work the miracle of feeding the 4000 million, was the question asked some 40 years ago. Today we must think firmly of twice that number, incomprehensible though this may be, and we must add to this enormous need for food crops the exponentially growing requirements for many other plants.

What are our chances of achieving the seemingly impossible? Only two rational strategies, alone or together, can be put forward to answer this challenge. Either we find more land on which to grow the additional crops required by man or we discover ways and means of raising productivity per unit area still further. Let us examine these alternatives in turn.

The present world area of cultivated land is estimated to be 1.4×10^9 ha, about 10% of the total land surface, but some authorities consider that about 3.2×10^9 ha are available for arable crops. However, before we get too excited about a possible doubling of the area under crops, we should remember that the extra land required is not by any means always in the same place where additional food is needed. The fastest growing population is not necessarily in a region in which big land development is possible, quite apart from the problems of finance and technical resources needed to bring these changes about. Many of the developed nations enjoy a high standard of living and either have adequate land to sustain themselves or are wealthy enough to be able to import food. For example, one may calculate that on average the population of Europe has available some 0.5 ha per person to provide food, but an additional 0.25 ha would be required to compensate for food imports and the supply of fish. By contrast, the large self-sufficient population of China can call on only 0.34 ha per person to sustain its demands. Even if it were possible to extend this modest land requirement to other countries, we must accept the fact that the areas already covered by crops are likely to be much more suitable for cultivation than any new ones, otherwise they would not have been settled in the first place. It must also be clear that, because the world consists of rich and poor countries, separated by political and economic divisions, it is dangerous and misleading to indulge in too much global arithmetic. If we add to these considerations the problems of climate, soil condition and topography which may limit land development, together with the knowledge that growing populations make their own demands on land for urbanisation, we are left with the conclusion that some scope for increasing the area under crops exists but that it is not likely to be as much as the doubling required by population growth. Furthermore, we must also take account of the competing land demands for the growing of industrial crops.

The second alternative involves considerable further increases in crop yield. There are certainly some prospects in this regard, although the problems are not inconsiderable. For example, in Japan the yield of brown rice has risen almost continuously since 1840 and in the last 30 years or so very substantial increases have been recorded in the face of a static or declining area devoted to this crop (Fig. 1.2). This most encouraging performance is matched by the analysis of world cereal production since 1950. During the following 20 years world output rose by 64%, whereas population growth amounted to only 43%. Much of the production increases were attributable to better yields per unit area. However, in this respect also there are regional differences which detract from the overall prospect. For example, during this period grain production per person increased by 1.5% annually in developed countries, but by only 0.4% on average in developing countries. In some of the latter, production measured in the same terms actually declined. Problems of distribution and international trade are inevitable. Another aspect to be considered is that increases in yield are often bought at the expense of more sophisticated

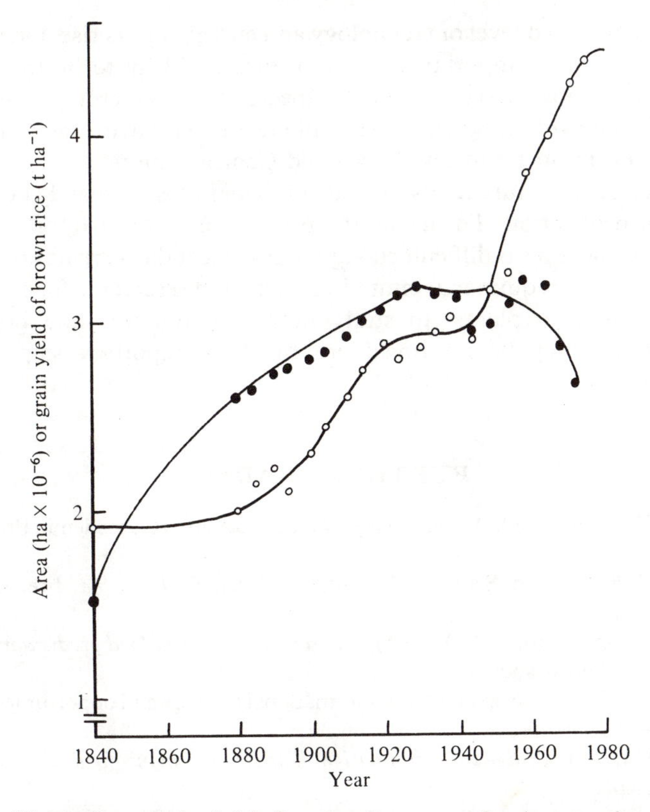

Fig. 1.2 Yield of brown rice (○) in Japan and area used for rice cropping (●) since 1840. (After Evans, L.T., 1978. Proceedings of Lincoln College Centennial Seminar, New Zealand.)

technology, and this requires sociological change, capital and a greater input of energy. Moreover, recent cultivars, which are largely responsible for the improved yields, require high levels of fertility and cultural skill to reach their potential, and it is largely at this end of the spectrum that yield increases have taken place. Nevertheless there is real scope for improvement, as shown for example in the Philippines in which the national average rice yield in 1987 was 2.53 t ha^{-1}. Field experiments located on farms and conducted with all necessary inputs of fertilizer, insecticide and herbicide have yielded 6.4 t ha^{-1} in the dry and 5.1 t ha^{-1} in the wet season, and even more significantly the same farmers themselves attained over 4 t ha^{-1} with identical cultural aids. It is reasonable to suggest that yield improvement per unit area is likely to be most easily attained at the lower end of the range and that, at the upper end, increases will be harder to come by, as the required level of technology and energy inputs rise. One further factor, gaining in importance in the future, is likely to be the gradual warming of global climates which is expected to favour crop production in the cool regions of the northern hemisphere but may have disastrous effects in Mediterranean countries, Africa and Central America.

Whatever the future holds, the fate of mankind is ultimately dependent on plant production. To sustain the present needs of food, oil, fibre and other raw materials is difficult enough, but to meet the demands of the next generation will require maximum ingenuity and research effort. To know the various crop plants, to study how they function and how their productivity may be improved are the first important steps in this direction.

FURTHER READING

Carlson, P.S. (ed.) (1980). *The biology of crop productivity*. Academic Press, New York.

Chrispeels, M.J. and Sadava, D. (1977). *Plants, food, people*. Freeman, San Francisco.

Cox, G.W. and Atkins, M.D. (1979). *An analysis of world food production systems*. Freeman, San Francisco.

Evans, L.T. (1975). Crop plants, an international heritage and opportunity. *Search* **6**, 272–80.

Heiser, C.B. (1973). *Seeds to civilization. The story of man's food*. Freeman, San Francisco.

Pirie, N.W. (1976). *Food resources, conventional and novel*. Penguin Books, Harmondsworth.

Robinson, D.W. and Mallan, R.C. (eds) (1982). *Energy management and agriculture*. Royal Dublin Society.

Seigler, D.S. (ed.) (1977). *Crop resources*. Academic Press, New York.

Scientific American (1976). *Food and agriculture*. Freeman, San Francisco.

Steele, F. and Bourne, A. (eds) (1975). *The man/food equation*. Academic Press, London.

GENERAL REFERENCES

Bailey, L.H. (1949). *Manual of cultivated plants*. Macmillan, New York.

Berrie, A.M.M. (1977). *An introduction to the botany of the major crop plants*. Heyden, London.

Bogdan, A.V. (1977). *Tropical pasture and fodder plants*. Longman, London.

Bronk, B. (1975). *Plants consumed by man*. Academic Press, London.

Brucher, H. (1989). *Useful plants of neotropical origin and their wild relatives*. Springer Verlag, Berlin.

Chapman, S.R. and Carter, L.P. (1976). *Crop production: principles and practices.* Freeman, San Francisco.

Cobley, L.S. (revised by W.M. Steele) (1976). *The botany of tropical crops.* Longman, London.

Frankel, O.H. and Hawkes, J.G. (eds) (1975). *Crop genetic resources for today and tomorrow.* Cambridge University Press.

Gill, N.T. and Vear, K.C. (1980). *Agricultural botany*, 3rd edn, Duckworth, London.

Jackson, D.L. and Jacobs, S.W.L. (1985). *Australian agricultural botany.* University Press, Sydney.

Janik, J., Schery, R.W., Woods, F.W. and Ruttan, V.W. (1974). *Plant science, an introduction to world crops.* Freeman, San Francisco.

Kochhar, S.L. (1981). *Economic botany in the tropics.* Macmillan India, Delhi.

Kowalchik, C. and Hylton, W.H. (1987). *Rodale's illustrated encyclopedia of herbs.* Rodale Press, Emmaus.

Lazenby, A. and Matheson, E.M. (eds) (1981). *Australian field crops*, vol 1, *Wheat and other temperate cereals.* Angus and Robertson, Sydney.

Lockhart, J.A.R. and Wiseman, A.J.L. (1978). *Introduction to crop husbandry.* Pergamon Press, Oxford.

Loewenfeld, C. and Back, P. (1974). *The complete book of herbs and spices.* David and Charles, Newton Abbot.

Lovett, J.V. and Lazenby, A. (eds) (1979). *Australian field crops*, vol. 2, *Tropical cereals, oilseeds, grain legumes and other crops.* Angus and Robertson, Sydney.

Mayer, A.M. and Poljakoff-Mayber, A. (1989). *The germination of seeds.* Pergamon, Oxford.

Molnar, I. (ed.) (1966). *A manual of Australian agriculture.* Heinemann, London.

Morton, J.F. (1977). *Major medicinal plants. Botany, culture and uses.* Thomas, Springfield.

National Academy of Sciences (1975). *Unexploited tropical plants with promising economic value.* Washington, DC.

Norman, M.J.T., Pearson, C.J. and Searle, P.G.E. (1984). *The ecology of tropical food crops.* Cambridge University Press, Cambridge.

Poehlman, J.M.C. (1987). *Breeding field crops.* AVI Publishing Company, Westport.

Purseglove, J.W. (1972). *Tropical crops, Monocotyledons.* Longman, London.

Purseglove, J.W. (1974). *Tropical crops, Dicotyledons.* Longman, London.

Simmonds, N.W. (ed.) (1976). *Evolution of crop plants.* Longman, London.

Simpson, B.B. and Ogorzaly, M.C. (1986). *Economic botany. Plants in our world.* McGraw-Hill, New York.

Sundararaj, D.D. and Thulasidas, G. (1976). *Botany of field crops.* Macmillan, Delhi.

Thomson, W.A.R. (ed.) (1978). *Medicines from the earth, a guide to healing plants.* McGraw-Hill, New York.

2 Plant structure

2.1 THE CELL

The basic unit of plant structure is the cell. All plants from the most simple to the most complex are composed of one or more cells. In plants that are of agricultural importance the cells are highly organised structures and are termed eukaryotic because each contains a well-defined nucleus, as distinct from prokaryotic cells found in plants such as bacteria, blue-green algae and mycoplasmas, which do not contain nuclei.

Study of a rapidly dividing stem or root apex (meristem) of a plant reveals a mass of newly formed cells surrounded by a box-shaped structure, the cell wall, which differentiates plant cells from those found in the animal kingdom (Fig. 2.1).

The cell contents

Within the cell wall is found the living content of the cell, called protoplasm. Even under a light microscope the protoplasm can be seen to contain a number of structures. The development of electron microscopy has substantially increased the number that can now be identified. However, because of their complexity of both micro-structure and function, only the major components of the cell will be discussed.

The protoplasm is separated into two major parts, the nucleus and the cytoplasm. The nucleus is usually spherical in shape and contains DNA which carries genetic information; this DNA is usually contained within the chromosomes. In a growing plant, when new cells are formed, the chromosomes reproduce exact copies of themselves in a process called mitosis. To facilitate the exchange of genetic information within populations of plants, the number of chromosomes within a cell is halved in

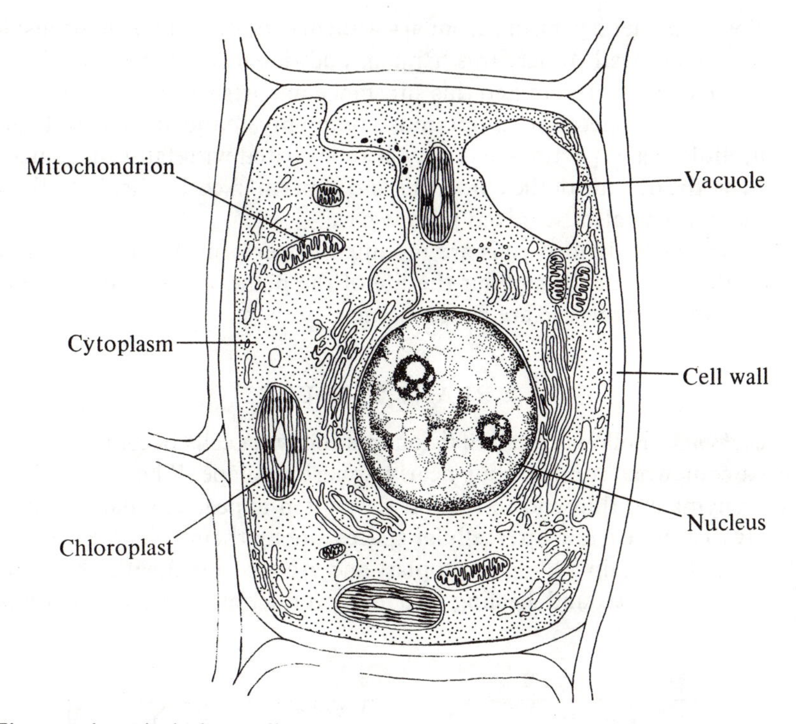

Fig. 2.1 A typical plant cell.

a process called meiosis to produce male and female reproductive cells. These cells, on their fusion, restore the normal chromosome number.

The rest of the living cell is filled with cytoplasm which is separated from the cell wall by a membrane called the plasmalemma. Within the cytoplasm are large clear areas called vacuoles. The vacuoles contain water and dissolved substances and are separated from the cytoplasm by another membrane, the tonoplast. Other easily recognisable structures within the cytoplasm are plastids which may be green (chloroplasts), red/yellow (chromoplasts) or colourless (leucoplasts).

Further structures contained in the cytoplasm, involved with cell function, are mitochondria, microbodies, and ribosomes. Within cells there may also be inert structures which can either be energy reserves or cellular wastes. These materials, known as ergastic substances, may be comprised of starch, protein, fats, tannins or other organic compounds.

The layer external to individual cells is the middle lamella. The oldest part of the cell wall is the primary wall which is on the outside of the cell and is composed mainly of chains of molecules of the polysaccharide, cellulose. Further layers, or secondary wall, may be deposited on the inside of the primary wall. Where this happens additional complex molecules of lignin are usually present and this substance imparts structural rigidity to the cell. Other chemical compounds commonly found in cell walls are hemicellulose and pectin. Cells may also be coated with fatty substances, cutin, on the outside of the epidermis and suberin in cork cells. Both these substances may also be mixed with waxes.

Young newly divided cells are modified to perform the functions that they will carry out throughout the life of the plant. Associations of cells with a similar structure are called tissues.

2.2 PLANT TISSUES

Parenchyma. Following cell division, parenchyma cells become enlarged into a somewhat complex but nearly spherical shape. The cell walls in parenchyma are thin and in section they often appear circular and have clearly distinguishable intercellular spaces among them (Fig. 2.2). Parenchyma cells contain a living protoplast and are associated with photosynthesis, storage, wound healing and the production of new structures.

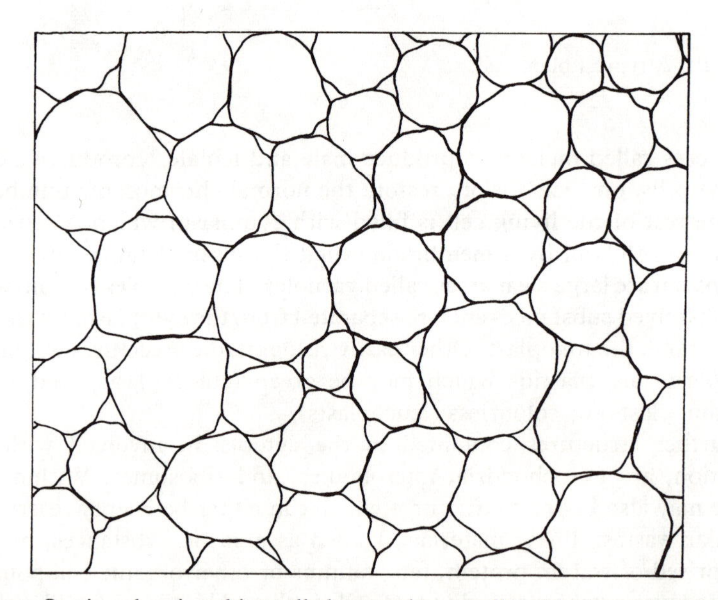

Fig. 2.2 Section showing thin-walled parenchyma cells with air spaces between them. Cells appear empty because contents are frequently lost in preparation of slides.

Parenchyma is found in the cortex of the stem and root, and comprises the majority of the tissue present in leaves.

Collenchyma. Because parenchyma cells have thin walls they have little structural rigidity. A modification of parenchyma is collenchyma. In this tissue the primary cell walls are thickened irregularly (Fig. 2.3). Collenchyma is a supportive tissue in young plants or structures and is found near the surface of stems, in petioles and around the veins in leaves. Like parenchyma collenchyma is a living tissue, and the cells are variable in shape.

Sclerenchyma. In mature plants structural rigidity is provided by the deposition of lignin in the secondary wall which forms following the end of growth of the primary cell wall (Fig. 2.4). The tissue thus formed is called sclerenchyma. There are two basic forms of sclerenchyma cells, sclereids which are of irregular shape and have highly lignified walls, and 'fibres' which are long thin cells also with thick secondary walls. Fibre cells are collected together into long strands which are the basis of a number of

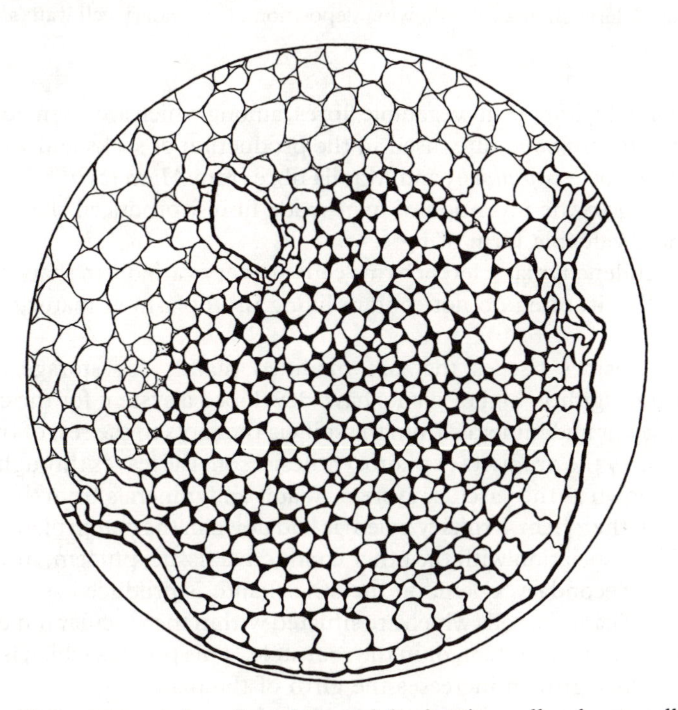

Fig. 2.3 Transverse section of a celery petiole showing collenchyma cells with irregularly thickened cell walls.

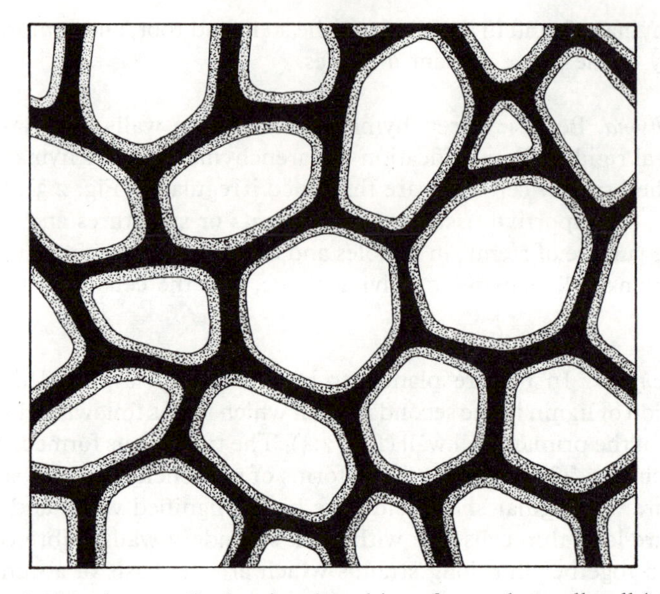

Fig. 2.4 Sclerenchyma cells showing deposition of secondary cell wall (stippled).

commercially important vegetable fibres, among which are linen fibre from *Linum usitatissimum*, jute, used in the production of sacks and woolbales from *Corchorus capsularis*, and Manila fibre from *Musa textilis*. The latter fibre, prior to the invention of man-made fibres, produced the strongest and most valuable form of rope.

Like collenchyma, sclerenchyma cells are very variable in shape and size but, unlike it, often do not contain living protoplasm at maturity.

Xylem. Besides tissues that are modified mainly for storage and for structural rigidity, higher plants have developed a system for the conduction of water and minerals from the soil via the roots to the rest of the plant and to carry the products of photosynthesis from the leaves throughout the plant, known as the vascular system. Water and minerals from the soil are carried in the xylem. Primary xylem is laid down in the young plant and it is found in association with the other conductive tissue, phloem, in vascular bundles. Secondary xylem, on the other hand, is produced as a result of division of cambial cells which are situated within the vascular bundles and in a ring around the stem, or in the root stele. This process which is known as secondary growth increases the girth of the plant.

The main cells in xylem are tracheids and vessel members. Xylem vessel members at maturity contain no living protoplasm and the walls are highly

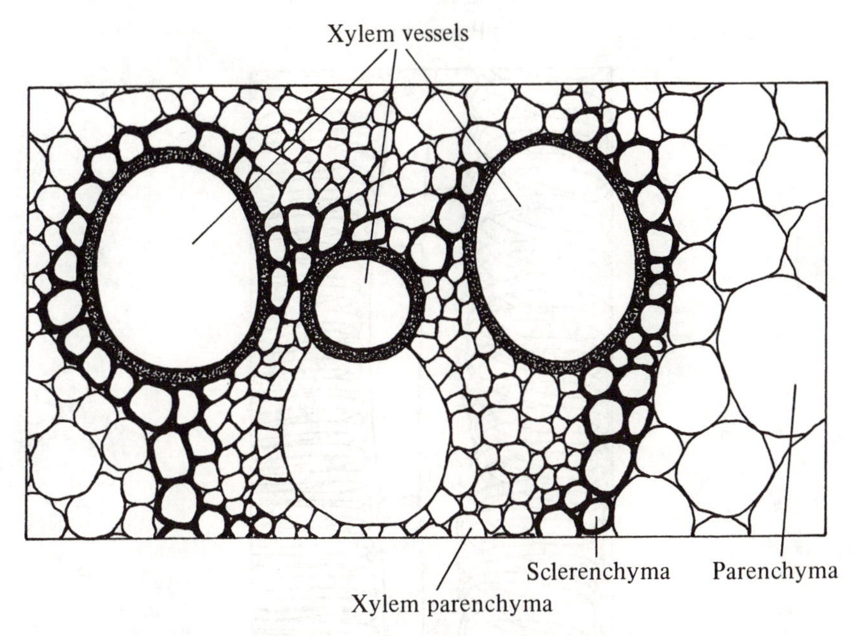

Fig. 2.5 Transverse section of a vascular bundle in a maize stem showing large xylem vessel members with thick secondary walls.

lignified (Fig. 2.5). The secondary wall of the cell is often deposited in ornate patterns (Fig. 2.6). Numbers of vessel members joined together form vessels. Both types of xylem cell carry water and minerals. The main difference is that in tracheids end walls are still present and conduction is assumed to take place via pits in the cell walls. In vessel members on the other hand at maturity the end cell wall has disappeared and the vessel members joined together are not unlike a length of water pipe. These interconnected vessels carry water from the roots throughout the plant.

Phloem. The other vascular tissue which is involved in the transport of the products of photosynthesis throughout the plant is the phloem. In young plants it is found in close proximity to the xylem in vascular bundles or in the stele. As with xylem there is primary phloem (young plants) and secondary phloem (older plants). Phloem is always situated towards the outside of the plant stem or root. The main cell type in this tissue is the sieve element (Fig. 2.7), so named because, unlike xylem where the cell walls usually disappear between contiguous vessels, in phloem they are retained, even though they are highly perforated to form sieve plates. Sieve elements at maturity contain protoplasm but no nucleus. They are

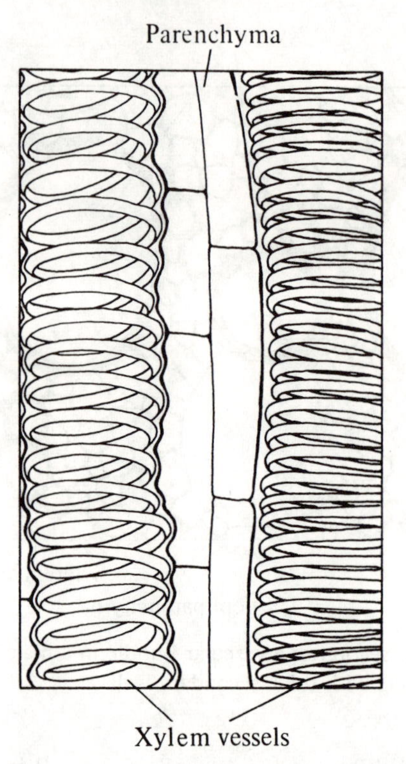

Fig. 2.6 Longitudinal section of a cucurbit stem showing wall patterning in xylem vessel members.

associated with living parenchyma cells called companion cells. The companion cells are thought to assist in the regulation of translocation.

Cambium. In all plants some cells remain which do not lose the ability to divide to form new cells. These are called cambial cells. There are basically two types of cambium. Vascular cambium is found in the vascular bundles and roots of young dicotyledonous plants and on division forms new xylem (secondary) cells towards the inside of the plant and new phloem cells towards the outside of the plant. By this process known as secondary growth the plant expands its girth. The other form of cambium is phellogen (cork cambium). Cork cambium is responsible for the formation of a secondary protective layer known as periderm on the outside of older plants. Phellogen divides off cells which eventually become suberised non-living cork cells, towards the outside of the plant, and living parenchyma cells or phelloderm towards the inside of the plant. In some trees such as

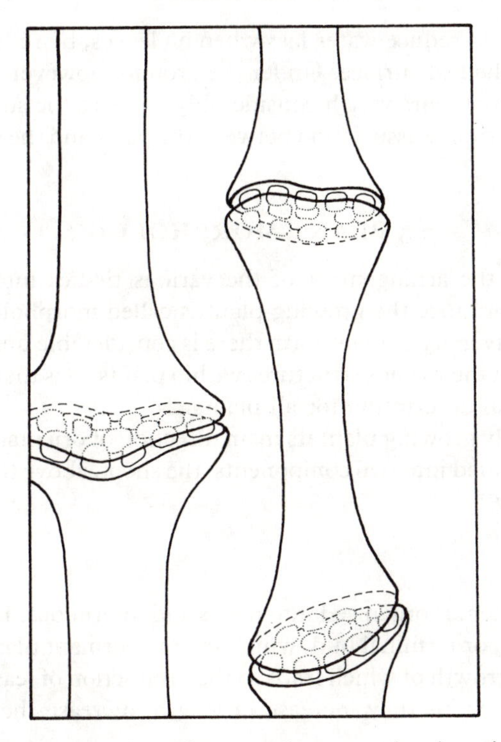

Fig. 2.7 Longitudinal section of a cucurbit stem showing sieve elements of phloem.

the cork oak, *Quercus suber*, the cork layers can reach considerable thickness and are stripped off to provide cork which is used in the production of corks for wine bottles and for flooring and insulation.

Epidermis. The epidermis of a plant consists of a single layer of cells which covers and protects the plant. Protection is obtained in two ways. The outer walls of above-ground epidermal cells contain the fatty substance cutin. Over the surface is a waxy layer, the cuticle. These two together restrict the loss of moisture from the plant to the environment. Also in the epidermis, especially in leaves, are cells which are specialised in the regulation of moisture loss from within the plant tissues. Within the epidermis occur guard cells which surround stomata (Fig. 2.14). The stomata give access to the tissues beneath the epidermis and are most common in leaves. Variation in turgor of the guard cells causes variation in the size of the stomatal aperture and therefore allows the plant to regulate the rate of water loss. Epidermal cells may also have appendages such as hairs, which help reduce water loss when on leaves, by reduction of wind velocity over the leaf surface. Under the ground, however, the epidermis produces the root hairs which considerably increase the surface root area and allow an intimate association between the plant and the soil in which it grows.

2.3 PLANT MORPHOLOGY

The study of the arrangement of the various tissues into systems and structures to produce the growing plant is called morphology. Notwithstanding the diversity of plant form there is considerable uniformity in the arrangement of the various structures within plants. It is therefore possible to use a general description for all plants.

In a vertically growing plant its main direction of orientation is the axis. The axis is divided into two components: the shoot, above the ground, and the root below.

The shoot

The shoot or stem is organised into nodes and internodes. It terminates at the shoot apex, or terminal bud, which in non-dormant plants is an apical meristem the growth of which leads to the production of leaves, nodes and internodes which, as they increase in length, increase the height of the plant.

At each node on the stem one or more leaves are borne and within the angle between the leaf and the stem, the axil, can be found an axillary bud. Like the shoot apex axillary buds may be either active or dormant. When a leaf falls off the stem, its former position at the node can be detected by a leaf scar below the bud.

Buds are short immature stems bearing many leaves and are usually enclosed in protective scales. When axillary buds elongate they form branches which are like the stem in structure. The branches will themselves bear leaves and have axillary buds. Buds may be vegetative in which case they will contain only leaf primordia, or reproductive when they may have both leaf and flower primordia.

The internal structure of a dicotyledon stem differs from that of monocotyledons. In a transverse section of a primary dicotyledon stem, vascular bundles can be seen arranged in a circle under the epidermis and the cortex (mainly parenchyma) and surrounding the pith (also parenchyma) (Fig. 2.8). When secondary growth occurs the distinct pattern of separate vascular bundles is lost as the vascular cambium divides and the plant increases in diameter by the deposition of secondary xylem towards the inside of the stem and secondary phloem towards the outside (Fig. 2.9).

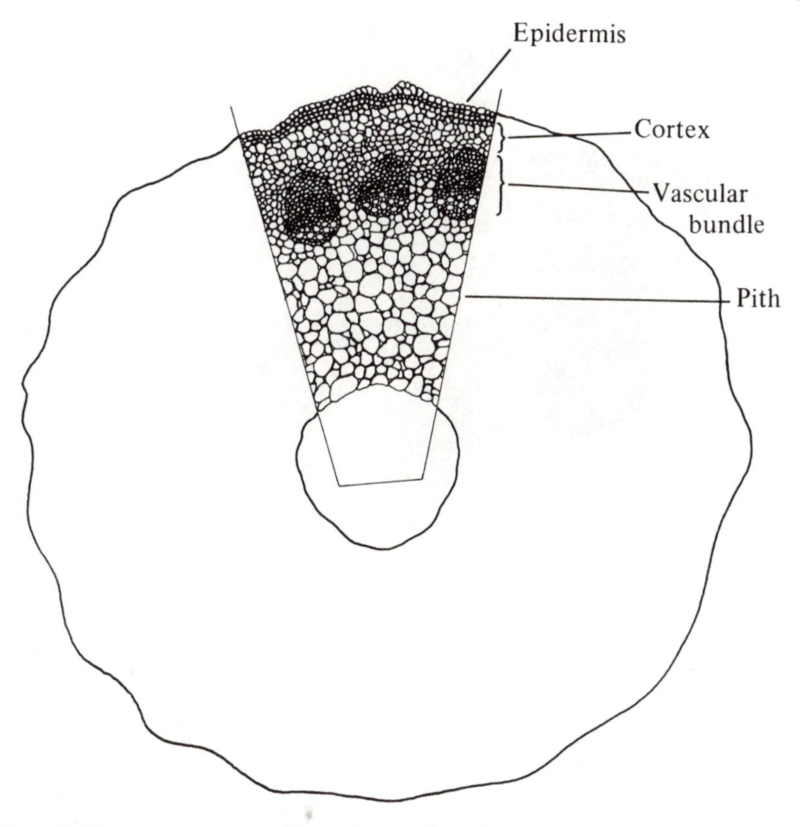

Fig. 2.8 Transverse section of a primary dicotyledon stem.

In a monocotyledon stem the arrangement of vascular bundles in the parenchyma is more random, though bundles are still concentrated towards the outside of the stem. Further, the bundles are still arranged with the phloem cells towards the outside and the xylem towards the centre of the stem (Fig. 2.10). In most monocotyledons there is no secondary growth and any increase in stem diameter which takes place is usually the result of division and enlargement of individual parenchyma cells. This process is known as diffuse secondary growth.

Stem modifications

Until now we have been discussing the type of stem structure that can be found in an erect stem such as tobacco or lucerne. However, there are many modifications of stem growth. Probably the most simple is where the stem

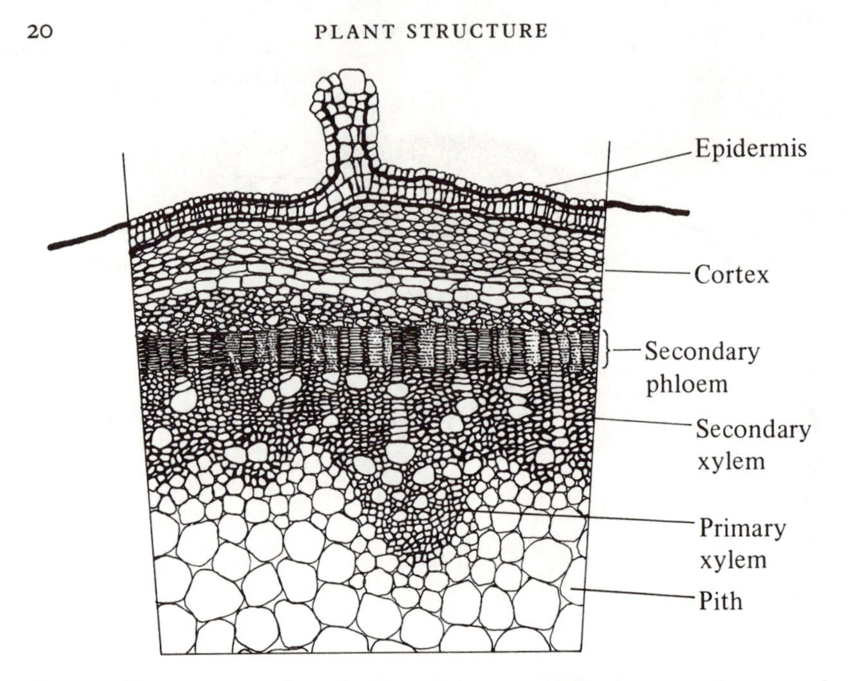

Fig. 2.9 Transverse section of a dicotyledon stem following secondary growth.

lacks structural rigidity and lies along the surface of the soil, i.e. is prostrate. Such a stem is a runner, as for example in plants of subterranean clover (Fig. 11.5). In some species, stems that lie along the surface of the soil form roots at each node and thus attach firmly to the soil. These are called stolons, as for example in white clover (Fig. 11.2). Other stems grow under the ground and form both roots and scale leaves at each node, as in the common weed yarrow (*Achillia millefolium*) (Fig. 6.3) which has such stems known as rhizomes. If rhizomes are cut up, each node is capable of creating a new plant. Rhizomes in their turn may be modified for storage and thus become much enlarged. A potato is such a modified stem and is called a tuber. Its origin as a stem structure can be seen from the leaf scars on its surface and the eyes (buds) which are present in their axils (Fig. 16.3).

In some monocotyledons the stem is considerably reduced in size. In many grasses prior to the plant becoming reproductive the stem is little more than a disc-like structure at the base of the plant from which leaves emerge (Fig. 4.6). Similarly in the onion the stem is situated at the base of the bulb and is extremely short (Fig. 3.1). In other plants such as members of the genus *Oxalis* stems may be modified to form corms which are short stem structures containing food reserves. As with rhizomes, plants which

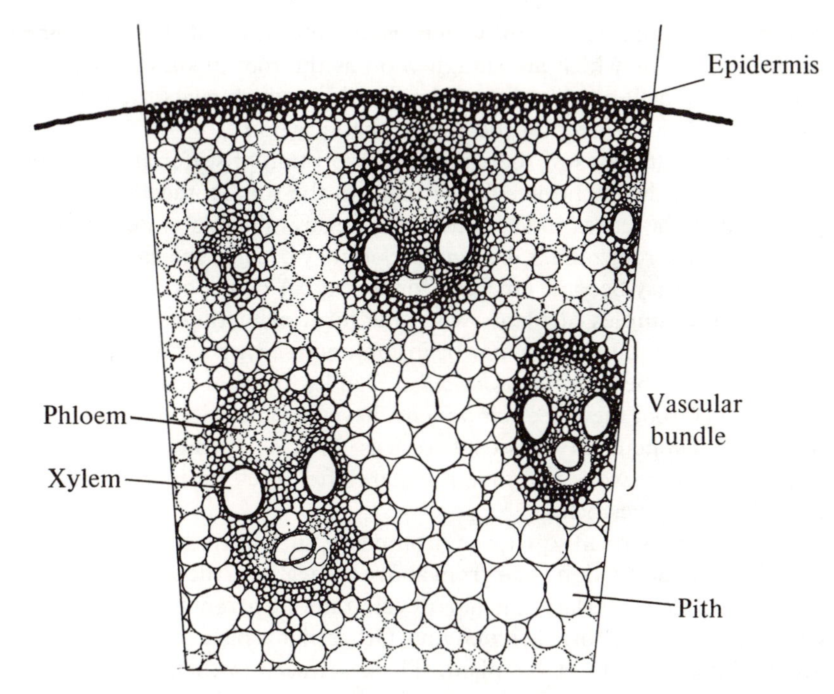

Fig. 2.10 Transverse section of a monocotyledon stem.

reproduce by means of corms may be difficult weeds to eradicate by cultivation or herbicides.

Stems which are modified for climbing are known as vines. Vines require structural support to climb and can either do so by tendrils as in the cucurbits (see section 10.2) or by twining as in certain members of the genus *Phaseolus* (see section 11.23).

Roots

The part of the plant axis that lies below the surface of the soil and absorbs water and minerals is known as the root. The first root to form on germination is known as the radicle. In dicotyledons it becomes elongated and like the stem produces lateral structures or secondary roots which in turn branch within the soil. In monocotyledons the first root which emerges from the seed usually does not survive for long and is replaced by a fibrous non-branching root system which arises at the base of the stem (Fig. 4.4). The root is terminated by the root apex which contains an active meristem by which the root increases its length. The end of the root is protected by the root cap which is a collection of loosely joined parenchyma cells which are sloughed off as the root pushes through the soil. The cap cells also secrete sticky mucilage which coats the tip of the root and aids in protecting the meristematic region.

The main functions of roots are the absorption of water and minerals from the soil, the anchorage of the plant in the soil, storage of nutrients produced by the leaves and conduction of sugar, minerals, water and gases to and from the root zone. Like stems, roots may become specialised, and in many plants they become enlarged for storage. A carrot or a sweet potato are typical examples of plants with this form of modification. In other plants the root modification may be in conjunction with a thickening of the hypocotyl and the stem. In *Beta vulgaris* in its forms of sugar and fodder beet and beetroot the three components have become thickened (Fig. 9.5). This also happens in various brassicas such as turnips, swedes and radishes. These swollen storage roots are often used by man for feeding himself or his livestock.

In plants growing in aquatic environments, such as rice, the root may be modified for conduction of air from above the water to the submerged root tissue. In these plants large air gaps develop among parenchyma cells in the cortex of the root to form aerenchyma. Other plants have prop roots which are modified to support the plant. These are seen in their most obvious form in tropical mangroves but are also found at the base of the stem of both maize and sorghum plants. In other plants the roots are modified to assist the plant in climbing. Thus plants such as ivy develop aerial roots which allow them to grow on vertical surfaces.

A few plants have the capacity to generate new plants directly from the root in the same way as rhizomes. Plants produced in this way are known as adventitious shoots and an example of a plant with this type of growth is the common agricultural weed sheep's sorrel (*Rumex acetosella*).

In a primary root (Fig. 2.11) there is not a great deal of difference in structure between monocotyledon and a dicotyledon. In both there is a central stele which contains the vascular tissues. The arrangement of phloem and xylem within the stele varies from plant to plant, and even among roots from the same plant. However, generally the xylem is towards the centre and the phloem towards the outside. Surrounding the stele is the pericycle which may comprise either parenchyma or sclerenchyma cells and the endodermis which comprises cells which have suberin deposited within their wall in a band known as the casparian strip. The presence of the casparian strip ensures that, although water and minerals can pass readily from the epidermis of the root through the cortex to the endodermis in the intercellular spaces, to enter the stele they must pass through the protoplasm of the endodermal cell. Outside the endodermis is the cortex of

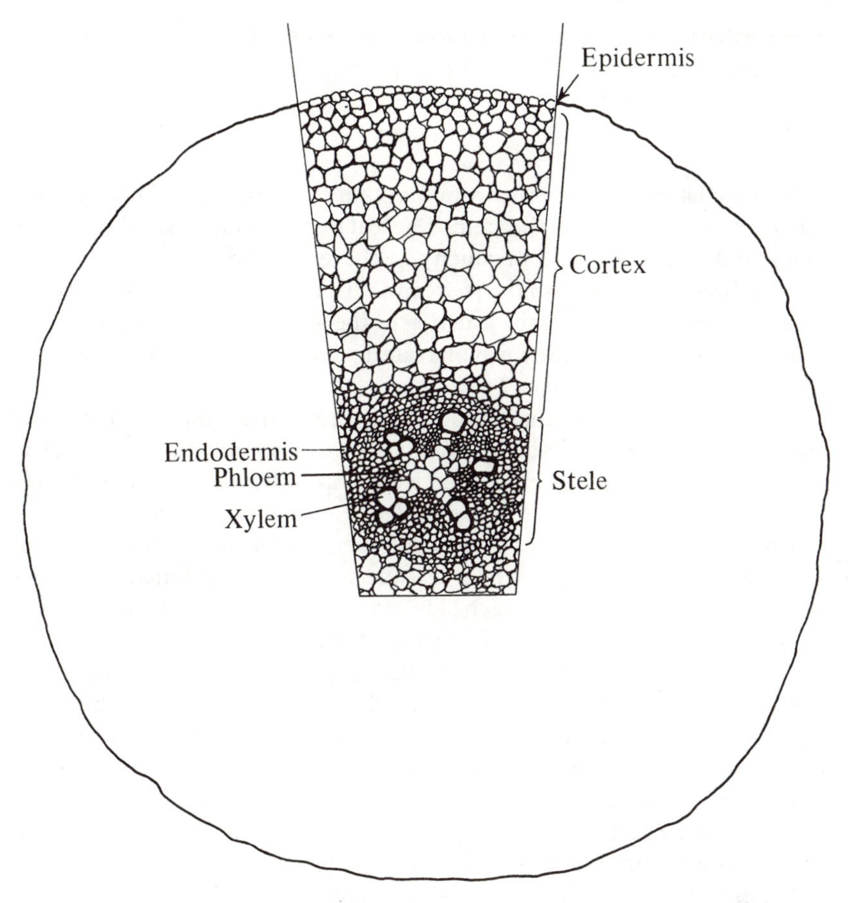

Fig. 2.11 Transverse section of a root.

the root which is usually composed of a wide layer of parenchyma cells. Cortical cells near the outside of the root may be differentiated into exodermis immediately below the epidermis. As already indicated, in roots the epidermal cells form lateral projections or root hairs which considerably increase the surface area of the roots and their absorptive capacity.

When roots in dicotyledons undergo secondary thickening, new xylem and phloem are laid down by cambium situated between the two tissues. The phloem is pushed towards the outside and eventually, as the root becomes large enough, the cortex and epidermis are sloughed off. The protection of the outside of the root is by the periderm which arises from the division of pericyclic cells which in turn may be replaced by phellogen.

The formation of lateral roots takes place from the stele and they are produced by division of pericyclic cells.

Leaves

Leaves are lateral outgrowths from the stem of young plants. Usually one leaf forms at each node on the stem. The leaf may be attached to the stem by a long stalk called a petiole, in which case the leaf is said to be petiolate, or it may be directly attached to the stem and is thus sessile. There may also be structures associated with the junction of the leaf petiole with the stem. In many plants these take the form of a lateral outgrowth of thin tissue along the petiole and are known as stipules.

Leaves are simple or compound (Fig. 2.12). In a simple leaf the leaf blade or lamina is entire as in tobacco. In compound leaves the leaf is divided into a number of individual leaflets as in the clovers. The leaflets may have small stalks called petiolules. Compound leaves are named depending on the number of leaflets and their arrangement on the midrib. Thus clovers and medicks with three leaflets per leaf are trifoliate. In plants with a number of leaflets on each side of the midrib such as broad bean or vetch the leaves are pinnate, while where the leaves arise in a whorl at the end of the petiole, as in lupins, the leaves are palmate. In some plants such as peas terminal leaflets may be replaced by tendrils to aid climbing.

Taxonomic descriptions of leaves also include the type of leaf margin and the shape of the leaf or leaflets. These terms are beyond the scope of a general text such as this. However, for plant identification it is generally necessary to consult a textbook of plant taxonomy as most keys are written on the assumption of a knowledge of these terms.

The arrangement of leaves around the plant stem or phyllotaxis is also of interest. The simplest arrangement is when the leaves are arranged alternately on either side, but they may also be opposite, or whorled or in varying combinations. Alternate leaf arrangement is common in grasses but can only be seen in species which elongate a stem from germination such as sorghum and maize. In other plants two sets of pairs of leaves may be produced along the stem one set being at 90° to the other. Thus when looked at from above there are four distinct rows of leaves visible, and this is known as alternate pairs. In some plants the leaves are arranged in a spiral around the stem to give the appearance of a three, five or seven bladed propeller when viewed from above. Finally, leaves may be arranged in whorls on the stem where a number of leaves arise at a single node. The common agricultural weed spurrey (*Spergula arvensis*) exhibits this form of phyllotaxis.

As with stems and roots, leaves may be modified. For example, scale

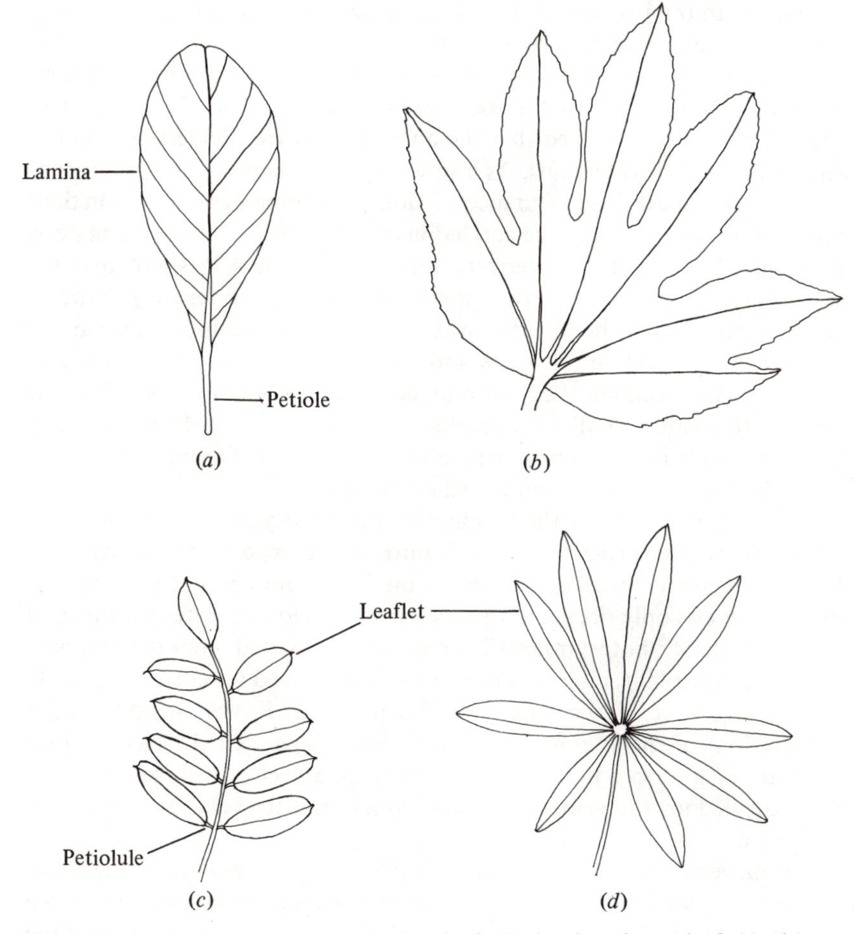

Fig. 2.12 Leaf types: (*a*) simple obovate leaf; (*b*) simple palmate leaf; (*c*) pinnate compound leaf; (*d*) palmate compound leaf.

leaves form on rhizomes under the surface of the soil, enclosing the subtending axillary bud and containing no chlorophyll. In some plants from very dry environments leaves may be much reduced in size and become folded on themselves to reduce water loss. This may be accompanied by growth of hairs on the leaf surface and the stomata being located in sunken pits. Some leaves become modified for food storage as in bulbs like onions and garlic where the below ground portion of the leaf or its axillary bud become enlarged with food reserves to carry the plant from one generation to the next (see Figs. 3.1 and 3.2). In other plants leaves may be normal in the juvenile plant but are modified to form spines in plants such as gorse or may be absent as in mature plants of broom. Modified leaves are also found within the seed of all flowering plants. The cotyledons are seed leaves modified for food storage and it is on the number of cotyledons within the seed that the angiosperms are divided into monocotyledons and dicotyledons. As implied by their names, monocotyledons contain a single cotyledon and dicotyledons contain two cotyledons in their seeds. If, on germination, the cotyledons remain within the soil, as in peas, they provide energy for the germinating embryo to push its shoot up to the soil surface. This form of germination, which occurs under the ground, is called hypogeal. In other plants, on germination, the cotyledons are carried up clear of the soil, as in beans and lupins, and become expanded and green. In this situation they not only act as an energy store but, because they contain chlorophyll, they produce photosynthate for the young plant while its main leaf system is becoming established. Germination when cotyledons rise above the soil is called epigeal.

The role of the leaf is the production of photosynthetic products. To maximise photosynthesis the leaf requires to be exposed to sunlight and leaf arrangement on plants tends to minimise mutual self shading. As photosynthesis only occurs in the presence of adequate water supply, and the products of photosynthesis have to be transported from the leaves to the rest of the plant, all leaves have a well-developed vascular system. By the arrangement of the vascular system within leaves the two major subdivisions of the angiosperms can again be separated. In dicotyledons the leaf veins form a branched network as in a beet or oak leaf, while in monocotyledons the veins are parallel to the midrib (vein) as in maize or sorghum.

A transverse section of a leaf reveals two layers of epidermis, one at the top and one at the bottom; one or both of these layers may contain stomata (Fig. 2.13). Vascular bundles run through the leaf and the remainder of the leaf is made up of parenchyma or mesophyll cells and intercellular spaces. The arrangement of parenchyma cells within the leaf varies from species to species. In classical descriptions, below the upper epidermis of the leaf the parenchyma is arranged in regular vertical parallel lines, and since under a microscope it looks not unlike the wall of a stockade, the cells have been named palisade parenchyma or mesophyll. Below this, the parenchyma cells are diffusely arranged and there are large intercellular spaces which allow exchange to take place with gases from the outside atmosphere. This tissue is known as spongy parenchyma or mesophyll. In some leaves, however, layers of palisade parenchyma may occur on both the upper and lower surface of the leaf and such leaves are called isobilateral.

Control of the rate of gaseous exchanges between the leaf interior and the

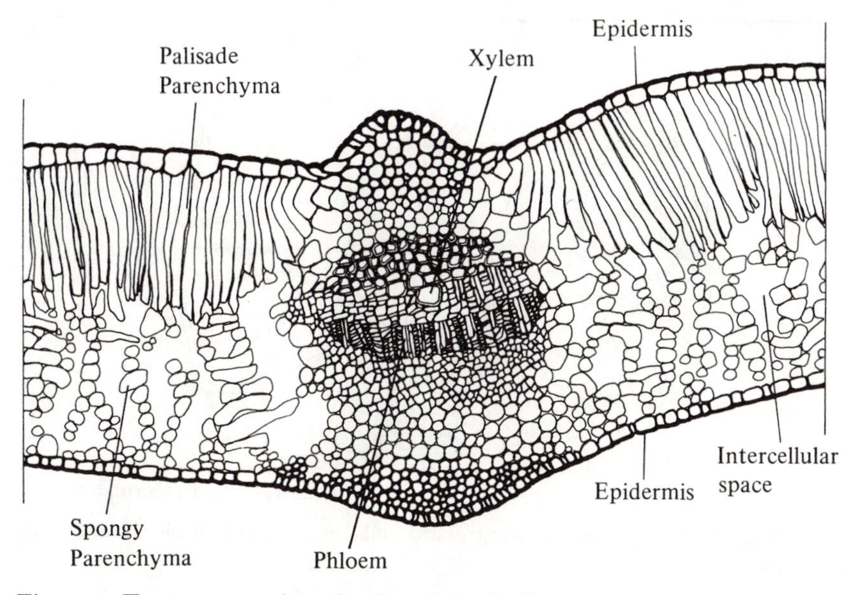

Fig. 2.13 Transverse section of a dicotyledon leaf.

outside environment is by the stomata (Fig. 2.14). When leaves are fully turgid, stomata are open and carbon dioxide moves freely into the leaf for photosynthesis to occur, and at the same time there is a rapid loss of moisture. If the rate of water loss becomes too great, stomata close due to loss of turgor in the guard cells. The water loss is therefore reduced but, at the same time, because of the reduction in gaseous exchange, so is the rate of photosynthesis.

The flower

In all flowering plants, both monocotyledons and dicotyledons, the seed is borne enclosed in an ovary contained within a flower. The parts of the flower usually are arranged in whorls and are inserted into the receptacle at the end of the flower stalk or peduncle. The first whorl is usually green in colour and is composed of sepals. A collective term for the sepals which may be fused, as in plants like the pea, is the calyx (Fig. 11.1). In plants that depend on pollination by insects or other animals, the next whorl is usually brightly coloured and the individual parts are known as petals, or collectively as the corolla. These two whorls of floral parts may not be differentiated in which case they are known as tepals and collectively as the perianth. In some plants there has been considerable modification of floral

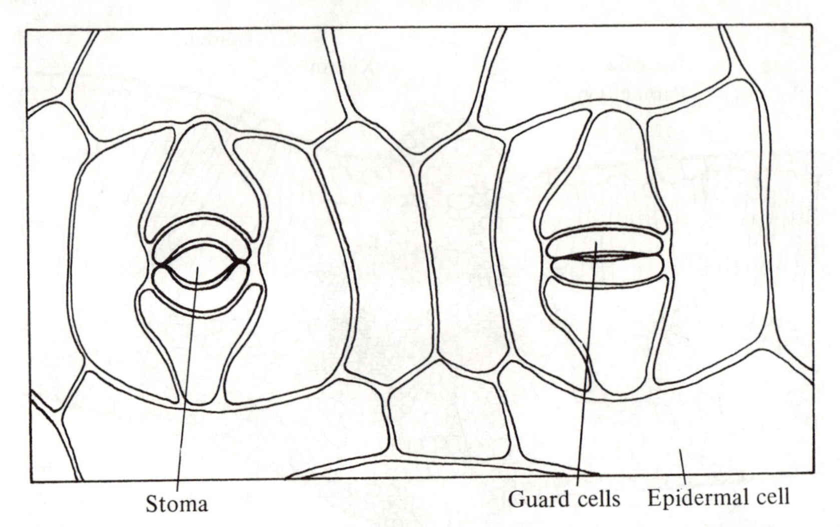

Fig. 2.14 Leaf epidermis showing guard cells and stomata in closed and open positions.

structure, and in the grasses which are specialised for wind pollination there are neither petals or sepals, as described in section 4.1.

The other parts of the flower are the reproductive or sexual structures. The male organs are known as stamens and each stamen comprises a long stalk or filament to the end of which is attached the anther. Within the anther the pollen cells produced by meiosis are formed. At maturity the anthers split releasing the pollen. Although many crop plants are self-pollinated, in many others complex mechanisms have been evolved to ensure that the plant does not cross with itself and that the pollen is carried to other plants.

The female reproductive structure consists of carpels. Flowers may contain one or more carpels which may be free as in the strawberry or joined as in the tomato. At the base of the carpel is the ovary, which encloses the ovules which are also produced by meiosis. Extending from the ovary is the style which ends in a zone receptive to pollen, called the stigma. The whole structural unit in a single combination of ovary, style and stigma is called a pistil (Fig. 11.1).

Not all flowers contain both male and female reproductive parts. Some flowers are therefore unisexual or imperfect. This arrangement can extend to male and female reproductive structures being borne on separate plants as in asparagus (see section 3.2) or the date palm, and such plants are said to be dioecious. In other species, although individual flowers are imperfect, they are borne on the same plant, as for example, maize in which the male flowers are carried at the top of the plant while the female flowers are borne on the cob in the leaf axils (see Fig. 4.38), or in the cucurbits where male and female flowers occur along the same vine (see Fig. 10.1). The more usual condition, however, is for flowers to be hermaphroditic, i.e. to contain both male and female reproductive organs together. When plants are hermaphrodites, or when both sexes are borne on the same plant, the species is monoecious.

Flowers may occur singly, that is they are solitary on a peduncle, or in groups in an inflorescence in which individual flowers are borne on pedicels. Inflorescences may be borne at the end of the axis (terminal inflorescence) or in the leaf axils (axillary inflorescence). Probably the simplest form of inflorescence is the raceme, in which the flowers are borne on short pedicels arising from the main axis, as, for example, in cultivated lupins (Fig. 11.9). In other species, pedicels may be absent and the flowers are sessile on the main axis, as in the spike of many grasses such as ryegrass or wheat (Fig. 4.11). At the other extreme the terminal inflorescence may be highly branched, with each branch bearing a number of flowers to form a panicle, common in other grasses such as oats (Fig. 4.15) and rice. Another characteristic inflorescence of a family of plants is the umbel which used to give its name to the Umbelliferae (now known as the Apiaceae). Umbels have the pedicels of the inflorescence arising at a common point as seen in the parsnip (Fig. 5.2) or the carrot flower. A further modification is the compound umbel, where a number of umbels arise from the same position. In other forms of inflorescence the pedicels grow from the main axis to produce a flat topped inflorescence. Thus, in the corymb, the oldest flowers arise from near the base of the stem, and thus flowers towards the outside of the inflorescence open first, while in a cyme flowering commences from the middle and proceeds towards the outside.

One other plant family has an inflorescence which is characteristic of that family alone and that is the capitulum or head of the Asteraceae. In this family the receptacle becomes greatly enlarged, is surrounded by bracts and has many sessile individual flowers inserted into it. These individual flowers which are known as florets (Fig. 6.1) may either have a conspicuous corolla as at the outside of a sunflower head (ray or ligulate florets) or may be quite inconspicuous as in safflower or thistles (disc florets). Receptacles in the Asteraceae may contain from a few to many hundreds of florets and may be all disc, all ray, or a combination of both types of florets (Figs. 6.1, 6.2).

Fruits

To a non-botanist a fruit is an apple, an orange or a banana. Botanically, however, a fruit is the structure that develops from the carpels and perhaps other associated structures which contains the seed. Although usually for formation of fruit both fertilization of the ovule and seed formation are necessary, in some plants fruit may be formed without pollination as in pumpkins and some varieties of citrus. In orchids such as vanilla, although the pollen takes no part in the production of the seed, fruit will not form until pollination of the flower has occurred. Where fruits are formed without sexual fusion of gametes, they are said to be parthenocarpic. In other species seedless fruit may form without fertilization, in which case ovules are fertilized but abort shortly afterwards. This can happen in cherries and grapes and thus a seedless fruit is produced, but strictly such fruits are not parthenocarpic.

The development of the fruit is an adaptation of the plant to assist in the seed dispersal. This adaptation can range from the formation of the pappus in the Asteraceae to aid in dispersal by wind, the formation of burrs in annual *Medicago* and *Trifolium* species to aid carriage in the wool and hides of animals, the growth of a large buoyant husk to assist water dispersal in the coconut, and the formation of sweet, sticky fruits, attractive to animals, which lead to the fruit being harvested and carried from place to place. Quite apart from dispersal by other agents, in many plants at maturity the fruit splits open and casts its seed out with explosive force. In some legumes in which this form of dispersal is a special feature, seeds can be found up to 3–4 m from the nearest parent plant.

Fruits may arise in a number of forms: they are called simple fruit if derived from a single pistil made up of one or more carpels, as in the pea or the tomato; an aggregate fruit if derived from non-fused carpels in which each carpel has retained its individual identity at maturity, as in the strawberry; or a multiple fruit when formed from the combined carpels of a number of flowers, as in a pineapple. Fruits are further subdivided into whether they are dry or fleshy at maturity, and the dry fruits on whether they burst open (dehiscent) or not (indehiscent). With a few exceptions, in most temperate agricultural crop plants the mature harvested fruit is dry. However, most horticultural fruit crops are fleshy, as are a number of tropical agricultural crops such as coffee and oil palm.

The development of fleshy fruit is considered to be more recently evolved than the dry fruit. Fleshy fruits may take the form of a berry which is juicy and in which seeds are immersed in a thick pulp comprised mainly of parenchyma cells. Berries may form a rind as in citrus, the cucurbits or the banana, or they may form without a rind as in fruits like the tomato and the avocado pear. A fleshy fruit that contains a large amount of extracarpellary tissue is a pome, the fruit type found in apples and pears. Both pomes and berries usually contain multiple seeds. A fleshy fruit that contains only a single seed and is enclosed in a strong endocarp is a drupe, as in coconuts, oil palm, cherry and plum.

Dehiscent dry fruits may be either single or multiseeded. The legume pod is generally a multiseeded dehiscent dry fruit derived from a single carpel. In other crops, such as members of the Brassicaceae or the Papaveraceae, the fruits are formed from one or more carpels into a capsule. At maturity, the capsule may either split open or disperse its seeds via pores or slits.

A single seeded indehiscent fruit in which the fruit wall tends to become fused with the seed is an achene, and this form of fruit is commonly found in the Asteraceae. Not dissimilar in structure, again with fruit wall fused to the seed coat, is the caryopsis which is the single seeded fruit borne by the Poaceae (see Figs. 4.1 and 4.19).

A final type of fruit generally of limited agricultural importance is the nut, as in the hazel nut which is one-seeded, indehiscent and surrounded by a hard dry shell at maturity. The suitability for agriculture of many small-seeded crop species has depended on whether it has been possible to suppress seed dehiscence and thus increase the harvested yield of crop.

Seed

Following the fertilization of the ovule by pollen to form the zygote it enlarges to form the seed. Basically a seed contains two components, an embryo which contains the genetic information necessary to produce a new plant, and an energy source to allow the embryo to survive long enough once germination commences to establish an independent leaf and root system. The embryo and its food reserves are enclosed in a protective covering, the seed coat or testa. Because of the presence of the energy reserves in seeds they form a valuable source of food for humans and their livestock. The main form of storage may be carbohydrate as in the starch of cereals like wheat, maize and rice, protein as in peas and the soya bean, or lipids as in castor bean, linseed, the peanut and the sunflower.

As with other morphological features, seeds from monocotyledons and dicotyledons are distinctly different in structure. In the dicotyledons there are two seed leaves or cotyledons enclosing the young embryo, while in the monocotyledons there is only a single cotyledon in the seed. In some seeds, particularly those that form the Poaceae, energy is stored in a large endosperm, but in many dicotyledons at maturity the endosperm is represented by little more than a layer of crushed cells on the inside of the testa. In these plants, the energy reserves are contained in the cotyledons which fill the entire seed coat.

The embryo consists of a plumule at the end of a very short stem, the epicotyl. Below the point of attachment of the cotyledon(s) is the hypocotyl which terminates in the radicle or primary root of the plant. Further descriptions of the features of seed of the major families of agricultural crop plants are contained in the chapters on individual families.

FURTHER READING

Bold, H.C. (1973). *Morphology of plants*. Harper and Row, New York.
Cutter, E.C. (1978). *Plant anatomy*. Edward Arnold, London.
Esau, K. (1965). *Plant anatomy*. Wiley, New York.
Esau, K. (1977). *Anatomy of seed plants*. Wiley, New York.
Fahn, A. (1990). *Plant anatomy*. Pergamon Press, Oxford.
Gemmell, A.R. (1969). *Developmental plant anatomy*. Edward Arnold, London.
Mauseth, J.D. (1988). *Plant anatomy*. Benjamin Cummings, Menlo Park.
Stevenson, F. and Mertens, T.R. (1976). *Plant anatomy*. Wiley, New York.
Weier, T.E., Stocking, C.R. and Barbour, M.G. (1974). *Botany: an introduction to plant biology*. Wiley, New York.

3 Liliaceae

There are a number of minor crop plants derived from this plant family. Apart from asparagus they are mainly grown for the flavour that they impart to our food rather than as a main part of our diet. There has been some argument amongst taxonomists as to the correct placement of the genus *Allium* and it is sometimes placed among the Alliaceae; however here it will be considered in its more traditional location as a member of the Liliaceae.

3.1 *ALLIUM*

Three important temperate agricultural crops belong to this genus. They are cultivated mainly for their pungent flavour and the harvested crop usually comprises fleshy leaf bases which are formed into bulbs in the case of onion (*Allium cepa*) and garlic (*A. sativum*) or a tightly rolled pseudostem in the leek (*A. ameloprasum*). A commonly cultivated garden herb is *A. schoenoprasum*, or chives, whose fine mild-flavoured leaves are used in a wide range of food preparations. Some members of the genus are also weeds of pasture. The problem they cause is not their aggressive properties as plants but the sulphur compounds they contain which give them their pungent aroma. The milk produced by cows grazing these pastures becomes tainted with a garlic-like flavour and has to be sold at a lower price.

Onion (*Allium cepa*)

Origin

The common onion is thought to have originated in the mountain lands of central Asia where Afghanistan, Iran and Pakistan meet. The crop has a long history of domestication as onions were depicted in ancient Egyptian murals and have been found in coffins of mummies. They were also referred to in the Old Testament. The crop was known to Hippocrates, and by the Middle Ages it was a popular food in Europe.

The onion plant

Onions are a biennial herb. However, for the production of onion bulbs, seed setting is not required and commercial crops are usually grown as annuals.

Following seed germination primary roots arise from the seed but as in most monocotyledons these are soon replaced by adventitious roots which arise from the short stem. Onions have a typical monocotyledonous root system. Roots are unbranched and do not undergo secondary thickening, and some of the older roots may die before maturity of the crop. The stem is a disc-like structure at the base of the plant and is also the site from which the leaves arise (Fig. 3.1). Within the leaf axils buds may be formed. The inflorescence which is borne on a scape pushes through the bulb which is formed from the thickened leaf bases. The leaves are alternate, and arranged regularly on either side of the stem. As in grasses new leaves emerge through the leaf sheath of the older leaves. The leaves, which are cylindrical in section, are initially solid but become hollow as the plant matures. They have a distinct white to blueish or glaucous appearance and are waxy to touch. When conditions become favourable the plant deposits carbohydrates in the leaf bases which swell up to produce the onion bulb, while the outer leaf base layers become papery and provide protection for the developing bulb. This protective coat can be extremely variable in colour ranging from white through yellow and red to brown.

Following bulb formation the onion usually becomes dormant for a period and it is at this stage that commercial bulbs are harvested. If a seed crop is required the bulbs are retained, although some plants bolt and produce an inflorescence in the same year if temperatures have not been low (see also section 9.1).

The inflorescence which is a compound umbel is borne on a scape. Prior to its emergence it is protected by bracts forming a membranous spathe which splits at maturity to reveal the flowers. The umbel may contain as many as 2000 flowers. Flowers are set on a slender pedicel with inconspicuous white to greenish perianth members which are arranged in two whorls of three. There are six stamens which are also in two whorls and alternate with the larger perianth segments. The ovary is superior with three locules and nectaries at the base, and there are two ovules in each locule. The onion fruit is a capsule at maturity which contains small wrinkled black seeds.

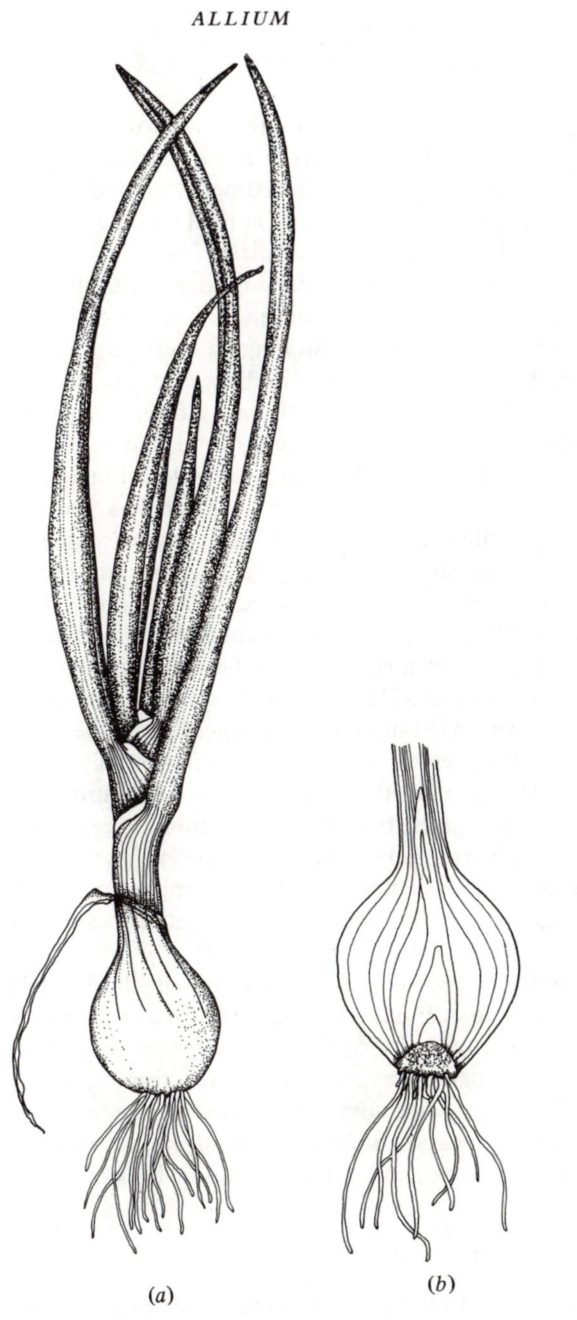

(*a*) (*b*)

Fig. 3.1. Onion (*Allium cepa*). (*a*) plant, (*b*) longitudinal section through bulb.

Cultivation and uses

Until quite recently, cultivated onions were usually established by hand from seedlings. With the sowing of crops on large areas, hand labour is not practical. The development of seed pelleting and precision seeding now allow the crop to be sown directly at final spacings. Young plants prior to bulbing may be harvested and sold as spring onions. Although onions grow best in cool climates with adequate moisture, for maturation, harvest and curing of the crop warm dry conditions are preferable. Bulb formation is controlled by photoperiod, and long days are required. Temperate onion cultivars do not therefore form bulbs in the tropics, although some lines have been selected that are able to form a bulb at high altitudes in tropical environments. At maturity, the tops of the plants commence to die off and fall to the ground, which is taken as an index of maturity when the plants are ready for harvest. American research has shown that a delay in harvesting until nearly all tops have fallen increases mean bulb weight and yield, although later harvesting may cause some decay in the crop. Following harvest the crop is cured. This is mainly to dry off the tops completely which, if allowed to remain moist, could act as a site of entry for rot fungi and bacteria. In America where machine harvesting is practised, the roots are separated from the bulbs, the tops are cut off and harvested onions are carried in bulk bins to storage. Onions are piled on slatted floors and hot air is blown through the pile to thoroughly dry the neck. In most countries the crop has to be stored to ensure a uniform supply of onions throughout the year. Factors that prolong storage life of the crop are suitable cultivars, careful handling during harvesting and curing to prevent damage and entry of rot fungi, temperature and humidity control in the store and the prevention of dry-matter loss by sprouting and root growth.

The onion is probably the major culinary herb of the world, and all cultures seem to utilise these pungent bulbs to enhance the flavour of their food. The vast majority of the crop is harvested and utilised fresh, and there are considerable exports of the crop from certain countries such as Egypt. However, with advances in food processing, a proportion of the crop is now processed into dehydrated onion, which allows long-term storage and a considerable reduction in the bulk of the crop for transport.

Garlic (*Allium sativum*)

Garlic is next in importance after onions amongst the species of the genus *Allium*. It was well known to the ancient Egyptians and is considered to have come from central Asia. For many years it was regarded with suspicion and disdain by northern Europeans as a food additive consumed

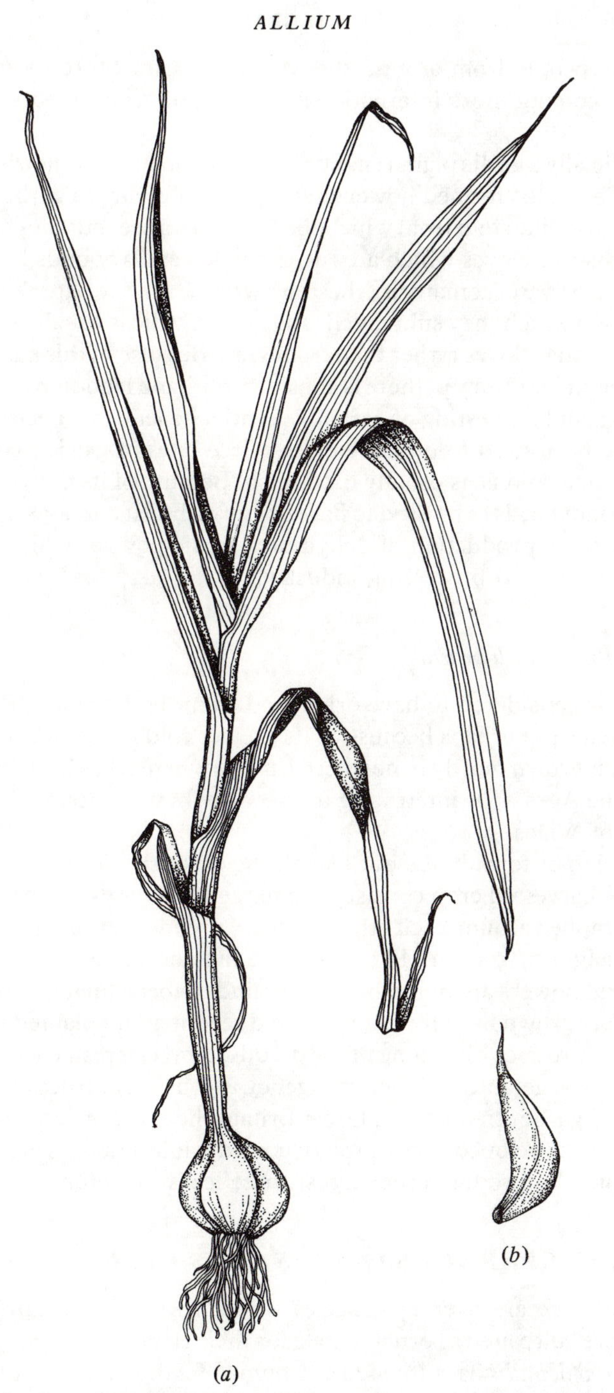

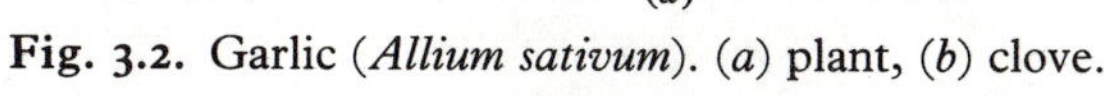

Fig. 3.2. Garlic (*Allium sativum*). (*a*) plant, (*b*) clove.

mainly by people from around the Mediterranean. More recently, with cuisine becoming more international, it is readily consumed throughout the world.

Superficially a garlic plant is not unlike an onion but is somewhat smaller in size. The bulbs formed however are quite different. In garlic there is a flattened disc-like stem from which the leaf bases arise, but the commercial crop consists of cloves which are the expanded axillary buds from within these leaf bases and at maturity these are wrapped in the papery remains of the leaf base which they subtended (Fig. 3.2). Garlic is usually established from individual cloves rather than from seed. Because of this and the small total demand for the crop, there is considerable hand labour involved in the cultivation and harvesting of garlic. Recently, genetic engineering techniques have been used to produce disease free lines of garlic plants.

Garlic, like onions, is usually used fresh. Because of its pungency only a small amount needs to be used to flavour any dish that contains it. There is, however, some production of dehydrated garlic powder which is mostly utilised in the food processing industry.

Leek (*Allium ameloprasum*)

The leek is considered to have originated from the Levant in the eastern Mediterranean. Perhaps because of its greater cold resistance than that of onions, it is grown mostly in northern Europe where it has been in use since the Middle Ages. It is interesting to speculate how it became the national emblem of Wales.

Leeks do not form bulbs but like all the other species reviewed in this genus the harvested crop consists of enlarged leaf bases. In leek these are tightly wrapped around each other to form a pseudostem and the plant may reach a height of 100 cm. The leaves are flat and keeled, up to 75 cm in length, and flowers are borne in a compound umbel which is usually pink.

Leeks are grown initially from seed and seedlings are planted in holes or trenches to produce blanching of the pseudostem. Harvested leeks are used entirely as a vegetable and their pungency is considerably less than that of onions and garlic. In northern Great Britain the sport of competition leek growing has developed. Prizes worth considerable amounts are offered to the person who produces the largest three leeks by volume.

3.2 ASPARAGUS (*ASPARAGUS OFFICINALIS*)

Although there are over 150 species within the genus *Asparagus* and a number are cultivated as ornamentals for their attractive foliage, only one, *Asparagus officinalis*, is cultivated as a human food. Asparagus plants form an extensive root system. Plants are perennial with annual stems which if allowed to grow may reach a height of 3 m. Stems are strongly branched and bear scale leaves, but photosynthesis depends mainly on cladodes, thread-like branches about 1 cm long resembling foliage leaves. The flowers are inconspicuous and are borne in the axils of the cladodes in groups of one to four. Asparagus plants are dioecious and red berries form on the stems of female plants following fertilization.

Cultivated asparagus can be established either from seed or vegetatively but, when established from seed, it may take up to three years to come into production. The harvested crop consists of the newly emerging spring stems which are cut off below ground level. Considerable amounts of the crop are canned. The dioecious nature of the crop causes processing problems as male asparagus spears are thinner and thus more rapidly processed than female spears, and thus there is a lack of product uniformity. Following harvesting a number of the stems are allowed to grow to build up root reserves for the following season. At the end of the year the stems die back, and are then removed. A recent development has been the production of genetically modified asparagus plants which are resistant to the herbicide glyphosate. This allows the herbicide to be used at recommended rates to remove the weeds from this perennial crop without the need for cultivation or hand weeding.

FURTHER READING

Jones, H.A. and Mann, L.K. (1969). *Onions and their allies*. Leonard Hill, London.

Calderbank, D.A., Rodger, B. and Kirkness, J. (1986). *Growing onions and shallots*. Ross Anderson Publications, Bolton.

Kowalchik, C. and Hylton, W.H. (1987). *Rodale's illustrated encyclopedia of herbs*. Rodale Press, Emmaus.

Peirce, L.C. (1987). *Vegetables, characteristics, production and marketing*. John Wiley, New York.

Sutherland, J. (ed.) (1984). Growing garlic: the unforgiving crop. *Proceedings, 1st Australian Garlic Industry Workshop, Weribee, 1984*.

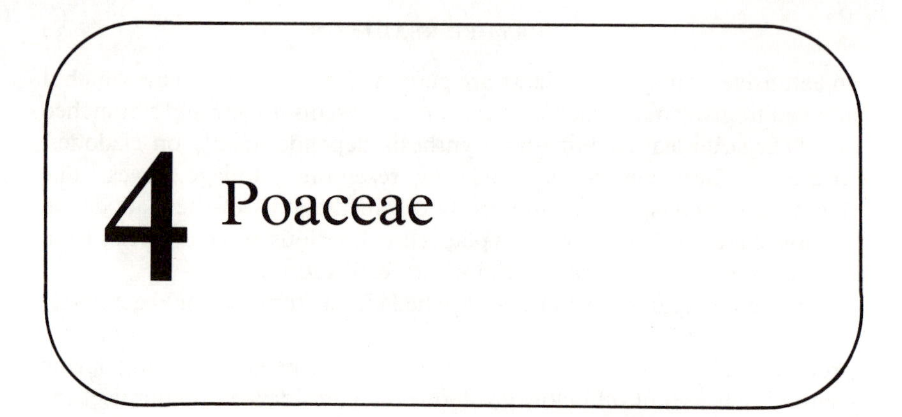

4 Poaceae

The Poaceae (Gramineae), the grasses, form one of the largest plant families consisting of some 610 genera and about 10 000 species. Members of this family occur abundantly in every climatic region. On the savannas, prairies, steppes and man-made pastures of the world grasses sustain grazing animals which provide us with milk, butter, cheese, meat, hides and wool. Rice, wheat, millet, and rye form the basis of the daily diet of many millions of people, while other cereals such as maize or barley are also extensively used for livestock production. Cereal grains form the raw material for such alcoholic fermentation products as beer, whisky or saké; maize is a valuable source of oil; sugar from cane is an important commodity on the international market, while other members of the Poaceae like the bamboos are widely used as building material. There can be no doubt that cereals and pasture grasses are economically the most important plants in the world, and it would be quite impossible to imagine how mankind could continue agriculture without them. We thus need to study this family in some detail, first by learning something of its general structure, and later by considering individual species of economic importance.

Grass systematics

In a group of plants with as many genera as the grasses and with such a wide distribution and long evolutionary history it is hardly surprising that classification is not an easy matter. The taxonomy of the Poaceae is a highly specialised study and hardly concerns us here. Although different schemes have been proposed, several sub-families or groups are generally recognised, of which the Festucoideae contain most of the temperate grasses, and the Panicoideae and Chloridoidae most of the tropical and sub-tropical grasses of economic importance. The Oryzoideae are represented by rice, the Bambusoideae by bamboo. We shall be dealing in some detail with species belonging to the first of these, especially the major cereals and main temperate pasture grasses. The Panicoideae include many highly productive grasses and cereals which follow C_4 type photosynthesis (see section 17.4), and several of these must also engage our attention. In fact, plant introduction and exchange, often fostered by international agencies, have been so active in recent times, that the climatic and geographic boundaries of species have become less rigid. We thus need to be acquainted with a wider range of plants, as we have the opportunity of making use of their potential in agricultural practice or, at least, in plant breeding. Taxonomically, the Poaceae are further subdivided into over 40 tribes, each containing many genera and species, but this will not engage our attention any further in this context.

4.1 GENERAL CHARACTERISTICS

The grass grain

What is usually called the seed or grain in the grasses is really a fruit, or caryopsis as it is properly termed (Fig. 4.1). The major part of this structure consists of a large store of carbohydrate, the endosperm, surrounded by the testa, the seed coat, and the pericarp, the fruit wall,

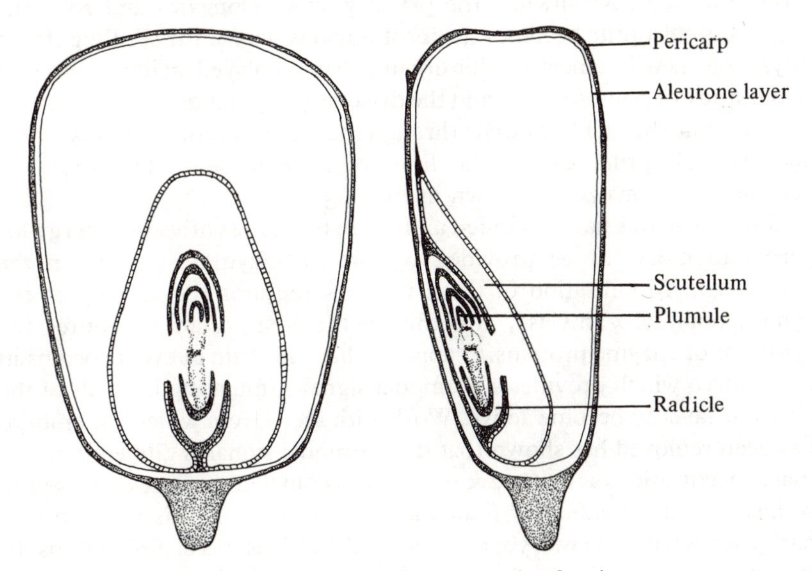

Fig. 4.1. Longitudinal sections through a caryopsis of maize.

fused together. Just inside this outer covering and before we get to the starch-filled cells of the endosperm, there is a distinctive region, most commonly only one cell thick, the aleurone layer, which is rich in protein and in certain species also in oil. This tissue plays an important part in the biochemical events of seed germination and it also influences the quality of cereal products.

Separated from the endosperm by a shield-like structure, the scutellum, lies the embryo. From the outside of the grain this looks like a small swelling but closer examination or preferably sectioning would reveal two separate parts: the radicle covered by a protective structure called the coleorhiza, and the plumule surrounded by a sheath, the coleoptile. The whole grass seed is either the naked caryopsis, as in wheat and rye, or it may be covered by fibrous structures which are loosely called husks or chaff representing parts of the inflorescence.

Germination

The dry grass seed normally contains some 14% of water, and at that level can be stored quite safely. During germination water is absorbed reaching 50–60% by weight, and the seed swells quite visibly. The first root appears bursting through the coleorhiza, and this is soon followed by two pairs of lateral roots, although precise details vary from species to species. Because these first few roots are attached to the seed, they are referred to as seminal roots (Fig. 4.2). Meanwhile the primary shoot elongates and soon the colourless coleoptile becomes the most obvious part of the seedling. Plant physiologists will remember this organ as having played an important part in the study of phototropism and the discovery of plant growth regulators. After a while the first leaf bursts through the coleoptile, other leaves follow and the coleoptile reaches the limit of its extension. The stages of germination in maize are shown in Fig. 4.3.

Until the leaves have unfolded and begin to photosynthesise, energy for germination has to be provided by the carbohydrates stored in the endosperm. Mobilisation of these reserves requires the activity of enzymes, notably α-amylase, and this in the first instance requires the synthesis of enzyme proteins. It appears that the chain of events begins in the embryo which provides a hormonal signal stimulating the cells of the aleurone layer to become active. Work with seeds from which the embryo has been removed has shown that the hormonal signal is gibberellin, and that concentrations as low as 2×10^{-11} M cause an increase in α-amylase and proteases and activation of β-amylase. So sensitive is this reaction that barley seeds without embryos are now used for bioassays of gibberellins. In the intact, germinating seed the enzymes so produced in the aleurone layer

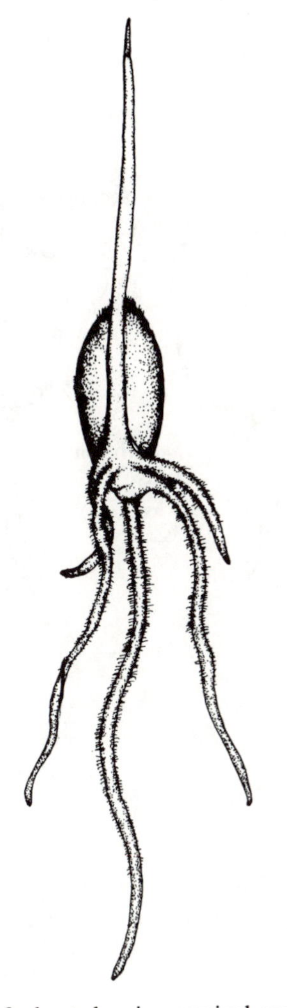

Fig. 4.2. Young seedling of wheat showing seminal roots.

break down the carbohydrates of the endosperm, and as a result simple sugars move across the scutellum to the embryo to sustain the growth of the seedling roots and shoot.

Grasses which have been in cultivation for a long time usually germinate quite readily once they have been sown. In wild species there may be varying periods of seed dormancy delaying germination, and these act as important mechanisms ensuring the survival of these grasses in adverse environments. Even cultivated species may well need a short period of

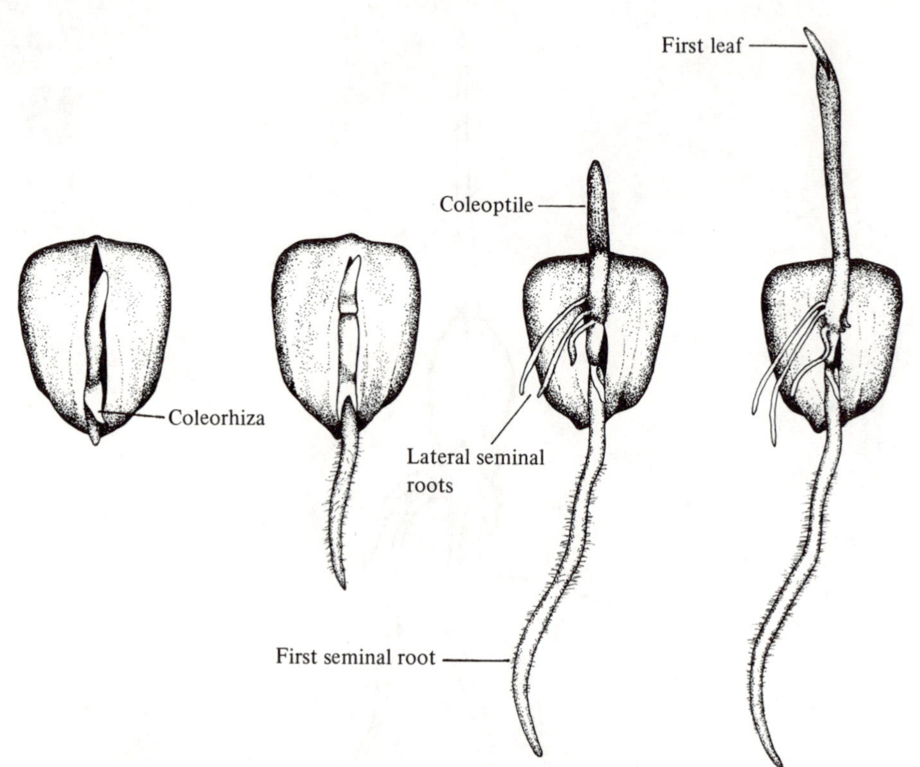

Fig. 4.3. Stages of germination in maize.

after-ripening. Where this is causing any difficulties, as in the malting industry where rapid and uniform germination of barley is required, gibberellins are used successfully.

The grass shoot

Although the seedling continues to produce new leaves, stem extension is initially confined to only part of the shoot axis, the mesocotyl, consisting normally of one internode. The degree of elongation tends to depend on the depth of sowing, because the net effect of this event is to bring the base of the young seedling close to soil level, and it is from there that further growth proceeds (Fig. 4.4).

If we cut a longitudinal section through the middle of the seedling or, better still, removed successive leaves carefully at their base with a dissecting needle, we would find right in the centre the stem apex (shoot

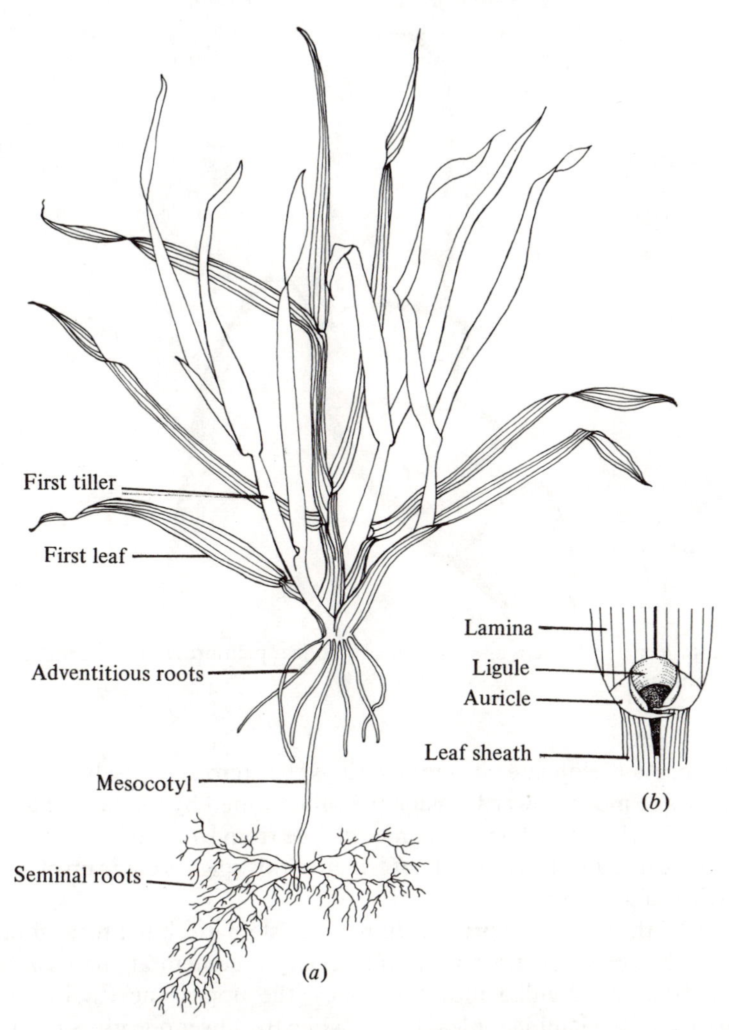

Fig. 4.4. (*a*) Grass seedling with seminal and adventitious roots, mesocotyl, and with tillers arising from the first three leaf axil positions. (*b*) Base of lamina and top of leaf sheath showing ligule and auricles.

apex, apical meristem, growing point). This intensely meristematic region looks white and glistening, and it resembles a small dome (Fig. 4.5). Along two flanks of this structure, opposite to one another, there occur small protuberances which mark the beginning of new leaves. Each leaf primordium, as it is called, grows not only in an upward direction but also around the stem apex, so that before long it forms a kind of tube encircling

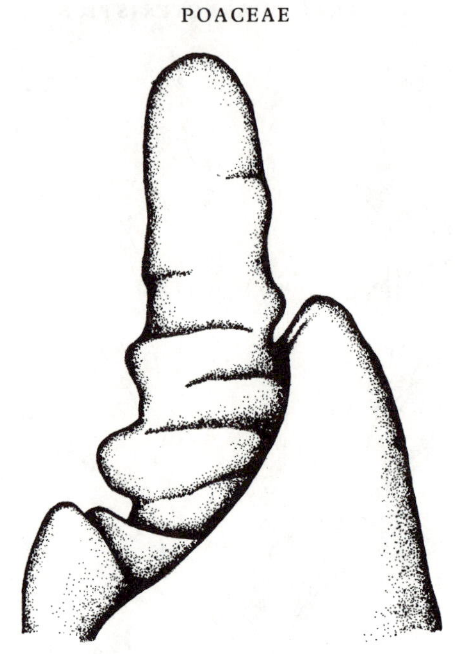

Fig. 4.5. Vegetative stem apex of wheat with leaf primordia in different stages of development.

the stem. As it elongates it soon overlaps the stem apex itself and then it continues to move upwards inside the tube formed by the next older leaf. As a result when a new leaf becomes visible from the outside, it is always seen in the middle of a young shoot which at this stage is made up of a series of such leaf structures.

Initially the whole leaf primordium is meristematic, but after a while cell division becomes localised at an intercalary meristem at the base. This region becomes divided into two zones, the upper one leading to the formation of the lamina, or leaf blade, while the lower one gives rise to the leaf sheath. Between these two structures there often occurs a small tongue-like outgrowth, the ligule, variation in which is one of the features by which different grass species may be recognised. Another feature aiding in recognition concerns two small projections at the base of the lamina, the auricles, which vary in size and may be present or absent depending on the species (Fig. 4.4).

It will be appreciated that each leaf is inserted at a node, and that at the base of the grass shoot we consequently find a succession of nodes separated by internodes. However, as long as the shoot remains vegetative and leaf formation continues, the internodes do not elongate and so the real stem of the grass plant is exceedingly short, even if the collection of leaf sheaths above give the appearance of much greater length. How important this arrangement is will soon become obvious.

At each node there occurs a complicated network of vascular bundles growing up and down the stem and connecting the leaves. Also arising from the nodes are root primordia which gradually produce the main supporting and absorbing organs of the grass plant, the adventitious roots. Because of the close proximity of the nodes, these roots appear to arise from the same region of the stem and thus form a very compact system. These adventitious roots continue to grow throughout the life of the plant, and their eventual size, distribution and longevity make them far more important than the seminal roots whose main function relates to the early growth of the seedling.

Tillering

A grass shoot is usually referred to as a tiller, and thus what we have been describing so far could well be called the first or main tiller. This is only the beginning of a much more complicated shoot system, because further tillers arise from buds in the axils of leaves (strictly speaking the leaf sheath). Each bud is a complete replica of the apical meristem except that it is protected by a prophyll which takes the place of the coleoptile. Active growth by this bud causes a tiller to be formed (Fig. 4.4), and this emerges from the encircling leaf sheath in one of two ways. In tufted or tussock-forming grasses the tiller grows upwards within the sheath and is first seen from the outside when the tip of its first leaf moves above the ligule of the subtending leaf. Alternatively, in other species, a tiller may break through the protecting leaf sheath and grow out horizontally to become a stolon or a rhizome.

Whatever the type of emergence, the important aspect of this form of branching is that within each leaf sheath there is an axillary bud and thus a potential site for a tiller. This is true not only of the leaves on the main shoot which in due course subtend the primary tillers, but also of the leaves of any other tiller, so that as time proceeds a complex system of tillers of different orders is built up. Each of them is potentially capable of independent existence, although on an intact plant they are in vascular connection so that some translocation is possible. Not all axillary buds develop any further, and those which form tillers do not appear at the same time nor survive for identical periods. The implications of these inter-relationships will become apparent a little later.

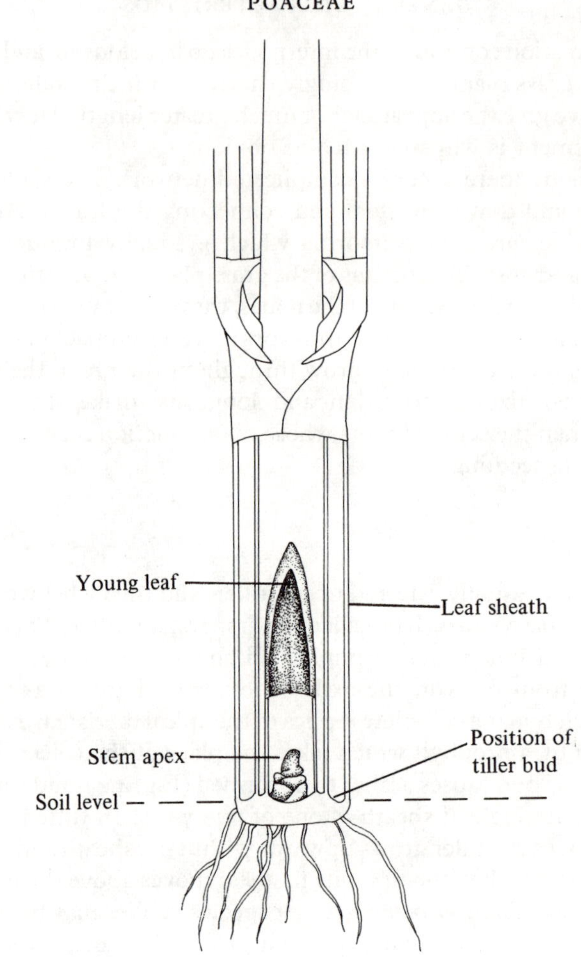

Fig. 4.6. Position of stem apex in the vegetative grass shoot shown after removing basal leaf sheaths.

Adaptability to grazing

Because grasses have evolved in the presence of grazing animals, they have developed adaptations enabling them to survive in the face of repeated defoliation. We have already seen that the stem apex while vegetative occurs at or close to soil level, usually well out of reach of grazing animals (Fig. 4.6). Not only can new leaf formation be resumed with a minimum of delay but, because the meristematic zone of each leaf also tends to escape damage, those leaves still in process of elongating at the time of defoliation continue growth even if their upper portions have been eaten. Axillary buds at the base of the plant are normally also below grazing level and, indeed, increased tillering activity is often the response to defoliation at certain stages of development. Because of these features new leaves appear quickly after grazing, thus restoring the productivity of the plant. A certain amount of hoof damage can also be tolerated, in so far that tillers partially separated by treading from the rest of the plant are able to strike more roots and survive independently.

Flowering

Very profound changes occur in the growth pattern of the grass plant when flowering begins. The first sign of the onset of reproduction is seen in the stem apex which elongates rapidly. Although still microscopic in size, it now produces many more primordia, and these soon look quite different from those which became leaves in the vegetative condition. Instead each primordium develops its axillary bud, while growth of the meristematic leaf cells becomes inhibited and no further leaves are produced. This is called the double-ridge stage. Once formed, the bud primordia develop rapidly and give rise to the various floral organs (Fig. 4.7). The whole stem apex thus transformed becomes the inflorescence of the grass plant, but equally important is another series of events which occur just below the

Fig. 4.7. Reproductive stem apex of wheat showing spikelet and floret primordia.

apex. In the vegetative condition successive internodes remained contracted, but now rapid elongation of the most recently formed internodes takes place, and as a result the stem apex is carried upwards inside the leaf sheaths to emerge eventually as the fully formed grass inflorescence. One of the consequences of these changes is that the stem apex is raised from the safety zone near the ground to become increasingly vulnerable to grazing animals or cutting implements. Grass tillers can now literally be nipped in the bud. This is a decided advantage if one wants to prevent inflorescence emergence and to encourage the plant to produce new tillers near the base. But if, on the other hand, the grass is grown for seed or grain production, grazing must be stopped immediately to avoid damage or loss of the young inflorescence.

Reproductive structures

The grass flower, or floret as it is commonly called, represents a highly reduced monocotyledonous type. The gynaecium is a single-seeded ovary usually bearing two styles terminating in a feathery stigma, which is well suited to pollination by wind (Fig. 4.8). In most grasses there are three stamens, although in the bamboos and in rice there are six. The floret is subtended by a pair of bracts, the lemma and, attached to its axis, the palea. Probably because these two structures protect the floral parts, what is considered to be equivalent to perianth members is restricted to two tiny protuberances, the lodicules, at the base of the floret. These membranous organs take up water readily at the time of anthesis, and by swelling they force the lemma and palea apart, thus exposing the stigmas and allowing the anthers to emerge. Cross-pollination by wind is further facilitated by rapid elongation of the filaments and the versatility of the anthers. As sufferers from hay fever will know, grass pollen escapes readily into the air. Despite these adaptations, not all grasses are cross-fertilized. In the common cereals, wheat, barley and oats, the stigmatic surfaces are commonly dusted with pollen before the florets open, and in some grasses, which are said to be cleistogamous, self-pollination is either the rule or it takes place under certain environmental conditions. Following fertilization, the ovule grows rapidly through the deposition of starch and in due course it occupies the whole cavity within lemma and palea. Eventually a new caryopsis is formed, and the life cycle is completed.

Florets are joined to one another by a short segment of stem tissue, called the rachilla, which usually persists and remains attached to the seed. Groups of florets, varying in number according to species, are referred to as spikelets, and these are normally subtended by two further bracts, the glumes (Fig. 4.8). In many grasses spikelets are sessile and opposite on a single or occasionally branched axis (rachis), in which case the in-

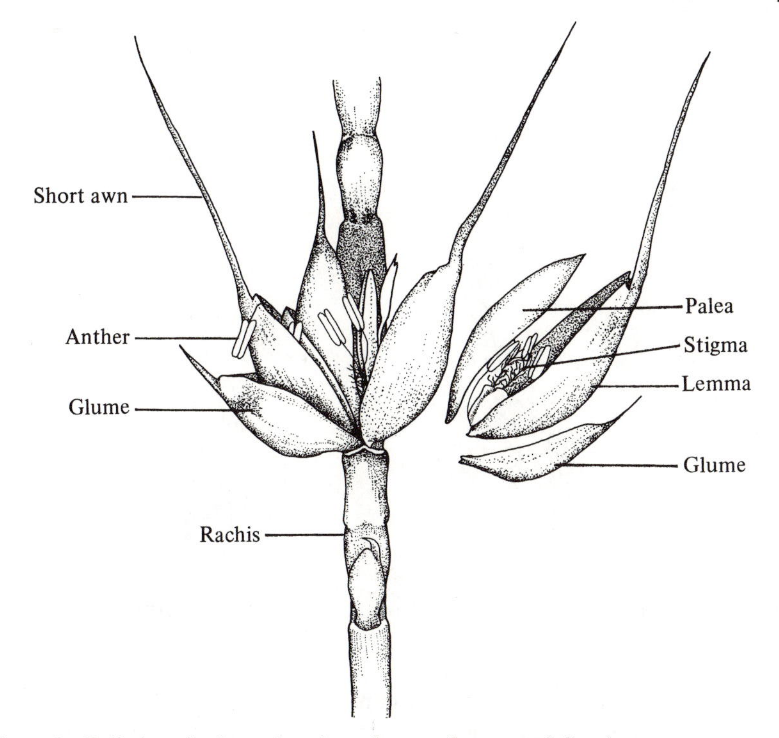

Fig. 4.8. Spikelet of wheat showing glumes, florets and floral parts.

florescence is termed a spike. In other species the spikelets occur on a highly branched stem structure, called a panicle. Still other species have a spike-like panicle, in which the spikelets are not opposite nor truly sessile but attached to an unbranched rachis by means of a short rachilla. Whatever the type of inflorescence they are all borne on a long, jointed stem called a culm.

In some species seed dispersal is aided by the occurrence of awns, bristle-like and often barbed structures usually attached to the lemma (Fig. 4.9). According to the point of their attachment, awns may be terminal, sub-terminal, dorsal or basal. Awned seeds are liable to be spread accidently by animals or man. In some plants, notably wild oat, the awn is bent and spirally thickened and, depending on moisture content, the spirals twist or untwist, thus helping the seed to be buried (Fig. 4.15).

Life cycle of the grass plant

The grass plant, as we have seen, is made up of a collection of tillers. Whether the plant as a whole behaves as an annual, biennial or perennial

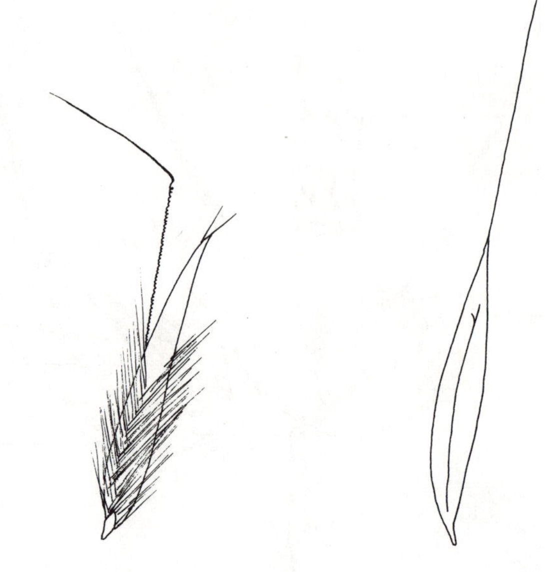

Fig. 4.9. Grass seed with kneed dorsal awn (*left*) and straight terminal awn (*right*).

depends on the life history of its constituent tillers. In annual cereals growing in the field a relatively small number of primary tillers arises on the main stem, and a few secondary tillers may also be formed. Once conditions become favourable for reproduction, a few of the major tillers will produce an inflorescence and flowering stem, while the remainder tends to die. Since the stem apex of the flowering tillers has become an inflorescence, no further leaves can be formed, and thus these tillers will also die when the grain is mature. The plant as a whole has thus completed its annual life cycle.

Some species are unable to flower in the same season in which they were sown but require a period of vernalisation in response to the low temperatures of the winter before reproduction can begin in the following spring. Such grasses describe a biennial life cycle, as their tillers either produce an inflorescence, or if remaining vegetative, die during the period of stem elongation in the flowering tillers. Winter cereals fall into this category as well as certain forage grasses, but in these it is possible to obtain greater longevity by removal of young reproductive stem apices through cutting or grazing at critical times.

However, very many grasses are perennials, especially those which are sown as pasture species and those which form the bulk of the natural grasslands of the world. Survival in these habitats depends on ability to withstand heavy grazing pressure, and this is achieved by almost continuous production of tillers throughout the growing season. Tillers that are present early enough to be exposed to environmental conditions conducive to flowering will tend to produce an inflorescence, set seed and then die, but many others will remain vegetative and keep on producing leaves. Although some of these tillers may flower in the following season and many others may die without ever having flowered, there should always be enough living shoots present to ensure the perenniality of the plants.

In a very real sense, therefore, a grass plant can be described as a collection of tillers, each with its own leaves and contributing to the adventitious root system as a whole. The performance of the plant will depend on the aggregate of individual tiller life histories. If the majority flower and the remainder die within the same season, we have an annual grass. Conversely, if only some of the tillers flower while others survive to continue to form leaves, the plant as a whole is perennial and, given a favourable environment, could keep on living almost indefinitely, even though the tiller population composing it will always be changing. This extreme plasticity, together with the position of the growing centres at the base of the plant, explains why grasses are so well adapted to being grazed.

Subfamily Festucoideae

Grasses belonging to this subfamily are adapted to cool climates and many require long days for flowering, often preceded by low temperatures. Photosynthesis follows the C_3 pathway (see section 17.4). The basic chromosome number is 7, and the mitotic chromosomes tend to be large.

Cereals

4.2 *TRITICUM*

Wheat (*Triticum* spp.) is one of the most important staple foods of mankind. About 37% of the world's population rely on it as their main cereal, it accounts for some 20% of the total food calories consumed by man, and world wheat production has risen to over 500 million tonnes, more than a third of total cereal output (Table 1.2).

Origin of wheat

Partly because man has been so intimately associated with wheat since the dawn of civilisation, and since it belongs to a tribe of plants which is sufficiently young to allow hybridisation between different genera, the origin of modern wheats is not simple. There are in fact several species of

Table 4.1. *Some of the main species of Triticum*

Ploidy	Rachis brittle	Rachis not brittle
Diploid (2n = 14)	*T. monococcum* *T. tauschii* *T. urartu*	
Tetraploid (2n = 28)	*T. dicoccum* *T. timopheevii*	*T. durum* *T. turgidum* *T. polonicum*
Hexaploid (2n = 42)	*T. spelta* *T. vavilovi*	*T. aestivum* *T. sphaerococcum* *T. compactum*

wheat which are closely related to one another (Table 4.1). *Triticum aestivum*, bread wheat, is a hexaploid with a chromosome number of 42. The basic number in the genus is 7, and in the diploid condition the genetic composition of bread wheat is usually given as AA BB DD, where A, B and D represent the complements of genetic material, or genomes, derived from one of its ancestors. The A genome originates from *T. monococcum*, einkorn wheat which derives its name from the fact that it produces only one grain per spikelet. This wheat still occurs as a wild plant in parts of Iraq, Iran and Turkey but its use as a cultivated cereal is limited by the brittleness of its rachis which disintegrates dispersing its separate spikelets. Wheats sharing this property are referred to as spelts.

The origin of the B genome is less clear. At one time it was thought to have been derived from a wild grass belonging to the genus *Aegilops*, but recent cytological and biochemical evidence has cast doubts on the validity of this theory. Present indications point to *Triticum* itself as the most likely source, but it remains uncertain whether the species involved is *T. urartu* which occurs, though not abundantly, in the same regions as wild diploid wheats, or whether directly or indirectly *T. timopheevii* or yet another tetraploid was involved. This cross, and there may well have been more than one, was an amphidiploid, in which hybridisation was followed by chromosome doubling. The donor of the D genome may have been another member of the genus *Aegilops* but more probably a wild diploid wheat, *T. tauschii*, and once again an amphidiploid was formed. Synthetic hexaploids have been obtained experimentally by crossing this wheat with the tetraploid *T. turgidum*, and it is thought that this mirrors natural hybridisation between these two species as the area of cultivation of the tetraploid encroached upon the distribution of the wild diploid. However, the great variability of hexaploid wheats suggests that similar hybridisations will have occurred more than once involving different genotypes.

Early history

Wheat is one of the first plants to have been domesticated by man, certainly at least as early as 7500 BC. Cultivation as a crop began in an area now occupied by northern Syria, south-eastern Turkey, and parts of Iran, close to what has been called the cradle of civilisation, and closely coinciding with the geographical distribution of the wild progenitors of wheat. Anatolia and part of the Balkans appear to have been another area of cultivation. Good evidence of the kind of wheat being grown comes from remains found near caves and other sites of early habitation, especially from fragments of pottery bearing the imprint of grain which must have got mixed up with the clay. These early settlements were in hill or mountain areas, and it was probably not until the fifth millennium that wheat growing spread to the Nile Valley and the plains of Mesopotamia.

The early wheats were very different from those grown today. In particular, the wild progenitors of wheat had evolved an effective method of seed dispersal and survival. The spike was very brittle and at maturity broke up into separate spikelets. These fragments were shaped like the head of an arrow, and this not only helped in their dispersal but enabled the seed to become buried in the soil and thus escape desiccation during the hot summer. This feature will have been a considerable disadvantage to the early farmer and food gatherer, and hence it is more likely that plants with non-brittle spikes would have been selected and their seed saved for subsequent sowing. Originally wild forms of *Triticum monococcum* were probably just collected by primitive man, but later einkorn wheat was cultivated and spread to many areas, especially along the Danube, Rhine and other rivers serving as trade routes. It still survives in some isolated mountain districts of Turkey and Yugoslavia.

Closely associated in space and time with the early diploid wheats was the development of tetraploids, including brittle forms of *T. timopheevii* but especially *T. turgidum*, a hulled non-brittle plant which gave rise to cultivated emmer wheat. Einkorn and emmer were evidently grown widely in many localities of the Near East, and cereal farming must soon have become as important to early man as the raising of sheep and goats. Emmer wheat spread rapidly from the more remote mountain areas to the fertile plains of Mesopotamia and Egypt, and to parts of Europe, India and Ethiopia. It is still grown on a limited scale in some areas. It is thought that by a series of mutations the toughness of the glumes of this wheat was reduced and that ultimately a freely threshing plant with a naked caropsis resulted. This is referred to as a variety of *T. turgidum* or as *T. durum*, depending on the taxonomic system used, a wheat which is still grown in hot, dry countries for the production of macaroni and similar pasta products.

Hexaploid wheats originated soon after diploids and tetraploids had become domesticated. Among the earliest archaeological finds, going back to about 7000 BC in Syria, were grains of *T. compactum*, and it was this and another related freely threshing wheat, *T. aestivum*, which became widely cultivated and spread into the irrigated plains of Mesopotamia and the Nile basin. *T. aestivum*, the bread wheat, is now one of the major crop plants of the world, which man has taken to every continent. The only other hexaploid still to be cultivated is the hulled *T. spelta*, spelt wheat, which is retained in some harsh environments such as the high plateau of central Iran and isolated areas in Europe. Although one might expect that wheats with a naked caryopsis evolved from those with hulled grains, there have so far been no archaeological finds of spelt grains dating back to before 2000 BC. It may well be that *T. spelta* existed at an earlier date but was not cultivated till later because tetraploid emmer with its naked grains was preferred.

Bread wheat (*Triticum aestivum*)

Wheat is an annual plant composed of rounded or rolled shoots. At the junction of sheath and lamina there is a small, rounded ligule and the leaves terminate in a pair of moderately prominent auricles, each supplied with a few bristle-like hairs (Fig. 4.10). The plant as a whole is glabrous. The extent of tillering depends on environmental conditions, but in a typical field situation three to four tillers and about the same number of secondary tillers may be produced, of which only one or two normally form an ear. The rest die during the period of stem elongation. Depending on the depth of sowing, a tiller may also appear from a bud in the axil of the coleoptile. The inflorescence is a spike bearing numerous spikelets, singly at nodes on opposite sides of a strongly jointed rachis. Each spikelet is subtended by a pair of stiff glumes and consists of up to nine florets, of which only some produce a grain. The lemma may terminate in an awn (Fig. 4.11), or a hooked awn-point (Fig. 4.10), or it may be awnless, depending on the cultivar. Probably because wheat used to be harvested largely by hand, there was until recently some preference for awnless cultivars which were easier to handle. This is now tending to change, partly because some selections with desirable yield characteristics happen to be awned, and partly because awns contain chlorophyll and photosynthesise, thus contributing to grain yield. The palea is smaller than the lemma and carries

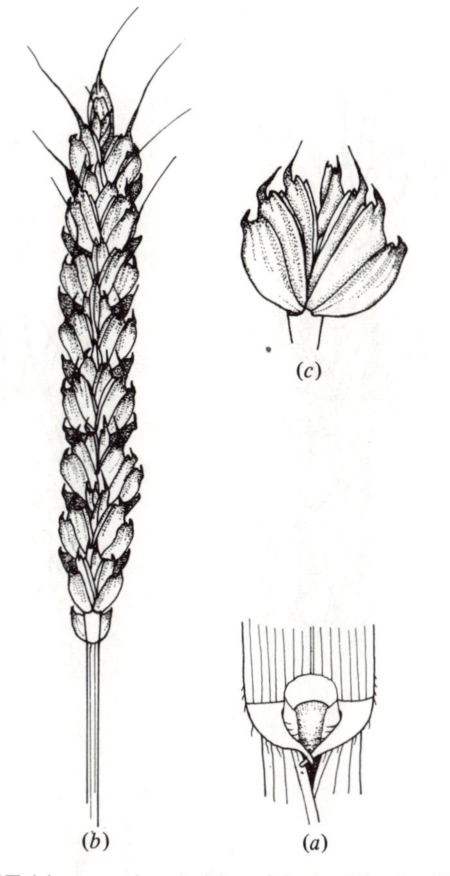

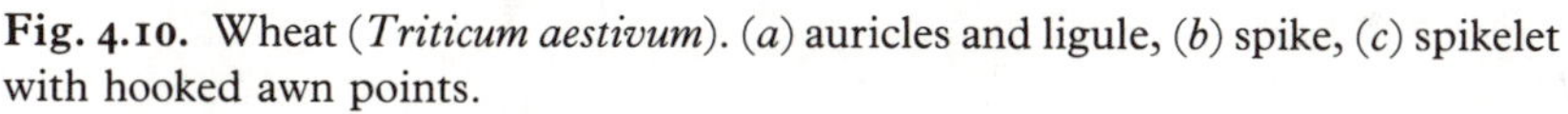
Fig. 4.10. Wheat (*Triticum aestivum*). (*a*) auricles and ligule, (*b*) spike, (*c*) spikelet with hooked awn points.

no awn. The naked caryopsis has a prominent crease on the ventral side and is hairy at the apex. Depending on cultivar, the glume, lemma and palea (collectively known as the chaff) and the grain itself may be a pale straw colour or they may be reddish brown. Wheat is largely self-pollinated, although the florets are seen to open allowing anthers and stigmas to emerge, as well as pollen to be shed. This stage is referred to as anthesis.

Physiology of yield

Wheat may be a winter annual or it may be sown in the spring. Cultivars needing to be sown in the autumn tend to have a requirement for the

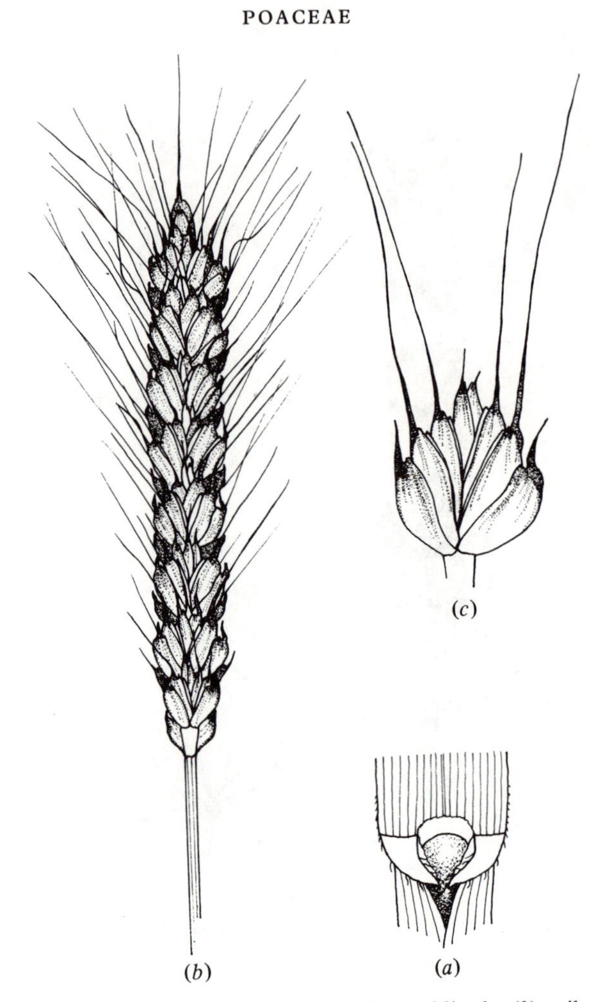

Fig. 4.11. Wheat (*Triticum aestivum*). (*a*) auricles and ligule, (*b*) spike, (*c*) spikelet with long terminal awns.

vernalising temperatures of the winter, before they can respond to the lengthening days of spring by initiating reproductive development. Other cultivars flower earlier if exposed to low temperatures, and in other cases it is agronomically advantageous to sow before the spring even if there is no physiological necessity to do so. However, there are many cultivars which are true spring wheats in being able to flower and yield satisfactorily following sowing in spring.

Grain yield of wheat per unit area has increased greatly in many countries since the beginning of accurate recording. This has been achieved by influencing agronomically and genetically one or more of the components that determine yield: the number of ears (spikes) per unit area of land, the number of spikelets per ear, the grain set as given by the number of florets bearing grain in each spikelet, and the weight of the individual grain. Each of these components is itself determined over a period of time by a range of physiological and environmental factors. For example, the number of ears will depend on the number of tillers produced, the time when they appeared, their position on the plant, and the extent of their survival which will itself depend on the density of the crop and environmental conditions. The time at which critical events occur is most important especially in relation to the time of fertilizer applications designed to increase yield components. Grain set, for instance, is known to be largely determined at the time of spikelet formation, several weeks before the ears emerge and florets become visible. The significance of these components to grain yield is discussed in section 17.3.

Physiologists have made considerable strides in the last 30 years or so in showing that the carbohydrates of the wheat grain are derived to a very large extent from current photosynthesis after the ear has emerged and not from storage organs. This means that the green parts of the plant after ears emerge are responsible for grain filling: the last leaf to appear (the flag leaf) and the one below it, the green culm, the ear itself, and the awns if present. Consequently the size of the flag leaf, the angle of inclination and its longevity have all been suggested as important selection criteria in plant breeding. In this connection it is interesting to note that the early diploid and tetraploid wheats probably had higher rates of photosynthesis per unit leaf area but that the flag leaves were smaller and senesced more quickly than in modern hexaploids. In fact, in the more primitive progenitors of bread wheat, there may have been very little export of photosynthates from the flag leaf, and assimilation in the ear itself accounted for most of the grain carbohydrate. The other major change has been a considerable increase in grain size and in the number of grain-producing florets in each spikelet. Consequently, modern wheats carry large numbers of big grains per ear, and this increase in the size of the 'sink' will have called for a greater 'source' of carbohydrates, with the flag leaf augmenting the supply from the ear (see also section 17.5).

Wheat breeding

It is probably true to say that more effort and resources have gone into the breeding of wheat than that of any other single crop. The rewards have been correspondingly high, and it is quite possible to show that part of the continuous increases in wheat yield can be attributed to improved cultivars, although even more effective has been the rising level of agronomic inputs. Resistance to diseases and pests has always been high on the list of breeding objectives, but there are also very many other requirements which have to be met. Some of these relate to local climatic conditions and involve such characteristics as lack of shattering to reduce seed loss in strong winds or presence of seed dormancy to prevent premature germination before harvest.

Perhaps the most widely discussed breeding achievement in recent times has been the introduction of dwarf or semi-dwarf cultivars. The first step was taken in the 1930s when Japanese breeders isolated a short-strawed wheat which became known as Norin 10. This cultivar was used for crossing after World War II by plant breeders in the State of Washington, where at high levels of fertilizer quite spectacular yields were claimed for a new cultivar called Gaines. This wheat together with others bearing similar characteristics became an important part of the genetic stock for a very large breeding programme set up in Mexico at the International Maize and Wheat Improvement Centre or CIMMYT for short. Many cultivars with one, two or three genes for dwarfness were released from CIMMYT and institutes in other parts of the world, and yields of 8–10 t ha^{-1} became widely reported. However, productivity at this level requires heavy applications of fertilizer and sophisticated agronomic techniques, without which the new cultivars may not necessarily be any advantage. The so-called green revolution is thus attributable as much to improved agricultural efficiency as it is to advances in plant breeding, but at the same time reliance on a high level of technology introduces a grave element of risk.

Dwarf and semi-dwarf wheats do not necessarily produce more dry matter but, because they have shorter straw, a higher percentage of the total is in the form of grain. Over the years there has been continuous improvement in what is called the harvest index: the proportion of total plant substance that is harvested as grain (see section 17.5). The great advantage of these cultivars is that they do not lodge, even in the presence of high nitrogen supply and irrigation. However, they have other and less obvious yield characteristics which are of great importance, notably better grain set, enabling more fertile florets in each spikelet to be developed. It will be interesting to see in which other ways it will be possible to improve wheat yield through breeding in the future, other than by selecting for disease resistance.

Another approach to wheat improvement has been inspired by the great success of hybrid maize. On the face of it wheat is not a promising plant for this type of breeding, because in a naturally self-pollinating species cultivars are usually bred as pure lines selected from segregating populations, following handmade crosses. However, the discovery not only of cytoplasmic male sterility but also of pollen fertility restoration has opened up new possibilities for further advances in wheat breeding. To what extent hybrid wheat will become important in practice will depend on a number of factors, notably how stable the male fertility restoring system can be shown to be, and whether hybrid vigour can lead to sufficiently large yield increases to pay for the extra costs of seed production. Present indications are that wheat breeding will continue along orthodox lines in the main, certainly for as long as plant breeders can claim continuing success.

Uses of wheat

Wheat grain is commonly ground into flour before further processing. This involves breaking open the grain, separating out the embryo and the bran (pericarp and testa), and crushing the endosperm to make flour. Commercial mills aim to obtain an extraction rate of 78%, but this depends on the amount of endosperm in the grain and also the ease with which the embryo and bran can be removed. Since the separation process makes considerable use of sieves, hard wheats giving 'gritty' fragments are preferred. Soft wheats yielding 'woolly' fragments that tend to clog up the sieves are less satisfactory.

Bread can be made in several ways. Traditionally, flour is mixed gently with water, salt and yeast are added, and the resulting bulk dough is allowed to rise in warm conditions for some hours through the respiratory activity of the yeast. Another short period of respiration follows after moulding of the loaves, or in tins before baking takes place. Modern methods involve rapid and intensive mixing at high speed, followed immediately by dividing the dough, only a brief fermentation period in tins and then baking. The work input in what is called mechanical dough development should reach a certain optimum figure for good quality bread to be made. Loaf volume and crumb texture are closely related to the protein (gluten) content of the grain and its biochemical composition. There are large genotypic differences in these respects, but environmental conditions also play a large part, notably soil nitrogen which will vary seasonally according to the nature of the soil, rainfall and fertilizer regime. Other factors also influence the baking performance of a wheat, as for example high amounts of α-amylase in the endosperm which may continue to degrade starch to sugars during the early part of the baking process, thus producing 'sticky crumb' which gives an unsatisfactory doughy consistency to the loaf. Attack by weevils during grain filling or premature germination during a wet harvest both have a disastrous effect on baking quality. It is thus clearly most important to the milling and baking industries to have all lines of wheat thoroughly tested before use, and the more closely the test resembles the industrial process the more reliable the results will be.

In warm, dry, continental climates like those of the Ukraine, parts of the United States, Canada and Australia so-called hard wheat is produced which is highly suitable for bread making. In cool temperate regions such as northern Europe or New Zealand, yields tend to be considerably higher but the grain is softer and quality less reliable. Much of this wheat is used for other products such as pastries, biscuits, breakfast cereals, noodles, starch, or for livestock rations. The best pastas such as spaghetti and macaroni are made from *Triticum durum* flour, and in fact none of the einkorn or emmer wheats is suitable for the production of leavened bread.

4.3 *HORDEUM*

Barley (*Hordeum vulgare*)

Barley, also known as *H. sativum*, is an annual cereal with a high potential for yield, adapted to a wide range of conditions in the cool temperate zone. It is grown extensively in Europe, the Near East, in parts of Asia, North America, Australia and New Zealand. Grain production throughout the world is about one-third that of wheat.

Origin of barley

Because barley is a diploid ($2n=14$) and tetraploids are uncommon, it is not possible to use cytological techniques to trace its ancestry as in wheat, and hence there have been even more conflicting views on its origin. It is now thought that cultivated barley probably arose from *Hordeum spontaneum*, an annual with a brittle rachis, which still occurs widely today as an aggregate of wild and weed races. Closely related and differing only by one of two linked genes is *Hordeum vulgare* with a non-brittle rachis and persistent grains, and it was this plant which gained ascendance during domestication. This occurred around 6000–7000 BC in an area stretching from present-day Israel and Syria to Iran and Turkey, although an even earlier date in the Nile valley has also been proposed.

Originally barley grew alongside emmer and einkorn wheat and leguminous grains but with the passage of time it became, in its own right, a highly important crop of early agriculture in the Middle East. It was grown under irrigation in ancient Egypt and Mesopotamia, especially after the rising salt content of the soils made wheat production increasingly difficult. Barley had an outstanding reputation as a human food, and whereas modern sportsmen are given a highly proteinaceous diet consisting of steak and eggs, ancient gladiators were fed on barley. It was not until the beginning of our era that wheat became the main cereal for human consumption in temperate climates and barley was relegated to an animal feed, apart from its use for the brewing of beer. Some of the races selected as food for man had naked grains, others were hooded through the provision of an inverted sterile floret, and others again had no awns at all.

The barley plant

In the vegetative condition the barley plant looks similar to wheat, except that the auricles are bigger but have no hairs (Fig. 4.12). The leaves which are rolled inside the shoot tend to have 15–20 veins as opposed to 11–13 in wheat. The inflorescence is a spike, and at each of its nodes there are three spikelets, each containing one floret. If all are fertile, one can see three rows of grains on each side of the rachis, and such a barley is referred to as six-rowed (Fig. 4.13). Some representatives of this type have rather lax spikes, in which case the lateral spikelets overlap slightly and give the appearance of having only four rows of grains, but morphologically they also belong to the six-rowed group. In the other and more common type of barley only the median of the three spikelets at each node contains a grain, while the other two sterile ones are represented only by their glumes and reduced lemma and palea. These are the so-called two-rowed barleys (Fig. 4.12). A single mutation involving two genes could change one type into the other.

The glumes are small, often with fine hairs on the edge and terminate in a point which could almost be called an awn. The lemma is a much more conspicuous structure, particularly as it carries a long, well-developed awn which gives the barley spike its characteristic appearance. The awn is serrated and clings to one's clothing, but it breaks off during threshing. The palea is smaller and fits like a lid on the curved lemma. Both are tightly fused to the pericarp, so that the barley grain typically consists of the caryopsis closely enveloped by lemma and palea, except that in some parts of the world there are naked or hull-less forms. Like wheat, barley is mostly self-pollinated before the florets are opened by the lodicules. Other slight differences from wheat are that on germination the coleoptile grows up inside the lemma and thus often appears at the tip of the grain, and also that the seedling can have up to eight seminal roots.

Physiology of yield

As in wheat, the great majority of the carbohydrate deposited in the grain can be attributed to photosynthesis after ear emergence. Reserve sub-

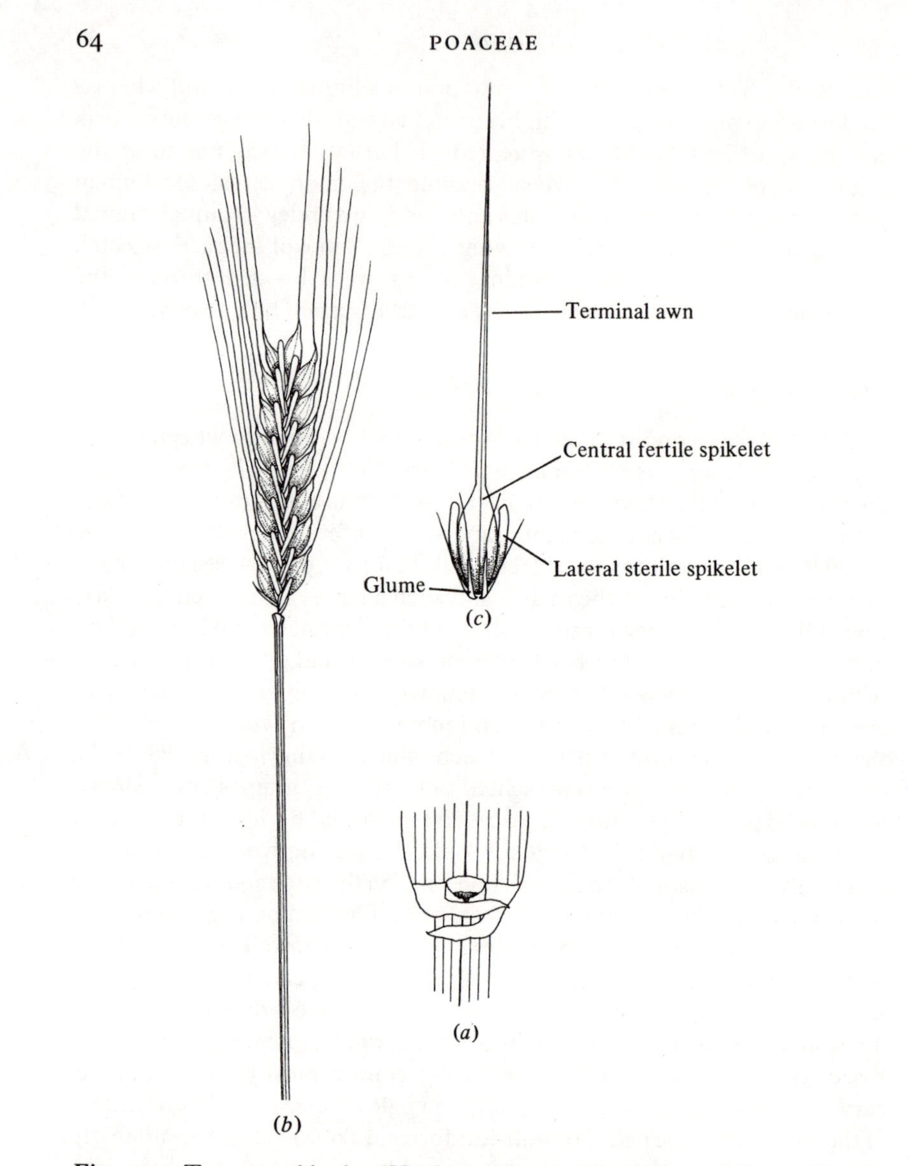

Fig. 4.12. Two-rowed barley (*Hordeum vulgare*). (*a*) auricles and ligule, (*b*) spike with fertile and sterile spikelets, (*c*) group of three spikelets.

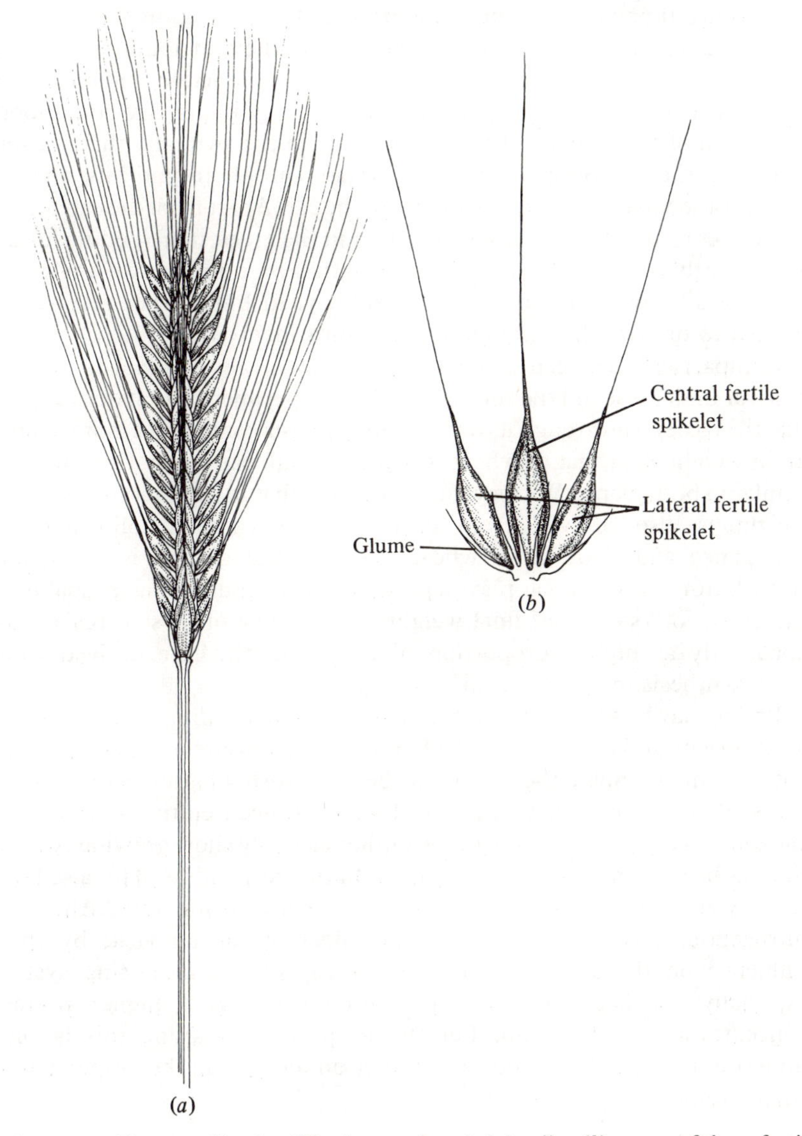

Fig. 4.13. Six-rowed barley (*Hordeum vulgare*). (*a*) spike, (*b*) group of three fertile spikelets.

stances are thus of only minor importance. Depending on the methods employed, varying estimates have been made of the contributions made by the different parts of the plant. The ear itself is very active, and moreover the awns contribute very significantly, especially under dry conditions. Nearly all of the stomata of a spikelet occur in the awn, and this in itself indicates the importance of these organs in photosynthesis and in transpiration in relation to other parts of the ear. Apart from the ear and its components, the flag leaves and sheaths and the green stem (peduncle) are also actively photosynthesising. Awn length, awn number, leaf angle and leaf size all have some influence on grain yield but further research is needed to unravel the complex relationships involved.

Comparisons between different cereals must always depend on the cultivar under test and the conditions of the experiment, but it does appear that the barley ear through its own photosynthesis contributes more to final grain weight than that of wheat. Losses through ear respiration are very similar in both plants. We may note in passing that in rice most of the grain weight is also derived from current photosynthesis following ear emergence and that, as in wheat, the flag leaf system is the major contributor. In rye the ear plays a positive part in grain filling, probably to the extent of 15–30% of final weight, but in none of these cereals is an apparently as high a proportion of grain carbohydrate derived from photosynthesis in the ear itself as in barley.

Barley may be sown in the autumn or spring, depending on cultivar and local agricultural practices. In the United Kingdom and in other countries with a similar climate the accent has been on spring barley, and over the years plant breeders have produced short-strawed cultivars which are capable of very high yields. In view of the relatively short growing season, this has been a remarkable achievement. Barley responds well to raised soil fertility, but care must be exercised in the use of nitrogen. Although nitrogenous fertilizers may have the effect of raising yield by their influence on the size and longevity of the photosynthesising system especially the flag leaf, they may also encourage a higher protein concentration in the grain. For the purposes of malting this is very undesirable (see below), and so would be an increase in the proportion of small grains.

Uses of barley

While still green and vegetative, barley can be used as forage for animals, either by direct grazing or as silage. At a later stage when the grain is still soft and immature and before the plant ceases to be green, barley may be cut and dried for hay. However, it is the grain itself which is the most important commodity for animal nutrition, mainly on account of its high starch content but also because the proteins are well balanced. Whole grain is not completely masticated by farm animals, and hence for improved utilisation and digestion it is advisable to crush or crack the grain, or to roll it into flakes after partial cooking. The performance of ruminants is improved through the addition of urea as a nitrogen supplement. Pigs and, in some situations, bovines are fed diets based mainly on such grain, as denoted by the term barley beef for example. Poultry can also be fed reasonable quantities of barley but only up to a certain proportion of the total diet. For direct human consumption the grain needs to have the husk (lemma and palea) removed but the grinding necessary to achieve this also takes away the aleurone layer and the embryo. The resulting barley flour and groats or pearl barley are used for the thickening of soups and stews. Breakfast cereals may also be made from the grain, but the main demand by the food industry is for a product made from barley, and that is malt.

The malting process is little more than controlled germination followed by kiln drying. For this the maltster requires clean, sound and plump grain with high germination capacity. A high content of carbohydrates is required, and hence small grains and those with more than 1.8% nitrogen may not be acceptable. After being cleaned and graded for uniform size, barley grain undergoes the first of three processes, called steeping, during which it is immersed in water for approximately two days or more, depending on temperature and other conditions. The water content of the grain rises rapidly from about 14% to 44%, and consequent activation of enzymes leads to active respiration. To avoid undue depletion of oxygen the steeping water is aerated and changed at intervals. If there is any likelihood of post-harvest dormancy, small concentrations of gibberellin may be added at this stage and so are other substances to control microbial activity and reduce rootlet growth. When steeping has proceeded long enough the grain is transferred to large drums or to rooms, referred to as germination floors or boxes, where at controlled temperature it is allowed to germinate during the next five days or more. To ensure uniformity of conditions, the germinating grains which are lying in layers some 75 cm deep are turned from time to time. As germination proceeds, six seminal roots appear and the coleoptile grows inside the lemma, but more importantly the endosperm is 'modified' by the enzymatic breakdown of starch to simpler carbohydrates. When this process has reached the required stage, it is terminated by the rapid transfer of the grain to a kiln where grain moisture is reduced to 8–15% at between 50–60 °C and thus enzyme action stopped. Lastly, grain moisture is reduced still further to some 4% by raising the temperature to 75–85 °C depending on the flavour and colour required of the final malt. The total time required for kilning is about 18–22 hours or longer, but the period and indeed all details of the processes involved vary greatly among different operators. As the dried grain, now called malt, is removed for despatch, the dried-up rootlets are also collected because they are rich in protein and thus a useful ingredient of concentrated animal feeds. Malt itself is used predominantly for the brewing of beer, although appreciable quantities are used as malt extract for the flavouring of biscuits, breakfast cereals, sweets and other products, as a meal for special types of bread, or for the production of malt vinegar.

There are several different techniques of brewing but essentially malt is first ground to produce a grist which is then mixed with warm water to give what is called mash. Malt sugar and other polysaccharides are now extracted, and eventually at a required level of specific gravity the liquid wort is separated from the spent grain. The sweet wort is next boiled with hops (see section 8.1), then cooled, and the resulting coagulum removed. Then follows fermentation by specific strains of yeast, after which the yeast is removed, and finally the end-product, beer, is filtered and bottled. There are many different techniques of brewing, and the outline of the process given here is intended as only a guide to indicate the steps involved. Similarly, there are very many types of distillation leading to the production of whisky, gin and vodka. Typically, Scotch whisky is made from malt grist, and the resulting unboiled wort is fermented with yeast. The alcohol level is allowed to rise to a point at which the yeast is killed, following which the liquid is distilled. Bourbon or grain whisky is usually made from maize with only minor additions of barley malt, while Irish whiskey tends to be based on a mixture of malt and barley grain. Vodka, a neutral spirit consisting essentially of ethanol, is obtained by distilling barley grain. Gin is produced along similar lines but is given its distinctive flavour through the addition of essential oils derived from juniper berries and other seeds. However, in all these processes a great variety of specific and often highly secret techniques are employed to produce a distinctive product. The spent grain following distillation is used as an animal feed. Until the advent of motor transport, well-fed Shire or Clydesdale horses used for the delivery of beer and liquor served as a telling advertisement for the industry and this by-product.

Barley grass (*Hordeum murinum*)

Barley grass is a small, rather hairy annual, with soft leaves bearing well-developed auricles and a short ligule. It is more easily recognised after flowering when it produces its characteristic bristly spikes (Fig. 4.14). At each node along the rachis there are three spikelets, of which only the middle one is fertile. The glumes and the lemmas all terminate in a long

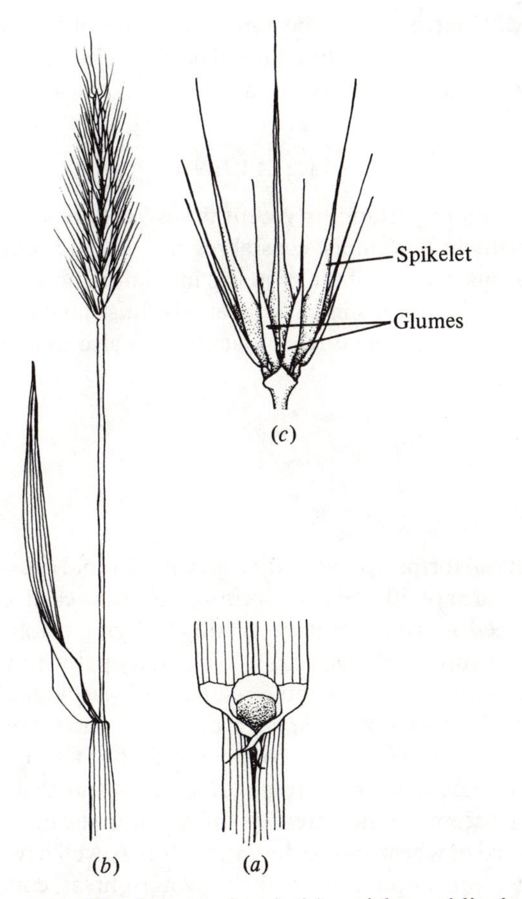

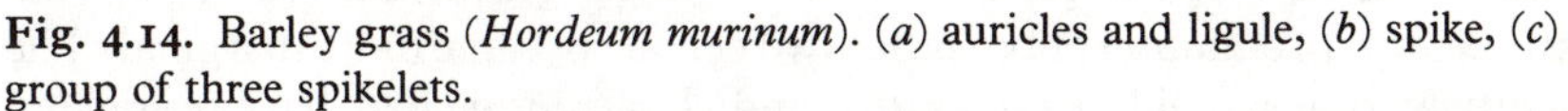

Fig. 4.14. Barley grass (*Hordeum murinum*). (*a*) auricles and ligule, (*b*) spike, (*c*) group of three spikelets.

bristle-like awn, and at maturity the whole spikelet falls to the ground and is disseminated. These so-called seeds cause problems, particularly on sheep farms, because the awns are sharp enough to work their way through the skin of an animal, quite apart from their nuisance value in causing eye inflammation and contaminating the wool. The spikelets frequently get in between the toes of sheep dogs and cause great irritation. Barley grass spreads readily and quickly colonises bare ground such as hedge rows, fence lines and gateways. Its abundance on stock camps, areas where animals congregate to shelter, indicates its preference for high fertility conditions. Geographically, barley grass is widely distributed in the British Isles and other European countries, south-west Asia, Australia and

New Zealand. Control should be aimed at preventing flowering, because the seeds remain viable for only a limited period. This is not easily achieved and continuous vigilance is required to prevent the spread of this weed.

4.4 *AVENA*

Despite declining importance in recent times, oats are still a widely grown cereal, especially in cool, moist coastal regions and in Mediterranean-type climates. The main value of the crop lies in its nutritious grain which may contain about 16% protein and 8% fat and is thus valued for direct human consumption or as an animal feed. Oat straw is also more nutritious than that of other cereals.

Cultivated oats

Origin and history

The genus *Avena* forms a polyploid series with diploid ($2n=14$), tetraploid ($2n=28$) and hexaploid ($2n=42$) representatives, each containing cultivated and weed species. Research is still clarifying cytological and other evidence, but it appears that several evolutionary pathways were involved, some independent and others inter-related. Archaeological remains have failed to show the presence of oats in areas where early wheats, barley and pulse crops were cultivated at around 6000–7000 BC. The fact that the earliest find of oats dates back to about 1000 BC and that it was made in central Europe seems to indicate that this plant first came to the attention of man as a weed of wheat and barley crops. It may well have happened that oats became more important in their own right as conditions for the successful production of the other two cereals became more marginal. The cultivated hexaploids *Avena sativa* and *A. byzantina* may have originated in this way, possibly from the weed species *A. sterilis* and *A. fatua*. In addition to these oats, diploid forms of *A. strigosa* appear to have spread northwards, possibly from the Iberian peninsula, to north-western Europe where they were found suitable for cultivation in cool, maritime climates. These so-called bristle-pointed oats are still used on calcareous soils in the Hebrides and Shetlands, although susceptibility to smut and the necessity to broadcast the seed have prevented their spread. An improved cultivar has been bred at the Welsh Plant Breeding Station, Aberystwyth, and also hybrids with *A. brevis*, the short oat. Another plant of interest is Pilcorn, a naked oat, which because of its importance in mediaeval British agriculture features in the *Herball* of John Gerard. Possibly because the embryo was vulnerable to damage and seed germination was poor, this oat disappeared again from Europe and its cultivation is now restricted to south-western Asia. It was the large grained *A. sativa* and *A. byzantina* hexaploids which became the main cultivated oats in central and northern Europe, from where they spread during colonisation to North America, South America and Australasia. Of the weed species, *A. fatua* has retained its important negative role as a widespread weed of cereal crops, as has *A. ludoviciana*, a variety or subspecies of *A. sterilis*.

The oat plant

Avena sativa, the common cultivated oat plant, differs from wheat, barley and rye by having no auricles and by the leaves being rolled in an anti-clockwise direction. The ligule is large and membranous. In distinction to wild oats which are commonly hairy, the cultivated oat plant is glabrous. Given favourable conditions and a low seeding rate, oats tiller more freely than other cereals and are thus frequently used as a winter greenfeed crop for grazing. The inflorescence is a much branched panicle bearing numerous spikelets (Fig. 4.15). Each of these is subtended by a pair of papery glumes and usually contains three florets. The two lower florets form a grain each, of which the larger comes from the basal position, but the uppermost floret either aborts or is represented by only a small lemma and palea. The grains consist of an elongated, hairy caryopsis, tightly surrounded by a tough lemma and palea. Although few modern cultivars have this character, cultivated and wild oats have a kneed awn inserted on the back of the lemma. The lower part of the awn is spirally coiled and hygroscopic, and as it picks up moisture it twists in one direction, only to uncoil in the other direction as it dries out. Possession of this mechanism bestows advantages in seed dispersal and burial on the wild species, but cultivated oats either lack an awn or lose it during threshing. On germination, the seedling usually forms only three seminal roots, the mesocotyl elongates almost immediately, and the coleoptile grows up inside the lemma and emerges at its tip. The oat coleoptile is well known to plant physiologists for the part it played in the discovery of growth hormones.

Uses and cultivation

Because of their relatively high protein and fat content, oats are valued for human and animal nutrition and, were it not for their lower yield, they would undoubtedly be grown more widely. Oats used to be and still are highly rated as a feed for horses, and the straw is also well regarded as a roughage for animal feeding. Whether this reputation is strong enough to

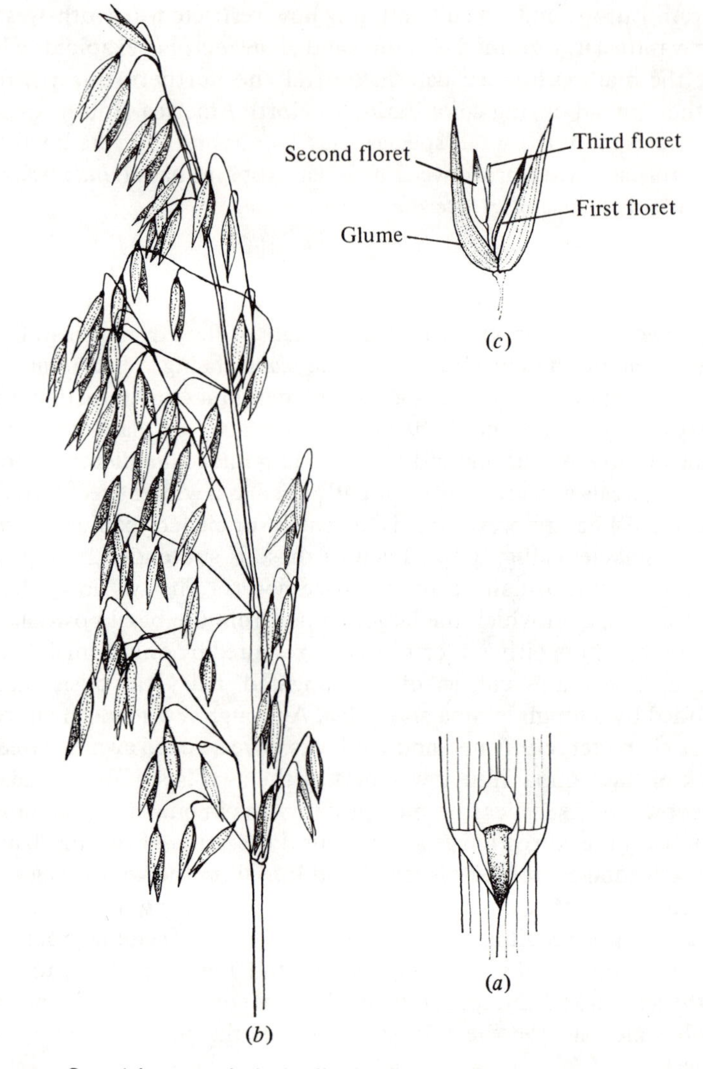

Fig. 4.15. Oats (*Avena sativa*). (*a*) ligule, (*b*) panicle, (*c*) spikelet.

justify harvesting by binder, maturing the sheaves in stacks, and threshing during the winter is somewhat questionable. We may also question the validity of Dr Johnson's judgement of oats as a grain 'which in England is generally given to horses, but in Scotland supports the people'. Although a high proportion of the oat crop is still fed to animals, it continues to hold its position in human nutrition on account of its nutritive value, and in fact is gaining in popularity on account of the reputed beneficial effects of oat bran on people with high blood pressure. Rolled oats and oatmeal are commonly employed in the food industry and in home cooking, and the use of porridge oats as a breakfast cereal in Scotland and elsewhere is strongly entrenched. For these purposes the caryopsis has first to be separated by milling. The residual husks can be distilled with acid and steam to give the aldehyde furfural which, among other uses, serves as a paint and lacquer remover.

Presence of the husks with its high fibre content reduces the feeding value of oats as compared with other cereals. Plant breeders are trying to produce naked cultivars, but so far these selections have not yielded well, naked caryopses are not formed invariably, and grain shattering is a problem. However, there has been progress in this direction and also in attempts to raise the protein content of the grain. Another character which it is hoped to improve is resistance to lodging, but so far the presence of a system of dwarfing genes as in wheat has not been established.

The leading countries producing oats are the USA, Canada, USSR, Germany and France, with appreciable areas also being grown in Scotland, Wales and Scandinavia. Although the crop is adaptable to a range of conditions, it prefers a cool and moist climate, especially during the summer. Despite this preference, oats are not as frost-hardy as some cultivars of wheat and barley. As long as moisture supply is not restricted, soil conditions are not highly critical, and the crop does not make heavy demands on fertilizer except nitrogen. On the other hand, growth is retarded by manganese deficiency, indicated by the symptoms of the so-called grey speck disease. Because of this, heavy liming should be avoided to keep the pH below 6.5, and it may be necessary to apply a dressing of manganese sulphate or to spray the foliage. Apart from being cultivated as a grain crop, oats are also commonly used as a winter greenfeed. Oats straw and chaff are fed to animals as a roughage and used to be highly regarded for this purpose in the days of the horse.

Wild oats

The common wild oat, *Avena fatua*, one of the worst weeds of cereal crops, is difficult to eradicate because of its short life cycle. It germinates mainly in the spring, and in late summer it matures well before the host plant, whereupon the spikelets separate and fall to the ground. The grain which is densely covered by dark-brown hairs, bears a twisted awn which helps to bury the seed (Fig. 4.16). However, immediate germination is prevented through dormancy controlled by inhibitors present in the lemma, palea and caryopsis. Although wild oats may remain viable in the soil for up to

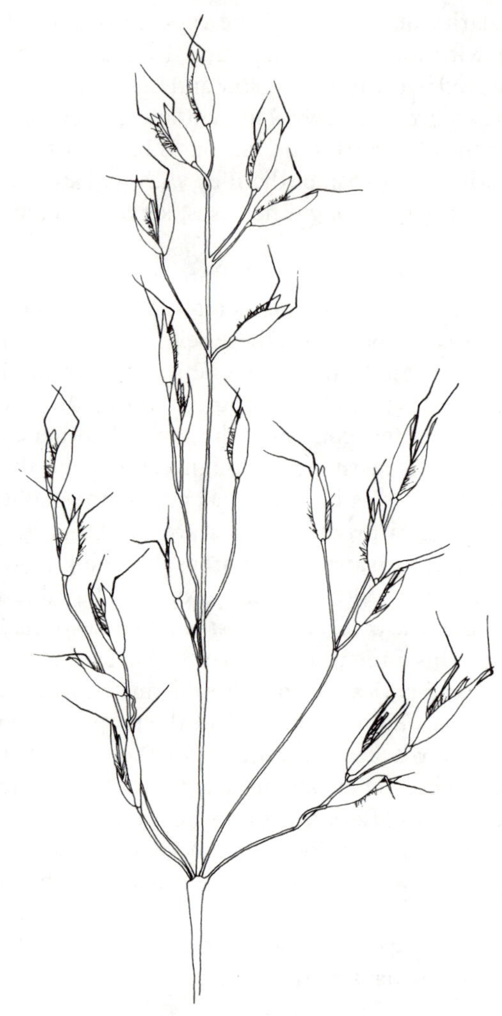

Fig. 4.16. Wild oats (*Avena fatua*) panicle. Note hairy lemma with kneed awn.

nine years, the majority of seeds do not survive for more than two or three years. Only some 10% are still viable after passage through an animal. The other widespread weed species is the so-called winter wild oat, *A. ludoviciana*, which germinates mainly in the winter. The florets of this species do not separate as easily as those of *A. fatua*. Control of wild oats is difficult. Several methods or combination of methods may have to be employed, including cultivation, lengthening the rotation, rogueing and selective herbicides such as carbamates, thiocarbamates and other chemicals.

4.5 RYE (*SECALE CEREALE*)

Rye is an annual crop which is grown widely in areas with cold winters and warm, dry summers, where its yield can be superior to that of other cereals. It is also capable of performing satisfactorily on acid soils naturally low in fertility. Although the importance of rye as a major bread cereal has declined in recent times, it is still grown on a big scale in Eastern Europe and in some regions of the USSR but there is also a small but persistent demand for it in many other countries in the cool-temperate zone.

Origin of rye

There is good evidence to suggest that rye first came to the attention of man as a weed in crops of wheat and barley. Domestication occurred around 3000 BC, considerably later than the date when the other two cereals began to be cultivated. The weedy races of rye, probably belonging to an aggregate perennial species *Secale montanum*, appear to have originated in eastern Turkey, north-western Iran and Armenia. The climate in these areas is sufficiently cold to give substance to the notion that crops of wheat and barley could easily have been contaminated by rye which was better adapted to these conditions. It appears that *S. cereale*, another aggregate species with many different forms, evolved from *S. montanum* either directly or through some intermediate species, and that eventually annual types with high grain-bearing capacity were selected for domestication.

The rye plant

Rye tends to tiller vigorously. The leaves are rolled within each tiller, and the young plant resembles wheat, except that the auricles are small, narrow and without hairs, and that they wither after a time. The culms are long, slender and wiry. The inflorescence is a spike bearing a single spikelet at each node (Fig. 4.17). Each spikelet is provided with a pair of rather narrow, pointed glumes and usually it contains three florets of which the third is sterile and does not develop. The fertile florets have a keeled lemma terminating in a strong awn and a smaller, somewhat boat-shaped palea. In distinction to wheat and barley, the stigma of rye remains receptive for some time after being exerted, and thus cross-pollination occurs. As the caryopsis swells, it forces the lemma and palea apart and thus becomes visible from the outside. When mature, it resembles a grain of wheat except

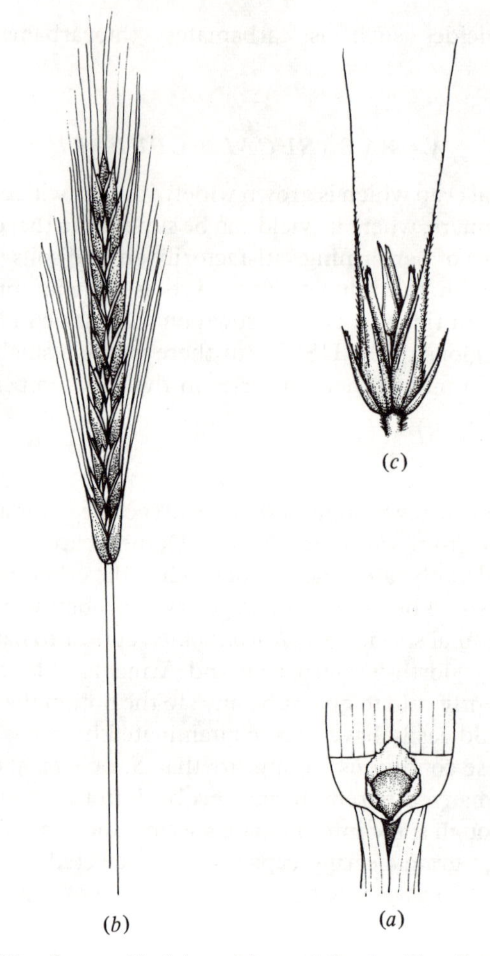

Fig. 4.17. Rye (*Secale cereale*). (*a*) auricles and ligule, (*b*) spike, (*c*) spikelet.

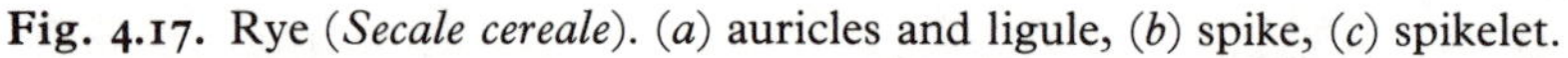

for being more slender and having a distinct dorsal ridge. Cultivated rye is a diploid with a chromosome number of fourteen. Depending on cultivar it may be sown in the autumn or spring.

Cultivation and uses

Because it tillers freely and can tolerate low temperature, rye is often grown as a winter forage crop. It establishes quickly and recovers well from grazing, even if temperatures are low. More commonly, however, it is cultivated for its grain and, before living standards were as high as they are now, dark rye bread was the staple diet of millions of people in central and eastern Europe. Compared with other cereals rye is greatly more adaptable to a wide range of cool temperate conditions, as shown by its distribution which includes northern Scandinavia, southern Chile, and extends from sea level to altitudes of over 4000 m in the Himalayas. It is drought resistant and tolerates soil acidity, although it is also capable of responding to favourable cultural conditions. Rye is also the most frost hardy of the cereals, which makes it suitable for cultivation in northern Canada or in eastern Europe. It is normally sown in the autumn, and in fact cultivars planted at that time require exposure to low temperature as a prerequisite to flowering before they can respond to long days in the spring. Petkus winter rye is well known to physiologists as the plant used in the 1930s at Imperial College, London to study the details of the process of vernalisation. As a practical corollary of these requirements it follows that special spring cultivars which do not need cold treatment must be used if sowing is delayed beyond the winter. Another physiological characteristic of rye is that photosynthesis in its ears contributes only between 15 and 30% to final grain weight compared with up to 45% in barley. Both cereals have awns that contain chlorophyll and photosynthesise when they are young, but it could be that the stomata of rye close more readily in response to water stress.

Rye grain is used in a variety of different ways. As an animal feed it is often mixed with other cereals, but by itself it tends to form a sticky mass in the mouth of livestock. For human consumption it is used predominantly as a bread cereal, although it lacks gluten, the proteins which give the dough of wheat its elasticity. Instead other substances, the pentosans, bind the water during mixing to produce a dough suitable for baking. Dark rye bread is consumed widely in eastern European and Scandinavian countries, although more often than not it is mixed with wheat flour in varying proportions. Several specialty products are also baked, such as Knaeckebrot, a crisp bread used in Sweden, or Pumpernickel in Germany, a very dark bread baked slowly for 18 to 36 hours. Rye flour has high nutritional value, largely because the protein is rich in the amino acid, lysine. The bread also contains good amounts of mineral and fibre, but the calorie content is lower than that of wheat, and for these and related reasons it has gained renewed and increasing popularity in the western world. Rye is also an important cereal for the distillation of alcohol. Special types of alcoholic drinks, such as Canadian whisky or Bourbon are based largely on rye. Another industrial use is for the manufacture of adhesives, because the flour has strong water binding properties. Rye straw was at one time greatly valued as material for thatching or as bedding for stall-fed animals. Its main use now is in the production of strawboards and paper.

4.6 TRITICALE

Triticale is a man-made name for a man-made plant. It refers to a cross between wheat and rye representing the genus *Triticum* and the genus *Secale* (Figs. 4.18 and 4.19). Intergeneric hybrids are very rare in nature, especially if chromosome number and size differ, and the progeny are almost certainly sterile. Isolated cases of spontaneous hybridisation between these two genera were first reported towards the end of last century but, because the plants produced were at best only partially fertile, they were considered to be academic curiosities rather than of any practical value. It was not until the advent of colchicine in 1937 which made chromosome doubling possible, that production of fertile wheat–rye hybrids became practical propositions. Considerable interest in the possible advantages of the cross culminated in 1964 in the adoption of a triticale programe by CIMMYT, the international maize and wheat research centre in Mexico. The aims are to enrich the high protein content of wheat with the lysine of rye, and to combine the high yields of one parent with the resilience of the other.

The first problem to be overcome was infertility caused by the difference in chromosome numbers. Two types of crosses have so far been employed. If one of the parents is hexaploid wheat (see section 4.2) and the other diploid rye, the resulting hybrid is treated with colchicine to become an octoploid. Alternatively, tetraploid durum wheat is used together with diploid rye, and this results in a hexaploid hybrid after colchicine treatment. Hexaploids are also produced by crossing tetraploid triticales with octoploid types.

In either case the hybrid can be backcrossed with wheat, rye or another triticale. A number of cultivars have been bred and, although considerable areas in over 50 countries are claimed to have been sown with triticale, some problems have still to be overcome. Despite the effects of colchicine, chromosome pairing at meiosis is often imperfect, largely because the rye chromosomes are larger and carry about half as much DNA again as those of wheat. As regards agronomic performance early selections were prone to lodge, since the long-day plants tested in Mexico grew very tall and were unable to support a heavy ear. Introduction of dwarfing genes has removed this problem but at the same time caused greater shrivelling of grains, attended by poor germination. Shrivelling could be associated with prolonged α-amylase activity and hence breakdown of starch in the endosperm. On the other hand, triticale grain can have a protein content approaching that of wheat but at the same time a higher biological value owing to the high lysine content inherited from the rye parent. Some advanced lines have been shown to be superior to high lysine maize and to

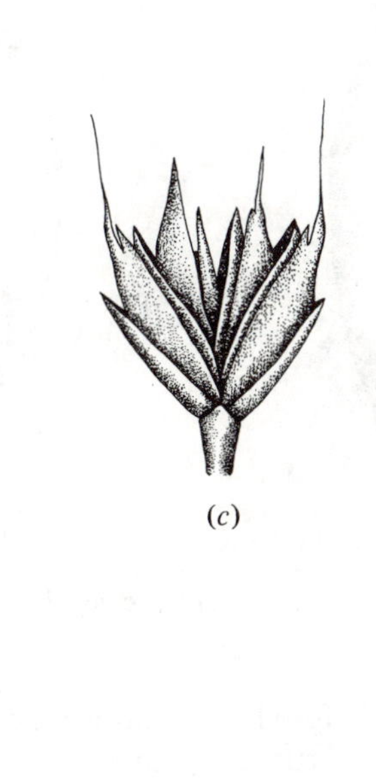

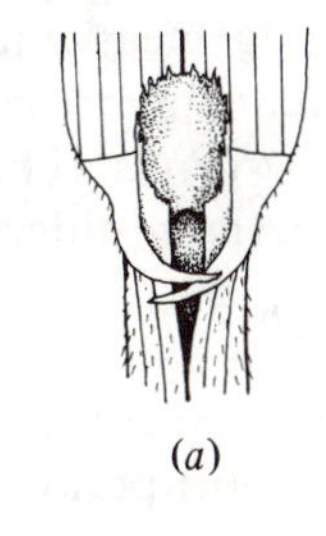

(b)

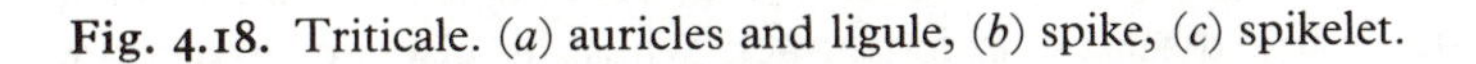

Fig. 4.18. Triticale. (*a*) auricles and ligule, (*b*) spike, (*c*) spikelet.

contain more protein. However, only a few lines of triticale have so far produced flour suitable for bread, because in general not enough gluten is present to enable the dough to rise. Mechanical dough development is helping to minimise this deficiency, but flour extraction rates still need to be improved.

Despite some of these problems, plant breeders are hopeful that triticale

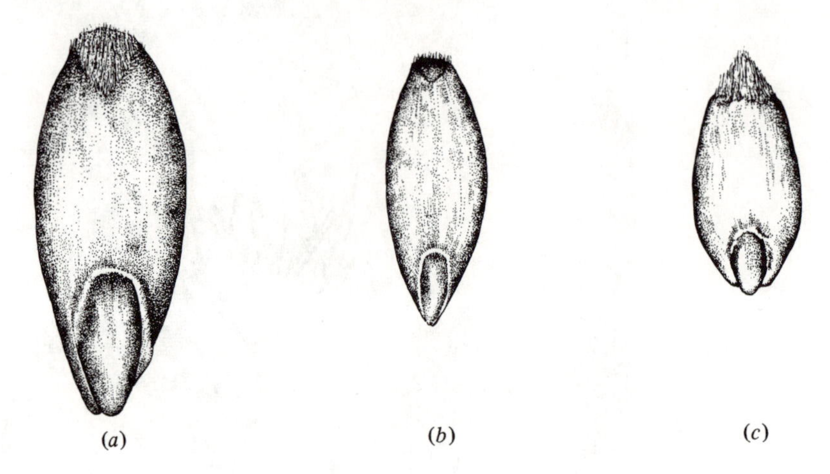

Fig. 4.19. Caryopsis of (*a*) triticale, (*b*) rye, (*c*) wheat.

cultivars will before long establish themselves as cereals superior to rye in yield and more rugged than wheat. It should be best adapted to cool climates, sandy soils of low natural fertility, and to high elevations. If experimental yields of 8.5 t ha^{-1} are repeatedly approached on a practical farm scale, and if acceptable amounts of protein of high biological value are obtained, this new cereal may well have an assured future. Perhaps we should suspend judgement until more research has been done. In the meantime let us remember that triticale is not a single species but the equivalent of a new genus. This means that there is bound to be great variability and hence large differences among the many lines that are under test will continue to exist for some time.

By 1985 more than 750 000 ha of triticale were grown throughout the world and in Poland alone the area was expected to reach one million ha within a relatively short period. This level of acceptance is based on the production of cultivars with reduced straw-length and well-filled grains, capable of growing on soils which are not suitable for wheat. A high level of resistance to many diseases which attack wheat is also an attraction, although this advantage may be short lived, as pathogens adapt physiologically to triticale as a host plant.

Festucoid pasture grasses

Most of the important temperate pasture grasses belong to the tribe Festuceae of the Festucoideae. This tribe contains both annual and perennial species. The inflorescence of these grasses is typically a panicle, although this may be reduced to a spike-like panicle and occasionally to a structure which is indistinguishable from a spike and is referred to as such. The spikelets typically contain several florets, and the plants concerned are either glabrous or only slightly hairy.

4.7 *AGROPYRON*

Alphabetically the first genus of the festucoid grasses used to contain a rather troublesome weed, couch grass (*Agropyron repens*), whose botanical name has, however, been changed recently to *Elytrigia (Elymus) repens*. This is a creeping perennial plant which persists for prolonged periods thanks to its extensive system of rhizomes which permeate the soil, forming adventitious roots and new tillers at their nodes (Fig. 4.20). The leaves are rolled in the bud, rather dull green in colour, usually with a variable cover of hairs. The auricle is fairly conspicuous but the ligule is short and rim-like. The inflorescence is a spike bearing numerous spikelets, each subtended by a pair of glumes. In distinction to ryegrass (Fig. 4.33) we note the presence of these two glumes but also the fact that the spikelets are facing the rachis with their broad side, and not edgewise, with the glumes and florets spreading out like the fingers of a hand. There are four to six florets per spikelet, each surrounded by a palea and a lemma capable of bearing a short awn. The seed, about 7 mm in length is the complete floret with a conspicuous rachilla, usually lying on its side because the lemma has a distinct keel. The seeds shatter easily, and many mature couch inflorescences have spikelets devoid of florets but with the glumes still attached. The economic significance of this grass is a negative one, because it is one of the worst weeds of gardens and arable land. It spreads by its seed, and indeed samples of crop seed must not be contaminated by it, but far more significant is the rhizome as a mechanism for spread and survival. Each segment, provided it has a node, is capable of generating a new plant which in its turn produces its own network of rhizomes. Cultural control is possible by repeated cultivation aimed at bringing the rhizomes close enough to the surface to let them dry out, but obviously success depends on a sufficiently long dry spell for this to be accomplished. Alternatively, translocated herbicides should be used, provided that there is a large enough leaf system to ensure translocation throughout the rhizomes. However, the fact that couch continues to be common suggests that control is not easy. A recent survey in the United Kingdom showed it to be present on between 68 and 92% of farms. *Agropyron* is closely related to *Triticum*, the genus to which wheat belongs, and there has been some interest in the performance of intergeneric hybrids which are capable of being produced. Perennial wheat is one of the

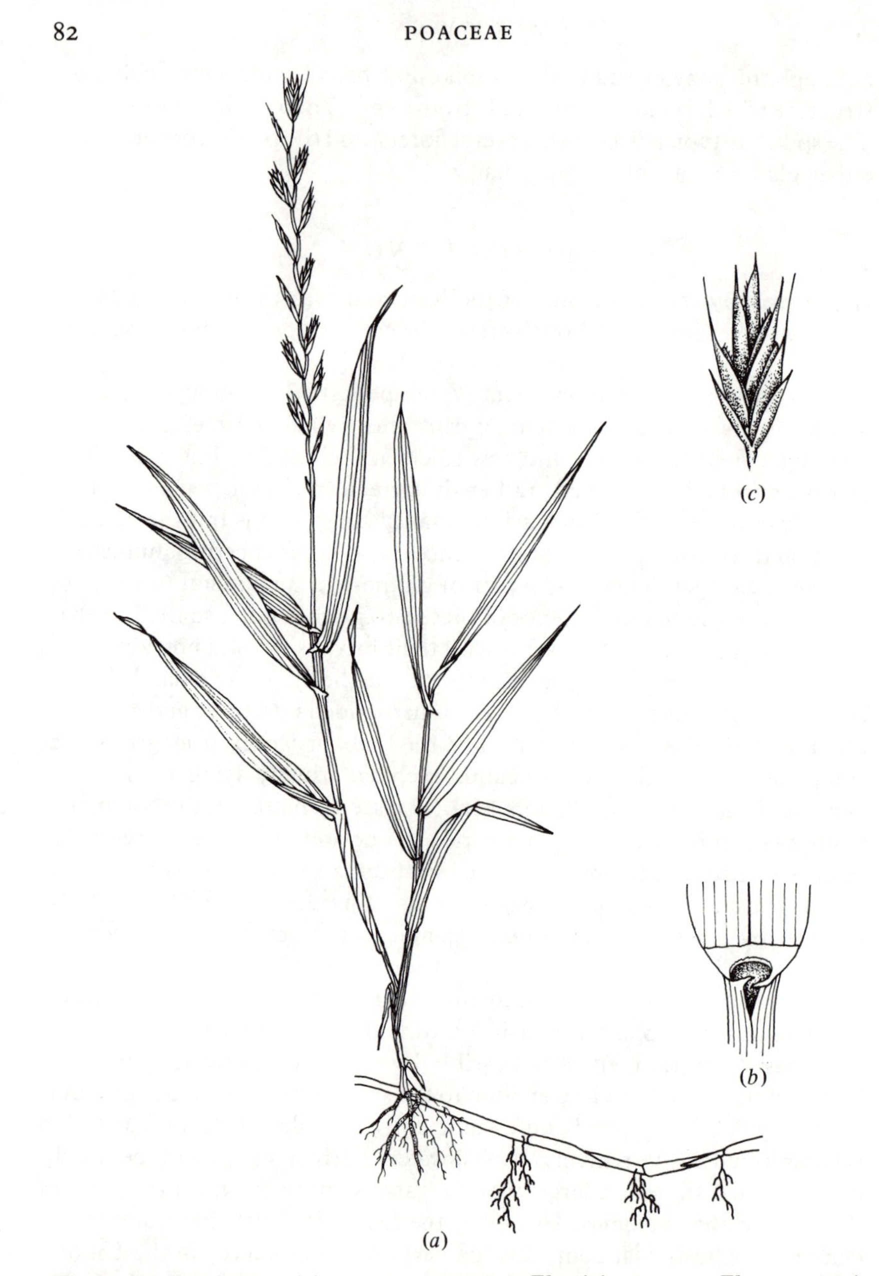

Fig. 4.20. Couch grass (*Agropyron repens* now *Elytrigia repens* or *Elymus repens*). (*a*) rhizomatous plant with spike, (*b*) auricles and ligule, (*c*) spikelet.

attractions of such a cross. The genus *Agropyron* is also represented by some wild species, for example *A. caninum*, bearded wheat grass, which has slender spikes with conspicuously awned lemmas and occurs in British woodlands.

The situation is very different in North America, for there some 120 million hectares of native grasslands are dominated by wheatgrasses that belong to this genus. Considerable areas have also been sown with improved species. It has been stated that no other group of grasses exceeds them in the western USA for the production of early, highly nutritious herbage, and that they are excellent for wind and water erosion control. On the central and northern Great Plains, the western wheat grass (*Agropyron smithii*) is valued as an early, nutritious grass which is also tolerant of alkaline soil conditions. Several improved cultivars are available. In intermontane areas it often occurs together with bluebunch wheatgrass (*Agropyron spicatum*), a long-lived, drought-resistant perennial. It is very palatable and nutritious but requires to be protected from too early or too persistent grazing. Extensively used in the Pacific North-west is slender wheatgrass (*Agropyron trachycaulum*), which has given rise to cultivars with high tolerance of saline or alkaline conditions, good forage quality and high yield. Crested wheatgrass (*Agropyron cristatum*), a native of eastern Russia, western Siberia, and central Asia has found wide use in North America for the regrassing of abandoned cropping areas, and low-producing rangeland. It grows well in low rainfall conditions, is drought resistant, and its persistence is enhanced by re-establishment from its own seed. This and the other wheatgrasses are cross-pollinated and thus greatly variable morphologically and agronomically.

4.8 *AGROSTIS*

Members of this genus are generally of low agronomic value, although they are widely distributed and at least one species is economically not unimportant. They are all very variable, and differences between species are not always very distinct. Browntop, common or fine bent (*Agrostis tenuis*) is generally a tufted plant with short rhizomes or stolons, although in moist and fertile conditions it is capable of spreading readily by means of these creeping stems (Fig. 4.21). The leaves are narrow and sharply pointed, typically forming an elongated triangle, with distinct, evenly developed ribs on the upper surface. There are no auricles, and the ligule is short and blunt. On flowering a highly branched panicle is produced, very slender and quite delicate, brownish in colour, and bearing very many small spikelets only about 3 mm long. Each spikelet is subtended by two

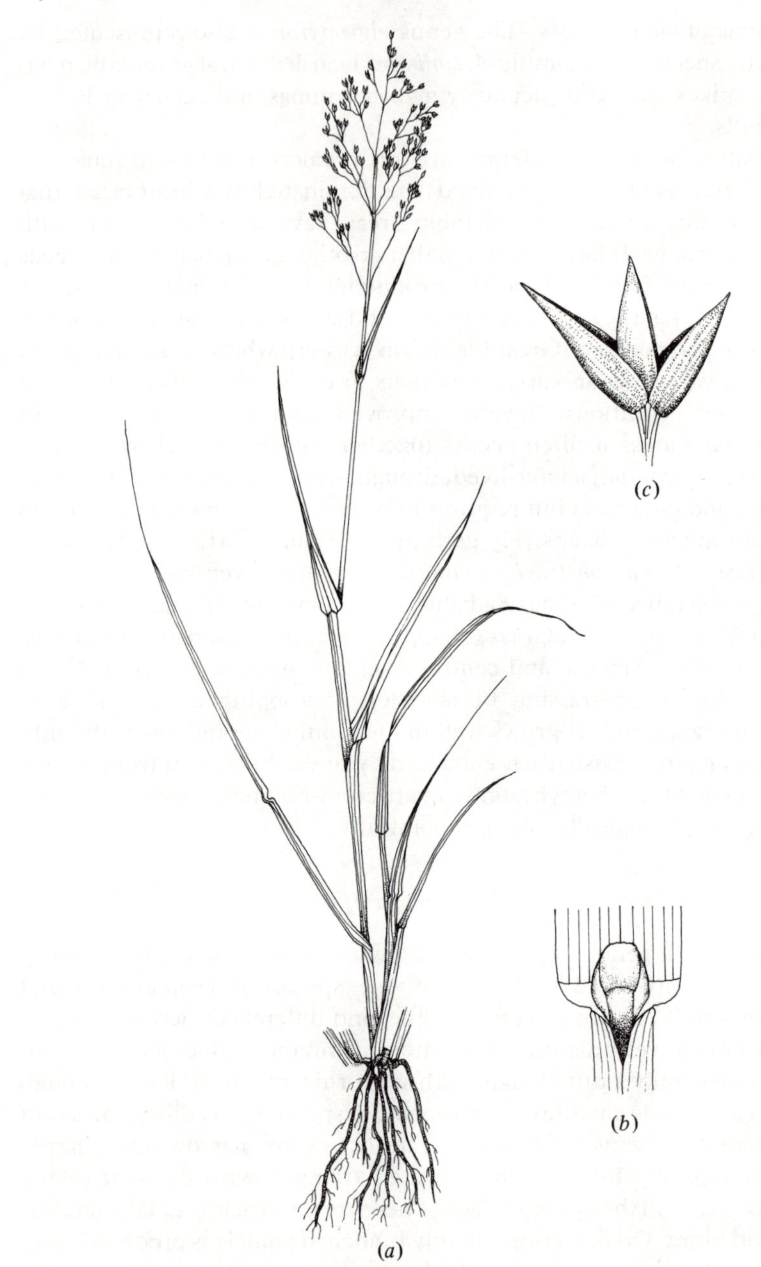

Fig. 4.21. Browntop or common bent (*Agrostis tenuis*). (*a*) plant with panicle, (*b*) ligule, (*c*) spikelet with single floret.

glumes and contains a single floret surrounded by a palea and an awnless lemma. Because there is only one floret per spikelet, the seed bears no rachilla. It is also very small with a 1000 seed weight of only about 0.1 g. Browntop is quite common as a constituent of poor grassland in low fertility situations, especially in hill and mountain areas. Its productivity is low, because it is late starting growth in the spring and little further leaf formation occurs after emergence of the inflorescence unless moisture and nutrient supply is good. Where soil conditions are so impoverished as to sustain only such grasses as *Molinia* and *Nardus* (see sections 4.22 and 4.23), as in several hill country areas of the British Isles, browntop or common bent does represent a preferable alternative, both on account of superior production and higher palatability. Similarly, it has some uses as an initial grass species following clearance of scrub or forest. However, wherever conditions are good enough to sustain growth of ryegrass and clover, there is no real justification for the widespread presence of *Agrostis*, and bad management would be suspected to have been responsible for its ingress. On the other hand, the species stands up extremely well to treading, and for this reason, and for its great persistence, it is held in high regard as a component of lawns and greens. In these situations repeated mowing removes reproductive stem apices before flowering can occur and thus continued leaf production is ensured. Furthermore, ability to form rhizomes and stolons is a decided asset in a species required to form a dense lawn. In this respect *Agrostis* combines well with the other main constituent of lawns in temperate regions, *Festuca rubra* (see section 4.18). On the other hand, the species is less welcome in an arable situation where it can be a weed which is not easily controlled. It follows that any area destined to become a ley or pasture needs to be thoroughly cleaned before highly producing grasses and legumes are sown.

A closely related species is creeping bent or fiorin (*Agrostis stolonifera*), which is also common in poor grassland. It can be distinguished by having a distinct, dome-shaped ligule, a compact panicle, and a slightly less pointed lemma. Black bent or redtop (*A. gigantea*) is an arable weed in the main, although in parts of the USA it is sown on wet and acid soils to which it is adapted. It is also a rhizomatous plant with an open panicle, slightly red in appearance, and a small, awnless seed. The leaves are much broader than in other members of the genus and the ligule is long and blunt. There are, however, two species of *Agrostis* which have awns: velvet bent (*A. canina* sub. sp. *canina*), and even more so bristle bent (*A. setacea*) which occurs on the acid moors and heaths of south-west England. Velvet bent may be used in lawns but bristle bent has no agronomic value.

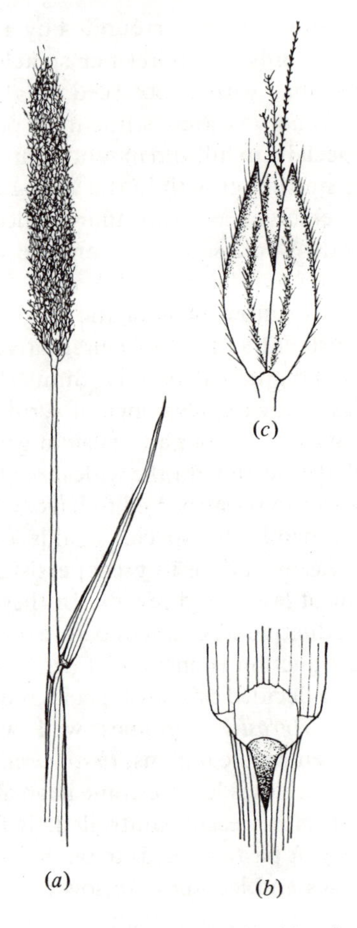

Fig. 4.22. Meadow foxtail (*Alopecurus pratensis*). (*a*) spike-like panicle, (*b*) ligule, (*c*) spikelet containing single floret with awn and stigmas showing.

4.9 *ALOPECURUS*

The most common member of this genus is meadow foxtail (*Alopecurus pratensis*), an early flowering species adapted to moist conditions and reasonably high fertility. This is a tufted perennial grass, capable of some spread through short rhizomes, entirely glabrous, with rounded tillers (Fig. 4.22). The leaves are dark green, with distinct ribs on the upper surface and of rather a dull appearance on the lower. There are no auricles, and the ligule is short and blunt. The base of the leaf sheath becomes brown with age. One of the distinctive features of this species is the inflorescence, one of the earliest to appear, which is a cylindrical spike-like panicle, presumably resembling the tail of a fox. On it are borne numerous spikelets on short stalks, each consisting of a keeled, hairy pair of glumes which surround a single floret. The lemma bears a long, curved, almost basal awn which extends well beyond the glumes. The whole floret drops to the ground when mature and becomes the seed. Although some 5 mm in length, it weighs very little and this, together with the hairiness of the glumes, makes seed cleaning very difficult, quite apart from the presence of many florets without a filled caryopsis. Satisfactory establishment is therefore difficult to obtain, and the resulting seedlings are slow to grow. Meadow foxtail is thus not often used in a temporary pasture or ley, although it is of some importance in parts of the USA, and a leafy cultivar was bred many years ago by the Welsh Plant Breeding Station, Aberystwyth. The importance of this species lies more in its ability to provide early spring growth in permanent grasslands under moist and fertile conditions, provided grazing pressure is not severe. It is perhaps better suited to contributing to swards destined to be cut for hay or silage, and it is also often seen growing in ditches and near hedgerows where moisture supply is adequate and the climate reasonably wet and cool. Even more demanding of wet conditions is a related species, floating foxtail (*Alopecurus geniculatus*) which derives its botanical name from the fact that the flowering culm is distinctly bent or kneed at the nodes. Of greater significance is slender foxtail or black grass (*Alopecurus myosuroides*), a locally abundant weed of heavy, arable land. Its success as a weed stems from its early flowering and seed production, well ahead of any cereal. Control is thus difficult, and in the autumn when the seeds germinate the soil may be too wet for any cultivation. Fortunately the seed does not persist for very long in the soil, but often long enough for slender foxtail to contaminate seed crops of improved grass species. A survey showed it to be present on 50% of the farms in England.

4.10 *AMMOPHILA*

Marram grass (*Ammophila arenaria*) has no agronomic value but serves an exceedingly useful purpose as a sand binding plant. It is a rhizomatous perennial with a strong, fibrous root system. The leaves are coarse and narrow, deeply ribbed on the upper surface and bearing hairs, and usually rolled inwardly as a device to prevent undue water loss. There are no auricles but the ligule is prominent and twin-pointed. The inflorescence is a compact panicle bearing one-flowered spikelets. The complete floret constitutes the seed but in practice marram grass is usually propagated by planting rooted tillers. It is so well adapted to dry coastal situations, its

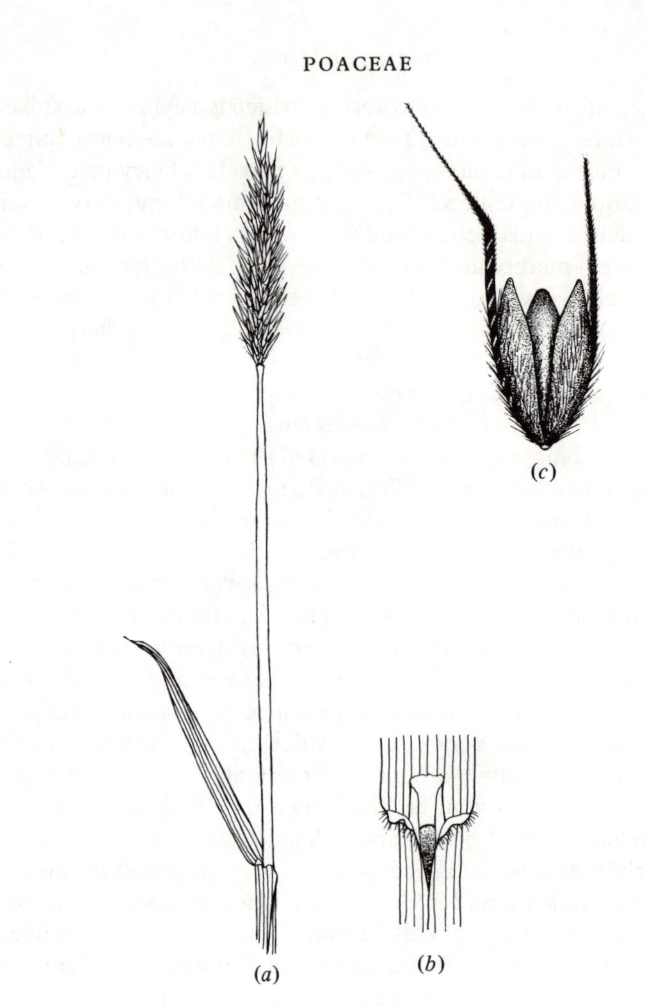

Fig. 4.23. Sweet vernal grass (*Anthoxanthum odoratum*). (*a*) spike-like panicle, (*b*) ligule and fringe of hairs in place of auricles, (*c*) spikelet with inner glumes and floret.

natural habitat, that it is extensively used for consolidation of sand dunes and other dry, unstable soils.

4.11 *ANTHOXANTHUM*

Sweet vernal grass (*Anthoxanthum odoratum*) derives its name from its characteristic sweet smell, usually associated with that of newly mown hay. The chemical substance responsible for this is the alkaloid coumarin which also occurs in the sweet clovers (genus *Melilotus*, see section 11.5). Although pleasant to the nose it imparts a bitter taste to the herbage and thus reduces its palatability. Sweet vernal grass is a tufted perennial with rounded tillers (Fig. 4.23). The leaves are dark-green, fairly hairy, with a distinct ligule but with the auricles replaced by a fringe of hairs along the basal leaf margin. The inflorescence is a somewhat lax, cylindrical spike-like panicle. The spikelets are subtended by a pair of thin, papery glumes which become yellow with age and are retained on the inflorescence to give it its characteristic appearance. Inside these two glumes is another pair of glumes (referred to by some authorities as empty lemmas representing a sterile floret), dark brown in colour and densely covered with hairs. The lower inner glume has a short dorsal awn, and the outer one a long basal awn which is twisted and bends distinctly about two-thirds along its length. Within these twin structures there occurs a single fertile floret, surrounded by a smooth, brownish lemma and palea. There are only two stamens, and the commercial seed is the complete floret surrounded by the inner set of glumes (or empty lemmas). Sweet vernal grass has a wide range of distribution throughout Europe, parts of Asia and the USA, New Zealand, and other countries. It is particularly well adapted to dry and low fertility soil conditions. However, despite its common occurrence it contributes little to herbage production and is not readily eaten by stock. Abundance of this species in areas capable of sustaining more productive grasses should be taken as a sign of pasture deterioration.

4.12 *ARRHENATHERUM*

Tall oat grass (*Arrhenatherum elatius*) bears, as the name implies, some resemblance to oats and was at one time classified as a member of that genus. It is a tall, somewhat coarse plant with rolled tillers bearing dark green, slightly hairy leaves. There is no auricle but the ligule is distinct and blunt. The panicle bears numerous spikelets, each containing two florets. The lower floret is staminate only and is subtended by a lemma bearing a twisted and bent awn which arises just above its base. The upper floret is complete and fertile, and its lemma carries a short, sub-terminal awn or no awn at all. The commercial seed consists of the whole spikelet without the glumes but including both the male and the fertile florets. Tall oat grass is not very palatable, and furthermore because of its uprightness does not stand up well to grazing. On the other hand, its ability to produce early spring growth and tolerance of dry conditions have given it a certain reputation as a hay crop in parts of Europe and the United States, but in Britain and elsewhere it is more common in hedgerows and waste areas than in sown grassland.

A closely related plant, *Arrhenatherum elatius*, var. *bulbosum*, is variously referred to as onion couch or onion-rooted twitch. These descriptive names refer to the lower internodes of the stems which are swollen to form

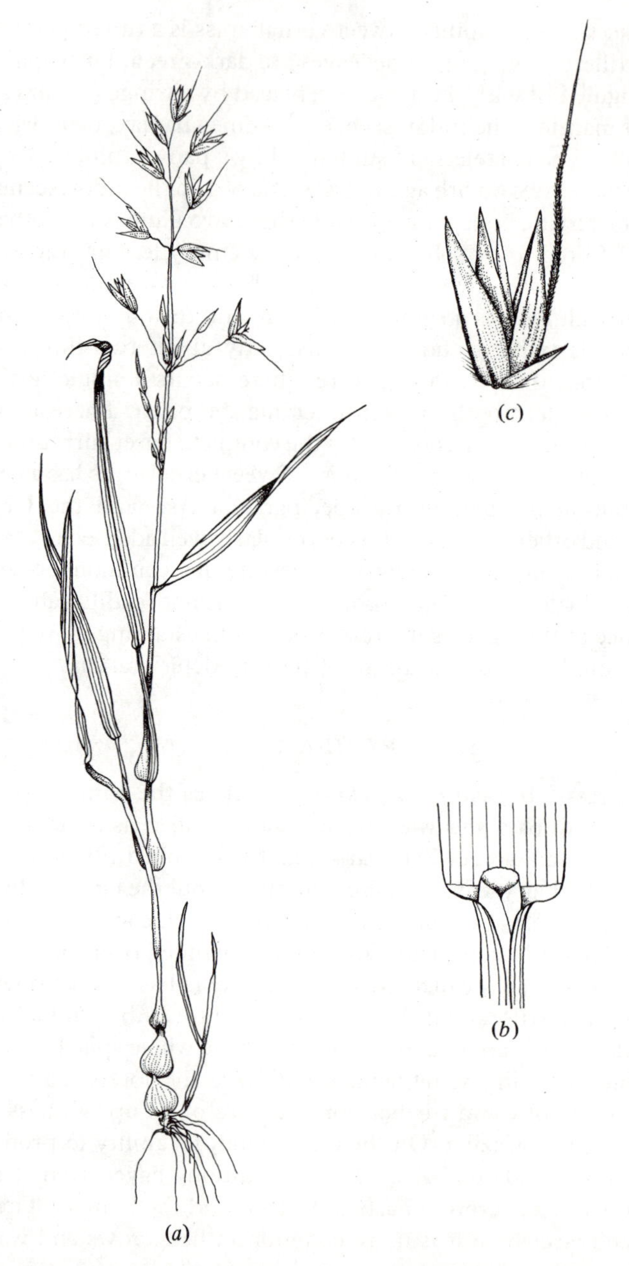

Fig. 4.24. Onion couch (*Arrhenatherum elatius* var. *bulbosum*). (*a*) plant with corms at base of culm and with panicle, (*b*) ligule, (*c*) spikelet with lower sterile floret bearing an awn and upper fertile floret.

a collection of rounded corms, not unlike a string of onions (Fig. 4.24). Each of these, when broken off and isolated is quite capable of producing a new plant, and this explains why this species is a serious weed of arable land and not easily controlled by mechanical means.

4.13 *BRIZA*

This genus deserves brief mention, not on account of its agronomic value but because its species are widely distributed and some are grown in gardens as ornamentals. Common quaking grass (*Briza media*) produces characteristic spreading panicles with slender, drooping branches. Suspended on them are the spikelets which hang down like small lanterns and wave or quake in the wind. Each spikelet contains about eight florets, and both the glumes and the lemmas are broad and without awns. Lesser quaking grass (*Briza minor*) is a similar but slightly smaller species, while *Briza maxima* is very much larger.

4.14 *BROMUS*

The genus *Bromus* contains several species of agricultural significance. Perhaps the most widely cultivated is smooth brome grass (*Bromus inermis*) which is used very widely in North America. This is a tufted perennial spreading vigorously by rhizomes (Fig. 4.25). The leaves are only slightly ribbed, almost without hairs. There is no auricle, but the ligule is distinct. The panicle is spreading and bears many spikelets, each containing eight to ten florets. In distinction to other members of the genus, the lemma has no awn or only a very short terminal one. Smooth brome-grass is particularly well represented in the corn belt of the USA and extends from there into Canada. Although it becomes dormant during periods of drought, it survives extreme temperatures and does best in cool, moist conditions. It is grown alone or with lucerne (alfalfa) or red clover and forms an excellent sward which should be grazed rotationally, cut for hay, or used in both ways combined.

One of the more controversial species in this genus is prairie grass or rescue grass (*B. wildenovii*). This applies not only to its taxonomic position, for it has at various times been known as *B. unioloides*, *B. catharticus* and even *Ceratochloa unioloides*, but also to its value and role as a pasture grass. It appears to have originated in South America but has since spread widely to parts of Europe, the United States, Australia and New Zealand. It is a tall, tufted perennial grass composed of flat or folded tillers (Fig. 4.26), rather like those of cocksfoot (see section 4.16). However, it can be distinguished from this species even before the

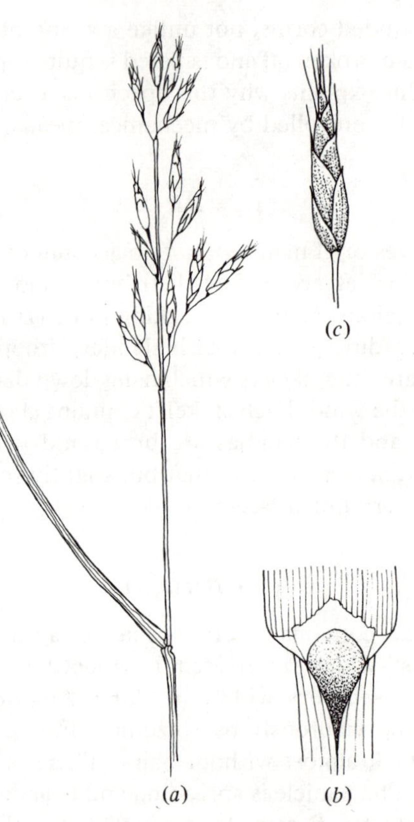

Fig. 4.25. Smooth brome grass (*Bromus inermis*). (*a*) panicle, (*b*) ligule, (*c*) spikelet.

inflorescence appears by the presence of a variable cover of fine hairs on the leaf sheath, and also by its light green colour. The ligule is conspicuous, but the auricles are absent. The inflorescence is an open panicle bearing flat spikelets, each containing six to twelve florets. There are two pointed glumes, the lemma may have a short terminal awn, and the commercial seed is the complete floret with a distinct rachilla. Flowering occurs over a prolonged period of time, the seeds shatter easily and may contribute to the persistence of the plant. Unlike many other grasses, prairie grass is highly self-fertile and, depending on environmental conditions, it may be self- or cross-pollinated. Agronomically its greatest value lies in its ability to provide winter and early spring growth, whereas its summer productivity is less desirable because of the presence of many stemmy inflorescences. Palatability is good but there is some doubt concerning the adequacy of its mineral content. Because of its upright habit it requires to be grazed

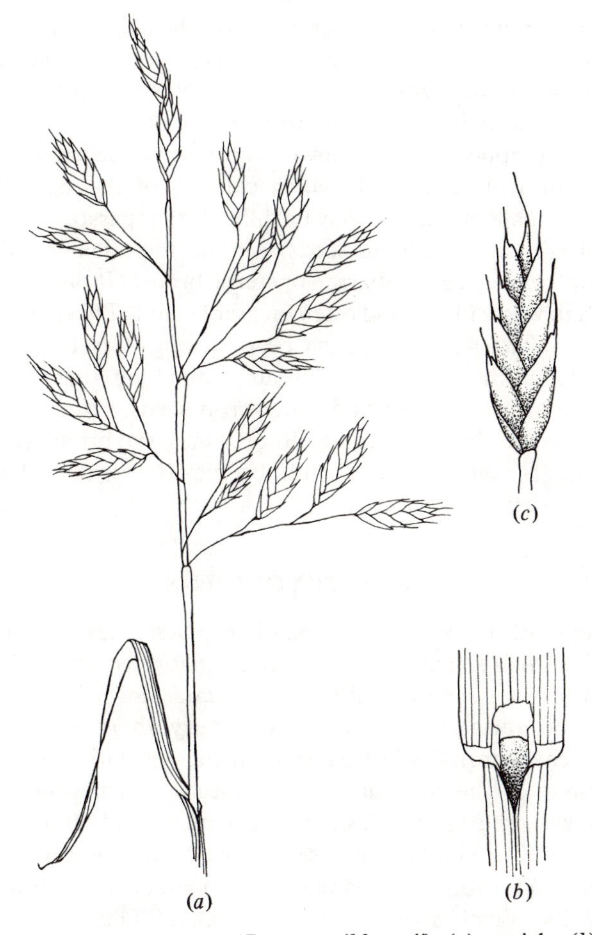

Fig. 4.26. Prairie or rescue grass (*Bromus wildenovii*). (*a*) panicle, (*b*) ligule, (*c*) spikelet.

rotationally, preferably by cattle rather than sheep. Several countries have now released improved and bred cultivars of this very productive species.

By contrast to the two species just described, soft brome (*Bromus mollis*) is a weed grass, common in waste places and sown grassland in Europe, parts of Asia and New Zealand. This is a tufted annual or biennial species with hairy leaves and sheaths. The ligule is short and serrated but there are no auricles. The panicle bears many large spikelets, each subtended by two hairy glumes which are unequal in size and containing eight to ten florets. The lemma is distinctly serrated along the apex, it carries a subterminal awn, and it is broad enough to overlap partly the neighbouring floret. The palea is also serrated along its upper portion but is much smaller in size. As long as it is not flowering soft brome is eaten by stock and thus it does not feature prominently in grazed pastures. In hay crops, especially those cut late, it has the opportunity to flower and scatter seed thus ensuring its continued survival. The seed, which consists of the entire floret with rachilla, is variable in size and may be difficult to separate completely from samples of ryegrass and other seeds. Among the other members of this genus is another weed grass, barren or sterile brome (*Bromus sterilis*) which is common in waste places and often on arable land. This plant has a large, spreading panicle, and the lemma bears a long dorsal awn. The seeds scatter easily but they are so large (about 20 mm long) that they are readily separated from cereal grain and other crop seeds. Other species in the genus *Bromus* are the field brome (*B. arvensis*), rye brome (*B. secalinus*), wood brome (*B. ramosus*) and upright brome (*B. erectus*); all wild grasses adapted to a variety of habitats.

4.15 *CYNOSURUS*

Crested dogstail (*Cynosurus cristatus*) is the best known member of this genus. It is a small tufted perennial entirely glabrous, with rolled to folded tillers and leaves which are ribbed above and glossy below (Fig. 4.27). There are no auricles and the ligule is very short. One of the more distinctive characteristics of this grass in the vegetative condition is the yellow colour of the base of the leaf sheath which should avoid any confusion with ryegrass which shows red to purple colouration in the same place. Any doubts will, however, be removed when flowering occurs, for crested dogstail produces characteristic one-sided spike-like panicles bearing at its branches two kinds of spikelets. The outer spikelets are completely sterile and consist only of the glumes and the empty lemmas of the florets. The inner spikelets contain three to five fertile florets subtended by a pair of glumes. The lemma is pointed or bears a short awn, and both it and the palea are brightly yellow. Crested dogstail seed, which consists of the entire floret with a rachilla, is thus easily recognised in any mixture. While it remains vegetative, this grass is palatable but not very high producing, although of some value during winter and early spring. However, the inflorescence is not eaten by stock and copious seed production allows the plant to persist, thus occupying space that should be taken up by more highly producing species. There are in fact no good reasons for sowing crested dogstail in preference to ryegrass or cocksfoot, even though some farmers continue to do so. Similarly, its inclusion in lawn mixtures cannot be recommended, especially as the culms are not

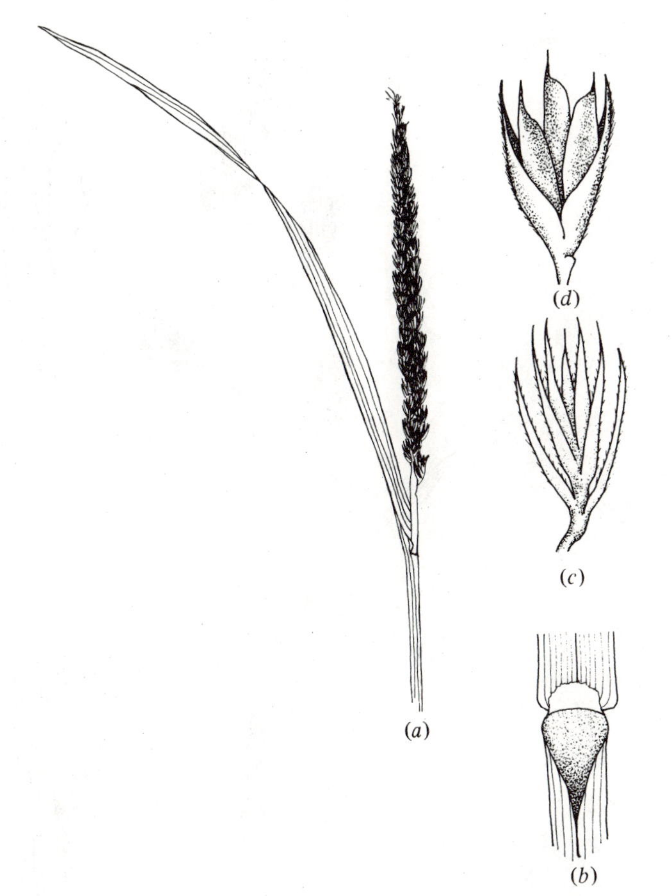

Fig. 4.27. Crested dogstail (*Cynosurus cristatus*). (*a*) spike-like panicle, (*b*) ligule, (*c*) outer sterile spikelet, (*d*) inner spikelet with fertile florets.

easily cut. Rough dogstail (*Cynosurus echinatus*) is less common although equally conspicuous through its broad, irregularly shaped inflorescence bearing distinct awns. It is a weed grass of no agronomic value.

4.16 *DACTYLIS*

This genus contains the highly important cocksfoot or, as it is known in North America, orchard grass (*Dactylis glomerata*). This is a strongly growing tufted perennial with highly flattened tillers (Fig. 4.28). The long leaves are dull- or blue-green in colour, with a prominent keel but with no

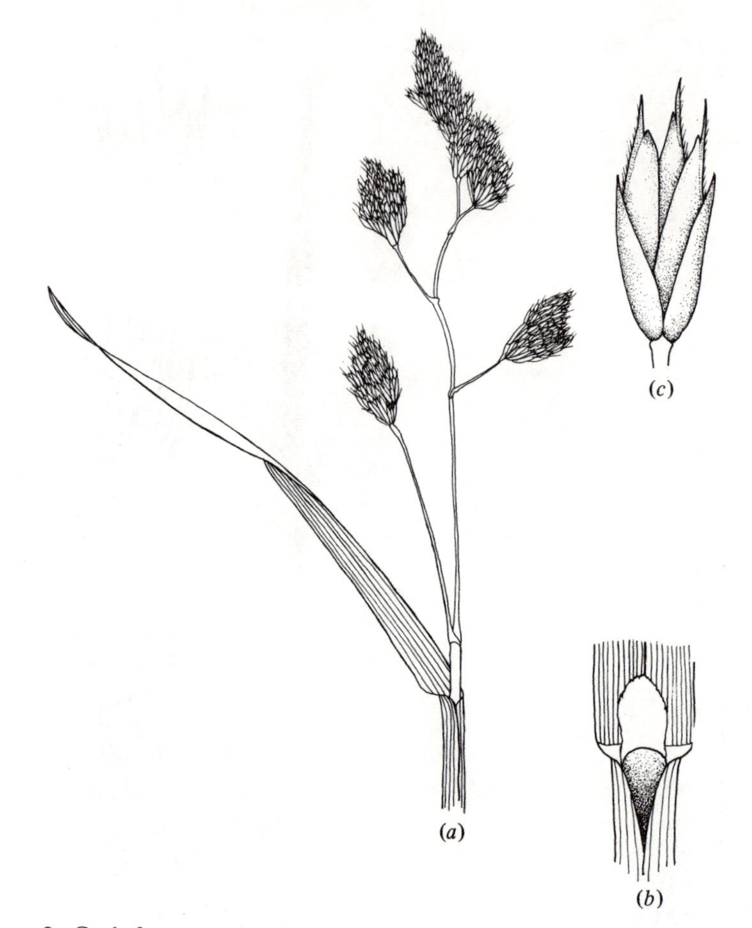

Fig. 4.28. Cocksfoot or orchard grass (*Dactylis glomerata*). (*a*) panicle, (*b*) ligule, (*c*) spikelet.

ribs showing on the upper surface. There are no auricles, but the ligule stands out conspicuously. The plant is entirely glabrous, and in this differs from prairie grass (see section 4.14). The inflorescence is a tall panicle, bearing spikelets crowded together in clumps, and this with some imagination may have given rise to the name of this grass. Each spikelet has two pointed, keeled and slightly hairy glumes, and contains three to five florets. The lemma is distinctly pointed and curved, and the palea is membranous. The commercial seed, consisting of the whole floret with rachilla, lies on its side when placed on a flat surface owing to the pronounced keel of the lemma.

Cocksfoot or orchard grass is a valuable perennial pasture species, well adapted to moderate to high fertility and to low soil moisture. If properly managed it is a productive component of a mixed sward together with ryegrass and clovers. Relatively slow establishment and sensitivity to harsh grazing are the main problems, and lenient rotational grazing by cattle is by far the best management. This ensures continued productivity and an acceptable level of palatability. On the other hand, unless it is regularly grazed or cut it has the tendency to become rather coarse and highly tufted, in which condition it is no longer acceptable to livestock. Some of these problems have been overcome or at least greatly reduced through the advent of a great range of cultivars with different characteristics. Leafiness has been improved, the pasture types are less prone to becoming coarse, and there are also more erect hay types which combine well with clovers. Cocksfoot seed production is a highly specialised enterprise requiring skilful manipulation of row and plant spacings and of fertilizer regimes, if yields are to be maintained after the first harvest. There is continued demand for this species, to a large extent because its tolerance of dry conditions makes it complementary to perennial ryegrass. Selections from the eastern Mediterranean are generally drought-resistant but at the same time they have a tendency towards summer dormancy.

4.17 *DESCHAMPSIA*

Tufted hair grass (*Deschampsia caespitosa*) is widely distributed in Europe over a wide range of latitudes, but is especially common in moist, shady habitats in Britain. It is a tall, coarse perennial, strongly tufted and forming distinct tussocks. The leaves are very harsh, distinctly ribbed, glabrous, with no auricles but a long ligule. The inflorescence is a spreading panicle with two to five florets per spikelet. The awned seeds, which are distinctive through the long and hairy rachilla, are produced in large quantities, and this creates the problem of curbing the spread of the plant. Tufted hairgrass is a troublesome weed on heavy and poorly drained soils. Because it is unpalatable, it cannot be controlled by intensive grazing, and control by cultivation is not often possible in the situations in which this grass thrives. Wavy hairgrass (*D. flexuosa*) is a smaller plant with a more slender, flexible panicle and with longer, twisted awns on the lemma. It is a tufted perennial with dark green, needle-like leaves which are permanently rolled inwards. The top of the fibrous roots is often red in colour. Wavy hairgrass occurs on heaths and acid hills. It is quite unpalatable and has no agronomic value.

4.18 *FESTUCA*

The genus *Festuca* contains productive pasture species suitable for intensive grassland farming and also fine-leaved species confined to hill country or employed as lawn grasses. It is closely related to the genus *Lolium*, and successful crosses between the two genera have been achieved. Meadow fescue (*Festuca pratensis*), a tufted, actively tillering perennial resembles Italian ryegrass (see section 4.21), in that the tillers are rounded, the leaves glabrous, ribbed above and shiny below, with conspicuous auricles and a short ligule. The leaf sheaths are also bright red at the base, and the only real differences from Italian ryegrass in the vegetative state are the persistent old leaf sheaths which are brown and the slightly less vigorous and tougher appearance of the sward. Once the inflorescence appears, there is no further cause for confusion, for in meadow fescue it is an open panicle, and not a spike, bearing spikelets which contain five to ten florets. The glumes are unequal in size, and the lemma bears a terminal point. The seed, consisting of the complete floret, is very similar to that of perennial ryegrass. It has a rachilla with a rounded end, not a flattened one as in *Lolium*. Meadow fescue is potentially a useful species, except that it does not compete successfully with the more aggressive perennial ryegrass. It can therefore not be grown in a mixed pasture containing ryegrass, although it combines well with timothy or cocksfoot. However, because of this restriction, meadow fescue has declined in popularity, even though its seasonal productivity is complementary to that of perennial ryegrass. Very well bred British, Dutch, Scandinavian and Canadian cultivars are available, and these continue to keep this grass in production.

Tall fescue (*Festuca arundinacea*) is a more robust and coarser version of meadow fescue resembling even more Italian ryegrass and its tetraploid derivatives, (Fig. 4.29). Apart from size and general agronomic considerations, the most reliable morphological difference is the presence of small, stiff hairs on the auricles and leaf bases of tall fescue in distinction to those of Italian rye-grass. The appearance of the large, open panicles of tall fescue will confirm the identification. Each spikelet contains three to ten florets, and the seed is the complete floret with a rounded rachilla. Both tall and meadow fescue could easily contaminate seed samples of perennial ryegrass, and great care must be taken to avoid any risk of this occurring. Tall fescue is a vigorous plant, well adapted to a wide range of conditions. United States selections have shown themselves useful in drought conditions as well as under irrigation, and winter greenness has been a feature of others. North African lines tested in Britain were superior in winter growth although they failed to withstand frost, while other strains have excelled in leafiness and a restricted flowering period. Tall fescue

Fig. 4.29. Tall fescue (*Festuca arundinacea*). (*a*) plant with panicle, (*b*) auricles and ligule, (*c*) spikelet.

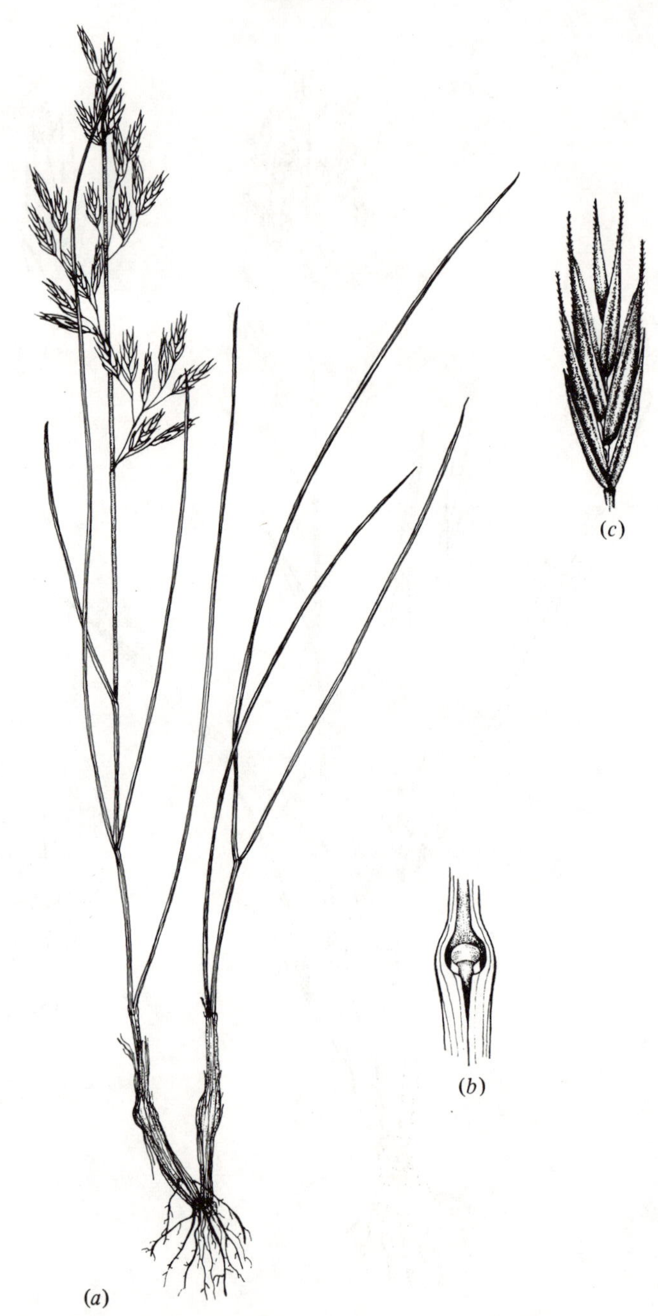

Fig. 4.30. Red fescue (*Festuca rubra*). (*a*) plant with panicle, (*b*) ligule, (*c*) spikelet.

prefers rotational grazing management, preferably by cattle, and in its establishment phase it is vulnerable to competition by other species. Unimproved tall fescue tends to be coarse and unacceptable to stock, and there have also been reports of it causing lameness through the presence of an alkaloid. However, none of these problems are of any consequence in improved strains.

Red fescue (*Festuca rubra*) is a very variable species with fine, needle-like leaves. It spreads by short, slender rhizomes, the tillers are rounded, and the foliage is dark green with some red colour at the base of the sheath. The open panicle, reddish in colour, contains five to eight florets per spikelet, and the lemma terminates in a fine awn (Fig. 4.30). Despite its hardy appearance, red fescue is quite palatable, and in low fertility hill country where other species could not thrive it plays a valuable role in sustaining animal production. It responds to raised fertility and, if given adequate water and fertilizer, it can become quite aggressive. At least two forms of *Festuca rubra* are known, of which one is *F. rubra* var. fallax (previously subsp. *commutata*) or Chewing's fescue, named after an early producer of seed in New Zealand. This plant is an important component of temperate lawns and sports turf, and for these purposes has always been a valuable article of commerce (see also section 4.8). A closely related species is sheep's fescue (*Festuca ovina*), another plant with needle-like leaves growing in hill country, and there are several other species and varieties which occupy similar habitats.

4.19 *GLYCERIA*

The species of this genus are well adapted to wet conditions and are thus normally found in low-lying, ill-drained pastures, along ditches and under heavy irrigation. Floating sweet grass (*Glyceria fluitans*) is a small, stoloniferous perennial with blue-green, flattened tillers, pointed leaves with a distinct keel, and a conspicuous ligule which is visible from a distance (Fig. 4.31). It has a slender, spreading panicle bearing cylindrical spikelets, each containing eight to sixteen florets. The lemma is very blunt and has a round back. Reed meadow grass (*Glyceria maxima*) forms tufts of upright shoots but relies on rhizomes for its spread. Both these species are palatable but have so far not been seriously considered for plant improvement.

4.20 *HOLCUS*

Yorkshire fog (*Holcus lanatus*) is the most important and widely distributed species in this genus. This is a tufted, sometimes slightly creeping

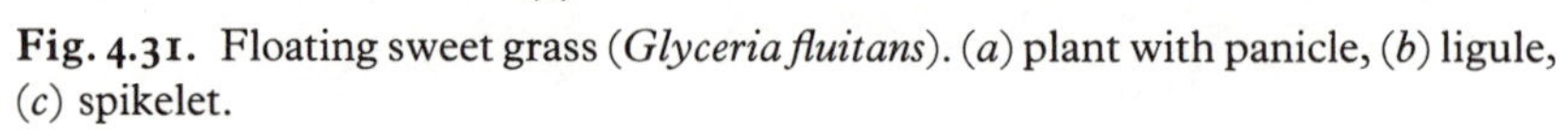

Fig. 4.31. Floating sweet grass (*Glyceria fluitans*). (*a*) plant with panicle, (*b*) ligule, (*c*) spikelet.

perennial with rounded tillers (Fig. 4.32). Both leaves and sheaths are softly hairy. The leaves are light- to grey-green in colour with no auricles but a prominent ligule. The sheaths are creamy in colour with longitudinal red stripes, rather like proverbial pyjama trousers. The panicle is compact when young but opens up later giving a pinky cream appearance. There are numerous spikelets, each subtended by a pair of large unequally sized hairy glumes. Within are two florets, the lower one perfect, the upper one borne on an extended rachilla is staminate only and has a lemma with a dorsal, hooked awn. The seed is either the complete spikelet or, after passing through a machine, it consists of the lower fertile floret surrounded by the shiny brown lemma and palea. Yorkshire fog is widely distributed, especially in humid and moderately fertile regions, but it seems to grow equally well in lowland dairy pastures or in hill country areas, provided that the annual rainfall is in the vicinity of 900 mm or more. Although it is palatable only when young and even though it is considered as an indicator of poor drainage and insufficiently intensive grazing management, its great adaptability to so many different conditions and its obvious ability to grow and persist have attracted attention. Improved selections with better palatability and resistance to leaf rust, one of the problem diseases of this plant, are under test and may lead to its systematic utilisation in grassland farming.

Creeping soft grass (*Holcus mollis*) has long rhizomes and prostrate stems. Its leaf sheaths are less conspicuously striped than those of Yorkshire fog, and a ring of long hairs at the nodes also aids in identification. At the flowering stage, the awn of the upper male floret projects well beyond the spikelet. This species is not uncommon in dry woodlands and may also be of some concern as a weed of arable land.

4.21 *LOLIUM*

Members of this genus are the most widely distributed and useful pasture grasses in many parts of Europe, New Zealand and other moist, temperate regions of the world. Although referred to as ryegrass, they are not related to rye and their present name is derived from the Old English term ray-grass. The genus contains annual, biennial and perennial species, all of them distinguished by being glabrous and bearing dark-green leaves which tend to be shiny on the under-surface. The base of the leaf sheath is often pigmented red or purple. The inflorescence is a spike with numerous spikelets pressed against the rachis. Each spikelet contains four to fifteen florets and is subtended by a single glume. The grain consists of the caryopsis surrounded by lemma and palea and bears a distinct rachilla. Several species of *Lolium* are of economic importance, particularly those

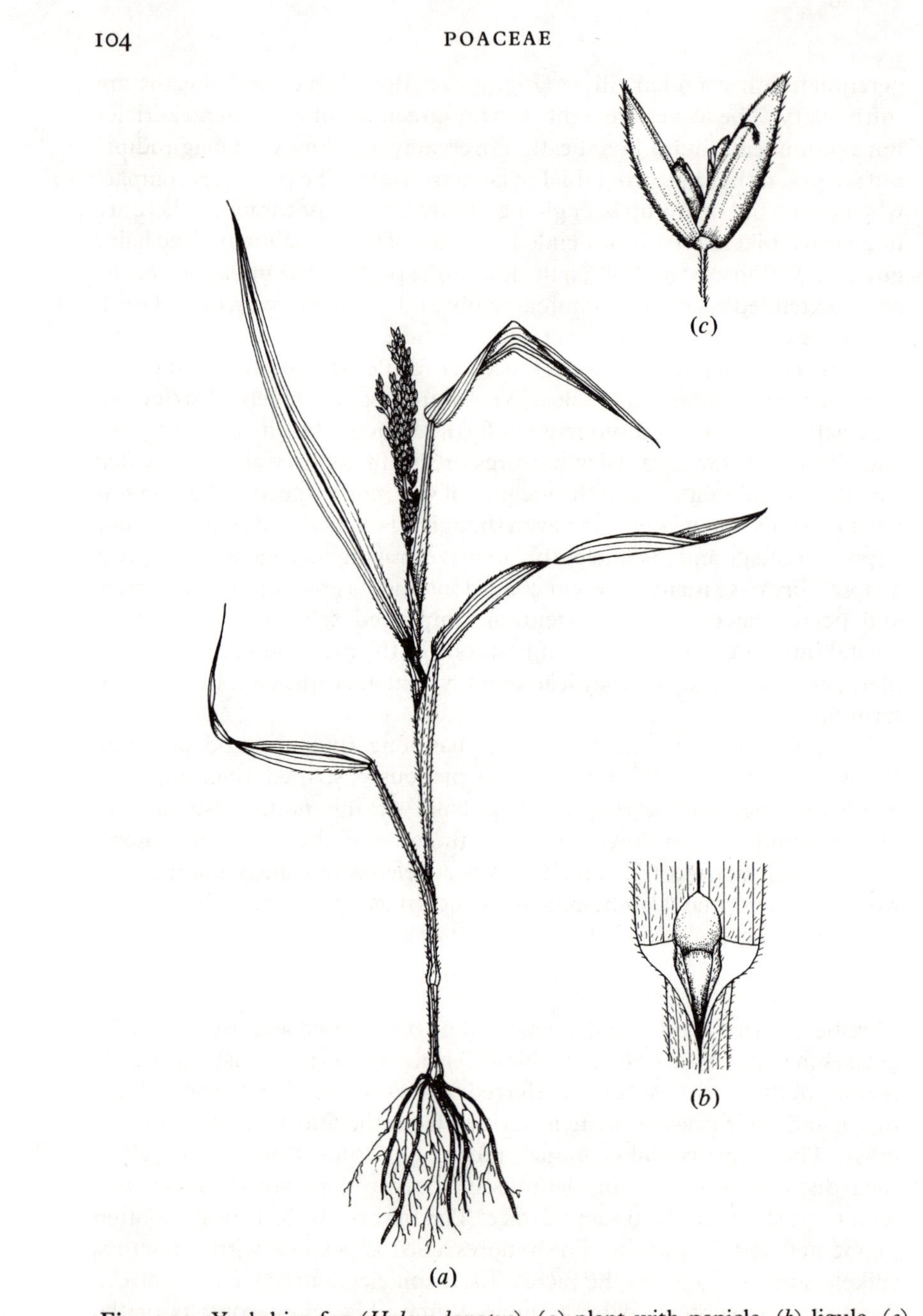

Fig. 4.32. Yorkshire fog (*Holcus lanatus*). (*a*) plant with panicle, (*b*) ligule, (*c*) spikelet with lower fertile floret and upper staminate floret.

which have become outstanding as pasture grasses for soils of medium to high fertility and with an adequate moisture supply. Intensive plant breeding has resulted not only in a profusion of cultivars, but also in the production of hybrids and polyploids.

Perennial ryegrass (*Lolium perenne*) deserves pride of place as the most common pasture grass in many temperate regions (Fig. 4.33). It is a long-lived perennial, capable of forming very many leafy tillers and thus a dense sward. The shoots are flattened and bear dark-green leaves which are distinctly ribbed above and very shiny below. The ligule is short and inconspicuous, and the auricles are small and often not easily distinguished from the slightly shelf-like base of the lamina. The base of the leaf sheath is bright red to purple in colour. The inflorescence is a spike with the spikelets closely appressed to the rachis and containing three to ten fertile florets. In distinction to other ryegrasses, the lemma bears no awn.

Perennial ryegrass is an extremely valuable pasture grass. By virtue of its vigorous tillering and rapid leaf production it recovers quickly from grazing, and the same characteristics together with its relatively prostrate growth habit enable it to withstand heavy animal hoof traffic. Even if given reasonably adequate but not necessarily good management, perennial ryegrass is extremely persistent. Its longevity is shown, for example, by the excellent permanent pastures in Kent and the English Midlands, which have been dominated by this species for centuries. There are so many tillers and both stem apex and axillary buds are so close to soil level that it is in fact quite difficult to eliminate perennial ryegrass by excessively harsh grazing management. This ability to stand up to hard grazing is probably one of the main reasons why it combines so well with white clover which requires a sward with a favourable light regime for optimum growth. Tightly grazed perennial ryegrass/white clover pastures are the mainstay of animal production in New Zealand which relies heavily on symbiotic nitrogen fixation by the legume and nutrient cycling through the animal. In other countries fertilizer nitrogen is preferred, especially during those times of the year when rhizobial activity is low, in which case somewhat less prostrate cultivars of ryegrass are used.

Perennial ryegrass, like most other species, is seasonal in its growth, and hence plant breeders have attempted to develop cultivars with different patterns of productivity. This has been achieved by making use of the wide climatic range of the species which extends from Scandinavia and the Baltic to the Mediterranean area, thus providing ample scope for plant introduction, selection and cross-breeding. One of the major factors determining seasonal distribution of leaf production is the onset of reproduction which leads to the formation of flowering stems. Tillering and leafiness are greatly reduced, and palatability declines sharply. Not

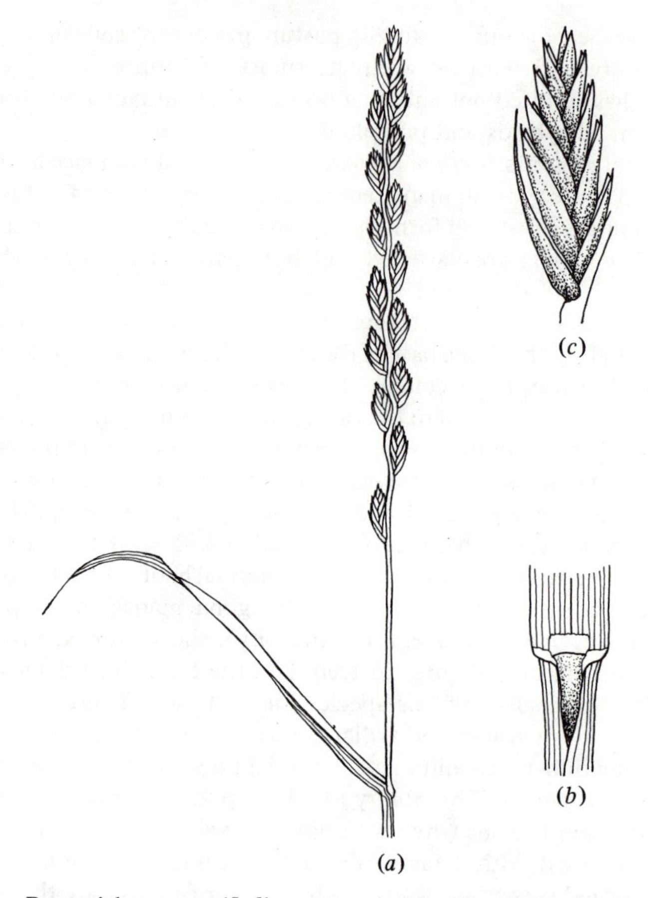

Fig. 4.33. Perennial ryegrass (*Lolium perenne*). (*a*) spike, (*b*) ligule, (*c*) spikelet

surprisingly, the time of flowering has been used very frequently by plant breeders as a selection criterion, and hence perennial ryegrass cultivars tend to vary widely in this respect. Not only that, but the length of the flowering period or more precisely, the ease of resumed flowering following cutting or grazing have been reduced quite skilfully by breeders to bring about improved productivity.

Perennial ryegrass is a long-day plant and requires a period of vernalisation before it will flower. Crops sown in the spring will not necessarily flower and set seed in the same year in cool, temperate countries, and if introduced into other climatic zones there could well be continued problems with seed production. On the other hand, because perennial ryegrass is an outcrossing species, there is perceptible genetic change as successive generations of the same ecotype adapt themselves to the prevailing environment. In consequence, many climatic races exist, and no single population remains unchanged over a period of time. Several examples could be cited to illustrate this considerable genetic variation. A case of practical significance concerns the cellulose content of the leaf and hence the proportion of soluble to insoluble carbohydrates which vary in different selections. These differences affect the weight gain of the grazing animal through the type of fermentation induced in the rumen and the rate of breakdown of the ingested grass. A simple test, the breaking strength of the leaf, can be used by the plant breeder to screen populations differing in these constituents. Improvements would be desirable, because perennial ryegrass leaves are rather tough in consistency through their high cellulose content. Even a sharp lawnmower does not necessarily cut its leaves cleanly.

Perennial ryegrass has a basic diploid chromosome number of 14, but it is possible to double it artificially by soaking germinating seeds in a solution of colchicine. Quite a number of tetraploid cultivars have been produced through the use of this chemical. Compared with diploid ryegrass, tetraploids tend to have larger seeds and bigger leaves. Tillering is often somewhat reduced and dry matter percentage can also be lower. On the other hand, improved soluble carbohydrate content appears to be associated with greater palatability and better intake by the animal, and total herbage production is often greater.

Italian ryegrass (*Lolium multiflorum*) is an annual or short-lived perennial species, depending on how it is managed. In distinction to perennial ryegrass, the leaves of this species are rolled in the bud and thus tillers are round rather than flattened (Fig. 4.34). The leaves are larger and broader, distinctly ribbed above and shiny below, and at the base of the lamina the auricles are well developed. The base of the leaf sheath is bright red in colour. The inflorescence is a long spike bearing numerous spikelets, each of which contains three to ten florets subtended by a single glume. The lemma terminates in a distinct awn, and the seeds are slightly larger than those of perennial ryegrass (about 1.9 g as compared with 1.6 g per 1000). Since it is possible for the awns to be broken off during seed harvesting and cleaning, and since seed size is not an absolutely reliable criterion, another method has been devised to distinguish between these two ryegrasses. When seedlings germinated on moist filter paper are examined under ultra-violet light, the roots of Italian ryegrass and its hybrids fluoresce as distinct to those of perennial ryegrass which do not.

Italian ryegrass establishes very rapidly from seed and quickly forms a dense, leafy pasture. It thus lends itself readily for use as a short-term ley or greenfeed crop, especially if advantage is taken of its ability to provide

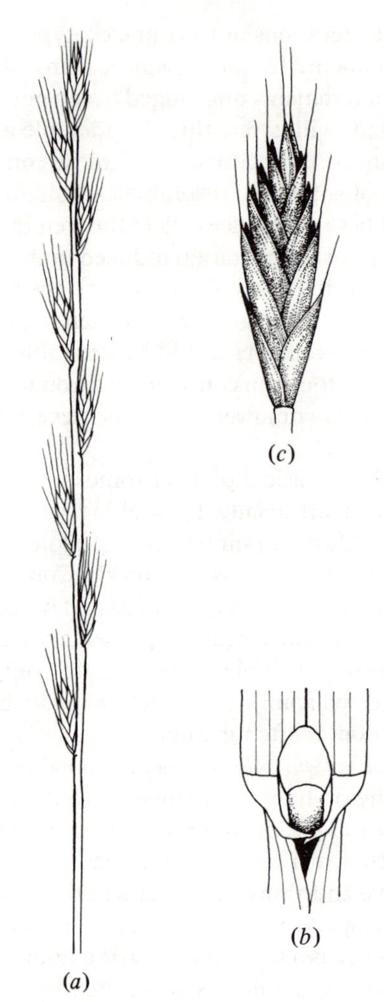

Fig. 4.34. Italian ryegrass (*Lolium multiflorum*). (*a*) spike, (*b*) auricles and ligule, (*c*) spikelet with single glume and awned florets.

good late winter to early spring growth. Alternatively Italian ryegrass may be included in a mixture to provide initial herbage production until more slowly establishing, perennial species take over. This device works well, provided grazing management is intensive enough to prevent Italian ryegrass from competing too strongly with the other pasture constituents. One of its advantages which assists in this strategy is its high feeding value, for it is readily acceptable to livestock at all stages of growth, even when it is close to the reproductive stage. In fact, by grazing at a time when the stem apices are sufficiently elevated above soil level to be removed by the animal, it is possible to reduce flowering and thereby to extend the life of the plant beyond a single season. If allowed to produce inflorescences, Italian ryegrass will flower as late as or later than most perennial ryegrass cultivars. It produces good seed yields, although time of harvesting is critical if undue seed shattering is to be avoided.

Italian ryegrass is a diploid with a chromosome number of 14 and thus lends itself to the production of tetraploids by use of colchicine. This has been particularly successful in extreme annual types, notably those derived from a region of the Netherlands called Westerwolds, and several of these cultivars have turned out to be extremely useful. These tetraploids tend to be vigorous, erect plants with thick stems and with broad leaves extending right into the base of the sward. Dry matter content is lower but soluble carbohydrates higher than in other ryegrasses. The inflorescences are also larger, and so are the awned seeds which weigh more than twice as much as those of the diploid (about 4.9 g per 1000). This requires sowing rates to be adjusted accordingly. Tetraploid Italian ryegrass establishes very rapidly and soon produces a dense sward of leafy material, especially if stimulated by nitrogen. Its main use is to provide winter and early spring production following sowing in the autumn, and in this respect it is superior to cereals used for the same purpose, largely because of its ability to recover quickly from cutting or grazing. The feeding value is high and leads to rapid animal weight gains, and in particular a high content of soluble carbohydrates is associated with good levels of solids-not-fat in the milk of animals grazing it. Westerwolds-type tetraploids are strictly annual and no vegetative tillers tend to survive the flowering season.

As members of the same genus with the same basic number of chromosomes, perennial and Italian ryegrass hybridise easily. Plant breeders, particularly those in New Zealand, have made use of this advantage by producing hybrid ryegrass cultivars which are formally referred to as *Lolium* × *hybridum*. These plants not unnaturally combine features of both parents, both botanically and agronomically. They tiller more freely than Italian ryegrass and are not as tall or upright, and on the other hand the auricles are better developed than in perennial ryegrass and the tillers are not as flattened. A high proportion, but not all, the florets and hence the seeds bear awns. One of the New Zealand bred hybrids used to be called short rotation ryegrass (now Manawa), and this should help to indicate its position in agriculture as a short-lived perennial capable of persisting three to four years. In this capacity it serves a useful purpose as the main or sole grass component of a pasture together with white and, in some areas, red clover. It establishes easily, though not as rapidly as its

Italian parent, and its early growth must be controlled to avoid suppression of clovers, unless nitrogen is supplied as a fertilizer. Grazing management should preferably be rotational. Productivity is best in the early part of the season, and long dry spells in the summer together with overgrazing will reduce persistence. On the other hand, not all cultivars have proved satisfactory under British winter conditions. Hybrid ryegrass has greater feeding value than perennial ryegrass and is acceptable to livestock at all times of the year, even when approaching ear emergence.

It is, of course, possible for the plant breeder to produce other hybrids apart from straight crosses of perennial with Italian ryegrass. For example, in New Zealand, Manawa (see above) has been crossed with perennial ryegrass to give a cultivar (Ariki) which is much closer in appearance and agronomic characteristics to its perennial than its Italian parent. Other combinations are equally possible and could well be of advantage when it comes to agronomic performance and pest or disease resistance.

Darnel (*Lolium temulentum*) is an annual weed species with a chequered history. It appears to be indigenous in southern Europe and was at one time quite common in Britain where its large seed contaminated wheat grain until seed cleaning machinery became sufficiently efficient to screen it out. Darnel seed was often the carrier of a poisonous fungus and thus rendered unscreened wheat dangerous. Another unusual feature of this species is that it served as a laboratory plant in physiological experiments concerning the induction of flowering, and it turned out to have very precise requirements for the reception of the photoperiodic stimulus. Darnel looks like a slightly coarser version of Italian ryegrass, but it is distinguished from it by its much larger inflorescence and particularly the size of the glume which is longer than the whole spikelet which it subtends.

Wimmera ryegrass (*Lolium rigidum*) is also an annual, but its rapid growth has been utilised in Victoria and other Australian states to take advantage of a short growing season. Very early flowering types have been selected. The aim is to develop winter growing annuals which together with subterranean clover can support sheep in dry areas.

4.22 *MOLINIA*

Purple moor grass or flying bent (*Molinia caerulea*) is very common in wet heaths, upland moors and generally peaty areas in Britain, although its distribution throughout Europe is wider than that. It is a large, coarse, tufted perennial with rolled tillers. The leaves are long, narrow and stiff with a slight cover of long white hairs on the upper surface. There are no auricles, and the ligule is replaced by a fringe of hairs. The inflorescence is a narrow panicle, green to purple in colour, bearing two to three flowered spikelets. The glumes are of unequal size and the lemma and palea also tend to be purple. The base of the culm is swollen. Although a frequent constituent of peatland vegetation, it is too unpalatable to be of any agronomic value.

4.23 *NARDUS*

Moor mat grass (*Nardus stricta*) is also common in Britain, but in dry mountain grasslands and moors. It is a tough, wiry plant with permanently folded needle-like leaves and stiffly persistent tillers. The inflorescence is a spike with the single-flowered spikelets arranged on one side of the rachis only, thus giving it a characteristic asymmetric appearance. The glumes are so highly reduced as to be virtually absent, but the lemma is long and ends in a fine awn point. The plant is not at all palatable but during the winter, when the grey-green leaves have turned almost white, it provides some standby sustenance for sheep.

4.24 *PHALARIS*

Phalaris (*Phalaris aquatica*), of Mediterranean origin, is widely grown in Australia and parts of the United States. It is a large perennial, with tillers rolled in the bud, glabrous and bearing blue-green to grey-green leaves. Auricles are absent but the ligule is conspicuous and serrated along its apical rim (Fig. 4.35). Phalaris is rhizomatous, and both rhizomes and leaf sheaths are showing red colour owing to the presence of anthocyanin. The inflorescence is a compact panicle bearing large numbers of single-flowered spikelets. A pair of winged glumes surrounds the floret with its shiny lemma and palea. Phalaris is used extensively in parts of Australia, where its Mediterranean growth rhythm fits the prevailing climate. It is at its best in mild winters and early spring but, once the summers become dry and hot, it ceases to be productive and becomes semi-dormant. It does not establish rapidly and may suffer from competition in the seedling phase. However, once established it can be very persistent and is capable of withstanding severe grazing. Palatability is not outstanding, and it is advisable to cut mature stands for hay rather than graze them. Fresh growth in autumn and early winter may cause phalaris staggers in sheep, but this can be prevented by dosing the animals with cobalt. In the United States a number of lines of phalaris have been produced under the name of Harding grass, and in the United Kingdom it has been crossed successfully with *Phalaris arundinacea*. This species, valued for its adaptability to a wide range of poorly drained soils and for its drought resistance, is grown widely in the northern parts of the United States and in southern Canada.

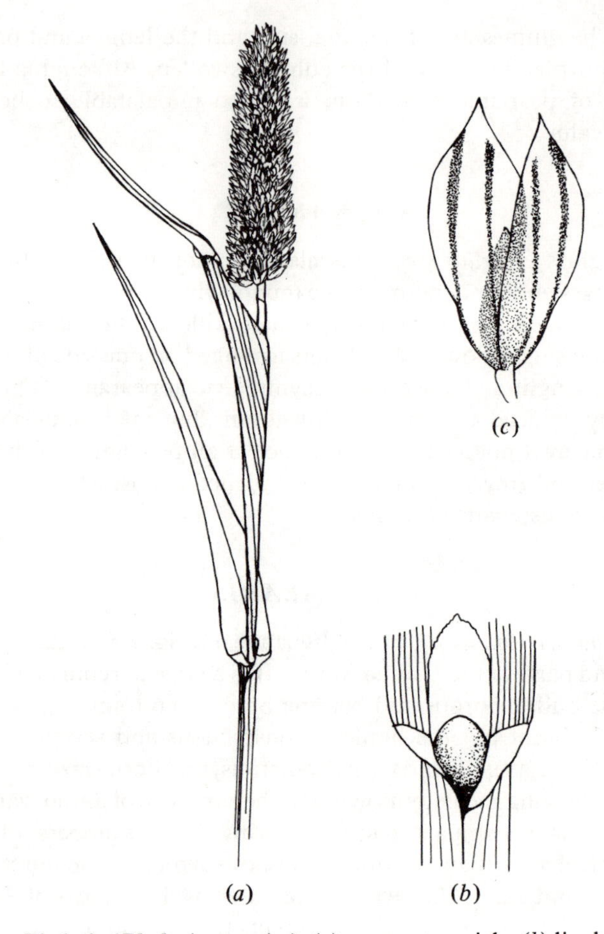

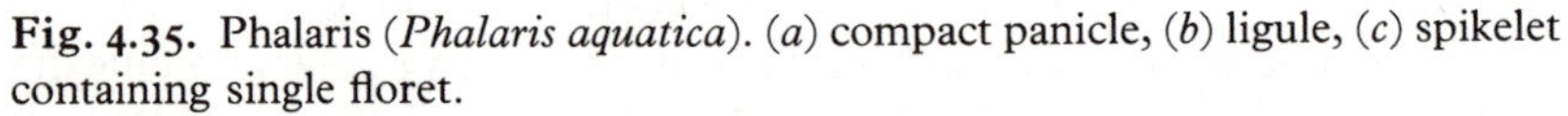
Fig. 4.35. Phalaris (*Phalaris aquatica*). (*a*) compact panicle, (*b*) ligule, (*c*) spikelet containing single floret.

It responds well to nitrogen fertilizer and is resistant to many pests and diseases. In Britain, *Phalaris arundinacea* or reed canary grass is not highly regarded, largely because it is unable to withstand the rigours of the winter. Although frost killing also reduces the length of the growing season in the USA, reed canary grass is used successfully for pasture, hay and silage. It begins growth early in spring and needs to be rotationally grazed at high stocking rate to avoid a build-up of coarse, stemmy material. Even on poorly drained land, this species quickly develops a sward firm enough to support cattle and sheep and to allow heavy machinery to be used. It is so vigorous that it is difficult to maintain legumes in association with it. Hay and silage yields and quality are good. Reed canary grass is used widely for soil conservation work, such as stream channel banks or grassed water ways, and it is one of the best species for irrigation with sewage effluents. Another member of the genus is Canary grass (*Phalaris canariensis*) which, as the name implies, is grown like a cereal for the production of bird seed. The oval spike-like panicle bears spikelets containing three florets. The glumes are green with white stripes and are large enough to surround the florets. In addition to being enclosed by a hairy lemma and palea, the seed is also subtended by a pair of small sterile lemmas.

4.25 *PHLEUM*

Timothy (*Phleum pratense*) is a tufted, glabrous perennial, light grey-green in colour, with rounded tillers which tend to produce swollen bases, so-called haplocorms, during periods of active growth which function as storage organs. The leaf veins are all equally prominent, auricles are absent, and the ligule though short is prominent and dome-shaped (Fig. 4.36). The inflorescence is a spike-like panicle with numerous spikelets forming a cylindrical structure which gave rise to the old name of the plant, catstail. Each spikelet contains a single floret surrounded by a pair of glumes. These are fringed with hairs along their keel and each terminates in a distinct short awn. Both lemma and palea are thin, silvery, membranous structures surrounding the brown caryopsis. Commercial seed tends to consist of the entire floret, unless machine dressing has removed lemma and palea, exposing the brown caryopsis. Timothy is highly palatable and nutritious, both as a pasture and as a hay plant. For vigorous growth it requires conditions of medium to high fertility and adequate soil moisture. It is thus mainly adapted to cool, moist climates, and some of the best strains come from northern Europe. North American selections tend to be early flowering and to be used predominantly for hay. Timothy is fairly slow to establish and may suffer from competition by other, more aggressive species. However, it is quite persistent, provided rotational grazing management is practised and moisture supply remains adequate. Conversely, it declines in vigour under close and continuous grazing, especially in dry seasons. Timothy is a hexaploid with 42 chromosomes, and it is thought that it originated as an allopolyploid following the natural crossing of a diploid and a tetraploid species. One of the diploid members of the genus is small timothy (*Phleum bertolonii*), a low growing, spreading plant which is smaller in all respects than the hexaploid species. Because of its greater persistence under grazing it has been used by plant breeders at the Welsh Plant Breeding Station, Aberystwyth and in the United States for the production of improved strains.

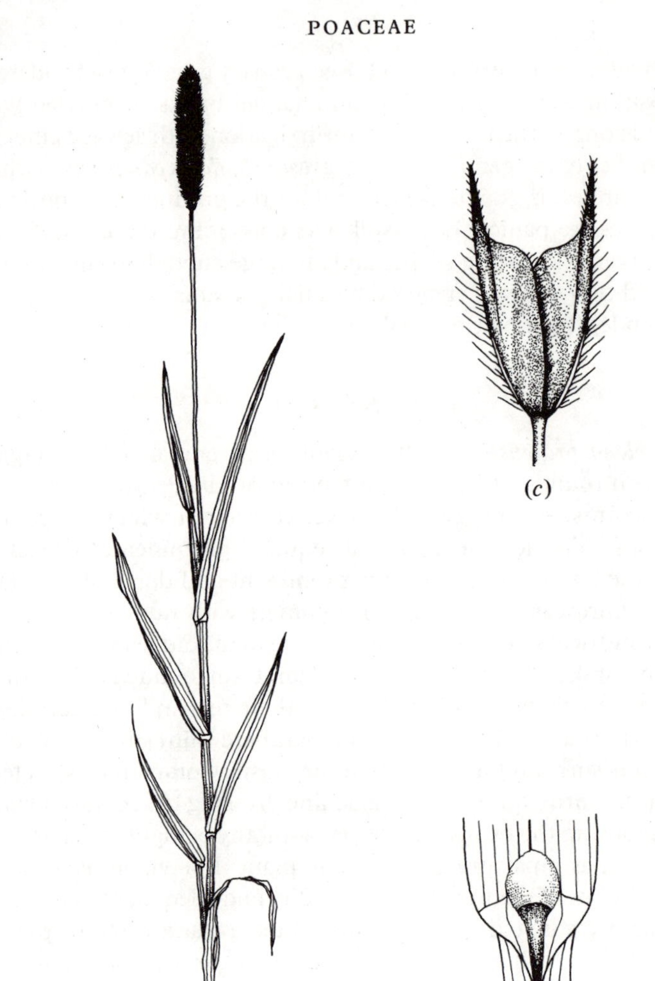

Fig. 4.36. Timothy (*Phleum pratense*). (*a*) plant with spike-like panicle, (*b*) ligule, (*c*) spikelet with two short-awned glumes, containing a single floret.

Some people find it difficult to distinguish the inflorescences of timothy and meadow foxtail (*Alopecurus pratensis*) described earlier (see section 4.9). Quite apart from the fact that the foxtail inflorescence is softer and more silvery, the main distinction is the presence of a single, soft awn as opposed to the two short and stiff awns on the glumes of timothy.

4.26 *POA*

Smooth-stalked meadow grass or, as it is known in North America, Kentucky bluegrass (*Poa pratensis*) is agriculturally the most useful member of this genus. It is a creeping perennial, spreading by rhizomes, and consists of flattened, glabrous tillers. The leaves are dark green and rather dull, folded in the bud, tapering only slightly from the base and terminating in a boat-shaped tip. As in other members of the genus, the central leaf vein is flanked by two lines running along the length of the lamina. Auricles are absent and the ligule is very small. The spreading panicle bears spikelets with three to five florets each. The glumes are unequal in size, and the lemma is keeled with hairs along its edge and at the base. The whole floret with rachilla forms the commercial seed. Although quite common in the United Kingdom and other European countries, smooth-stalked meadow grass is not regarded highly there as an agronomic prospect. It is low yielding and, although its rhizomes make it persistent and possibly even difficult to get rid of when the pasture is ploughed, it takes a long time to establish. Winter and spring growth are satisfactory but there is little activity in the summer. The species is very variable genetically with somatic chromosome numbers ranging from 28 to 154, partly because it may be apomictic (reproducing by seeds in the absence of fertilization). In the United States, Kentucky bluegrass has for long been one of the most important pasture grasses covering 20 to 40 million hectares in north-eastern USA and further areas in Canada. It is also considered to be the most important lawn grass. Although midsummer production is low, its contribution at other times and its nutritive value are highly regarded. Rotational grazing or cutting to maintain a sward height of 5–15 cm is considered optimal, together with regular liming and fertilizer applications to sustain the grass and its associated legumes.

Rough-stalked meadow grass (*Poa trivialis*) is a stoloniferous creeping perennial, also with flattened, glabrous tillers and folded leaves. The leaves taper distinctly from the base and have a pointed tip (Fig. 4.37). They are slightly shiny on the underside and also show the double line running longitudinally on either side of the central vein. There are no auricles, and the ligule of the lower leaves is short but on the upper leaves it is long. Reproductive structures are similar to those of *Poa pratensis*, except that

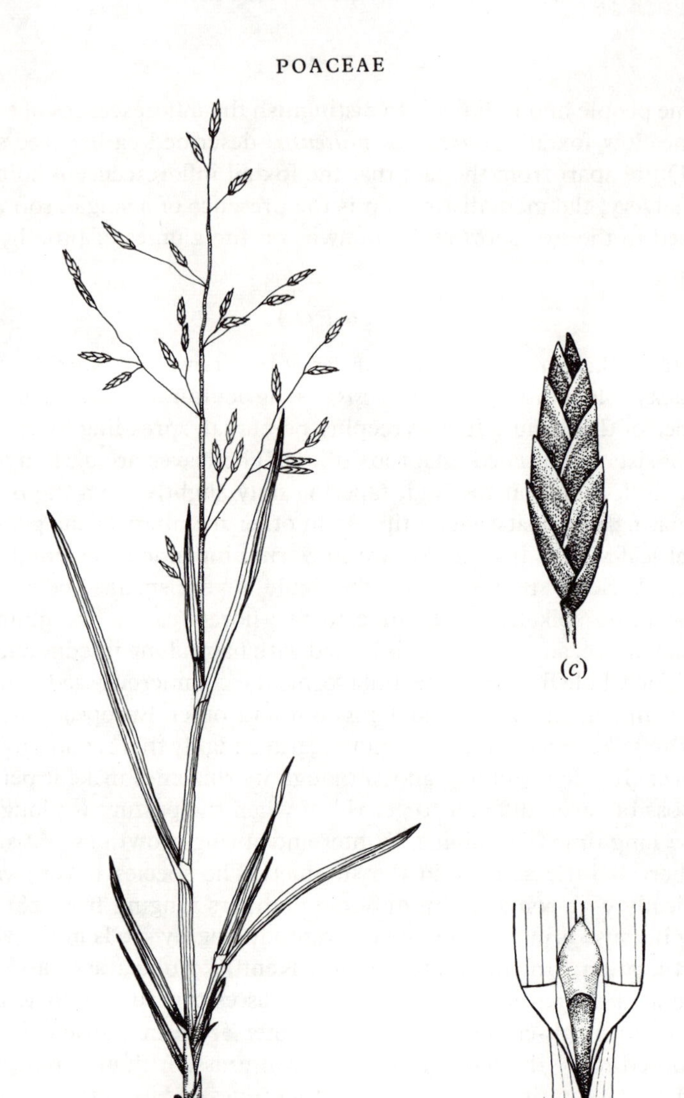

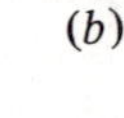

(a)

Fig. 4.37. Rough-stalked meadow grass (*Poa trivialis*). (*a*) plant with panicle, (*b*) ligule, (*c*) spikelet.

the lowest node of the panicle may be slightly rough through the presence of some very small spines and that only the lower half of the keeled lemma is hairy. Rough-stalked meadow grass grows well under wet conditions and is thus common in damp, low lying pastures. Its yield is low and confined mostly to spring and early summer. It was at one time included in pasture seed mixtures in Britain but is now out of favour. In the United States its use is usually confined to shady lawns.

A few other members of the genus *Poa* deserve to be mentioned. The first is Canada bluegrass (*Poa compressa*) which resembles Kentucky bluegrass except for a distinct blue-green foliage, fewer basal leaves below the flattened panicle and later maturity. It does not recover well from grazing. Annual meadow grass (*Poa annua*) will be known to all gardeners and arable farmers in temperate regions as a widespread weed. It has relatively broad and blunt leaves with a conspicuous ligule, but perhaps its most distinctive feature is that it flowers at any time of the year because it is a day-neutral plant. Control by cultivation is not difficult but copious seed production ensures survival of the species. Several other *Poa* species, mutton bluegrass, Texas bluegrass, and big bluegrass are native to North America, and *Poa nemoralis* occurs in British woodlands.

4.27 *TRISETUM*

Golden or yellow oatgrass (*Trisetum flavescens*) is a low-yielding perennial adapted to dry situations in England and Wales. It is a tufted perennial with rounded tillers and leaves which are ribbed above and hairy on both surfaces. The hairs on the sheath point downwards. Auricles are absent but the serrated ligule is prominent. The inflorescence is a panicle, smaller than but resembling that of tall oatgrass (see section 4.12), but the spikelets become distinctly yellow as they ripen. There are two to three florets per spikelet, and the lemmas are furnished with a twisted, bent dorsal awn. Golden oatgrass is fairly drought-resistant, it is readily eaten by stock, especially sheep, but its productivity is too low for it to be included in sown pastures.

4.28 *VULPIA*

Ratstail or squirreltail fescue (*Vulpia bromoides*) is an annual which occurs in Britain in dry grassland and as a weed contaminating ryegrass seed samples. The plant resembles a small fescue. It bears a panicle with numerous many-flowered spikelets. Each floret contains only one stamen. The lemma bears a long terminal awn which is useful in the dispersal of the seed. It flowers and scatters its seeds early and thus continues to reappear.

Another member of the genus, *Vulpia myuros*, occurs in waste places in Britain but was also introduced to Australia where it plays a minor part as an annual constituent of unsown grassland.

Subfamily Panicoideae

Members of this subfamily occur commonly in the tropics or subtropics, and if cultivated in more temperate climates, their growth is confined to the warm, frost-free period of the year. Most of the species are large, vigorous plants, and many have the C_4 type of photosynthetic mechanism (see section 17.4). Some stem formation tends to occur before the onset of the reproductive phase, and the mature stems are often so tall and heavy as to require the support of so-called prop or strut roots. These are strongly thickened adventitious roots arising from one or more nodes above soil level whose function is to provide mechanical strength rather than to absorb water or nutrients.

Two tribes are now recognised in making up this subfamily. The first of these tribes are the Andropogoneae, large grasses with solid internodes and with panicle-type inflorescences, which include such economically important plants as sugar cane, maize and sorghum. The second tribe, the Paniceae, contains quite a number of agricultural genera, among them various types of millet but also numerous forage grasses such as Guinea grass, Pangola grass or Kikuyu grass. The basic chromosome number in the subfamily Panicoideae is 9 to 10, and the mitotic chromosomes are small.

Tribe Andropogoneae

Most of the genera belonging to this tribe are large grasses with solid internodes. The spikelets contain one or two florets, and these may conform to the usual bisexual pattern or they may have either male or female organs only. Alternatively some genera have separate male and female inflorescences. Economically the most important plants in this tribe are maize, sorghum and sugar cane.

4.29 MAIZE (*ZEA MAYS*)

In terms of production throughout the world, maize is at present the third most important grain crop following closely upon wheat and rice. Although requiring a frost-free climate for growth and reproduction, maize is now increasingly grown in temperate regions, not only as a forage crop for silage but also for the production of grain. Vigorous growth and high yields are attributable to the C_4 pathway of photosynthesis of the plant (see section 17.4).

Origin of maize

Maize is one of several important crop plants which the early explorers of the fifteenth century discovered in the New World and brought back to Europe among their trophies. Although it is possible to pinpoint Central America as the early centre of cultivation of maize, the origin of the plant is more obscure because it does not apparently occur in the wild form. However, there occurs in Mexico a closely related grass, teosinte (*Euchlena mexicana*), which hybridises freely with *Zea*, and thus it has been suggested that modern maize is either of hybrid origin or that in some other way it is a derivative of teosinte. Maize pollen has been identified in excavations in Mexico City dating back some 80 000 years, and it is also well known that maize was originally a very much smaller plant than it is now, closer in stature to teosinte.

The maize plant

Maize is a tall annual grass with thick, solid stems usually supported by prop roots. The leaves are broad and smooth with a conspicuous midrib. Very few tillers are produced. The plant bears separate male and female inflorescences. The highly branched terminal panicle, or tassel as it is called (Fig. 4.38), contains only male florets consisting of a lemma and palea, two lodicules and three yellow, green or purple anthers. Two florets surrounded by glumes make up a spikelet, and these also occur in pairs, one almost sessile, the other on a short pedicel. Pollen is shed and carried by wind to the female inflorescences or cobs which occur in the axil of some leaves in the mid-region of the stem, in the same position in which tillers are normally found. Each cob consists of a number of compressed internodes, bearing at each basal node a modified leaf with little or no lamina and terminating in a ligule. This tightly packed structure is referred to as the husk and serves to protect the female inflorescence, which in its turn forms an unbranched axis analogous to a spike. On it the sessile spikelets occur in distinct longitudinal double rows. Each spikelet contains two florets, one sterile, the other fertile, but in both the lemma and palea are highly reduced in size. The same applies to the glumes, so that all that is visible in a mature cob (Fig. 4.39) are the rows of naked caryopses which need to be removed to display these other structures whose normal protective function has been taken over by the husk. On the other hand, the styles are highly conspicuous because of their great length. They extend

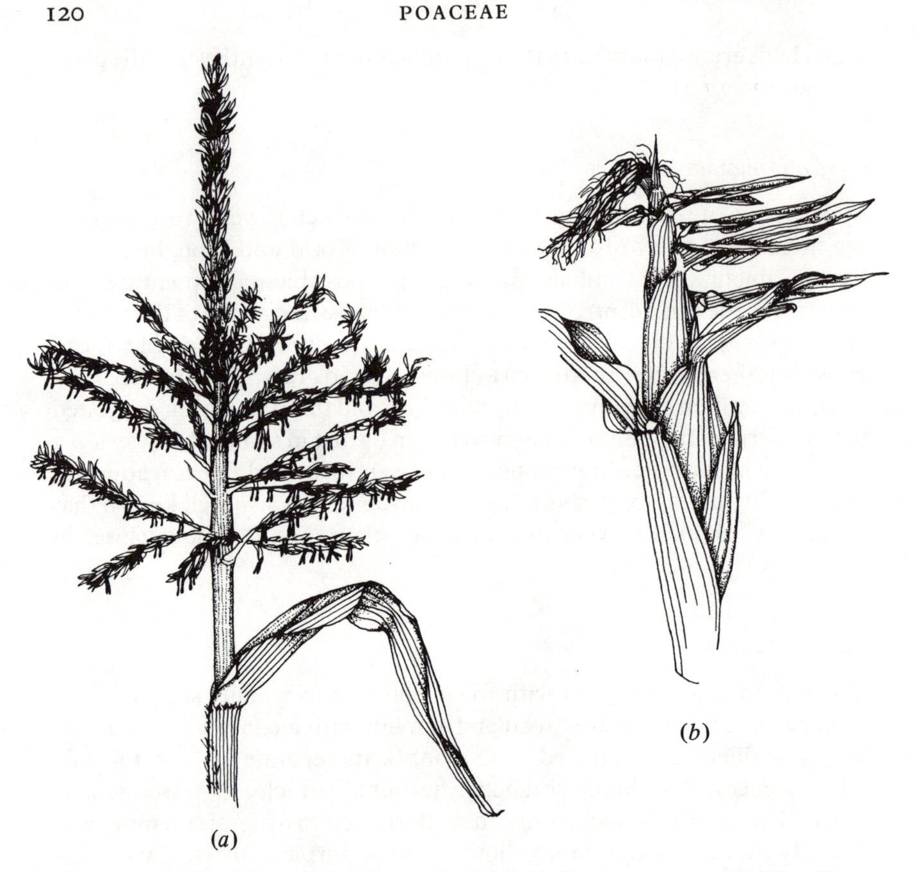

Fig. 4.38. Maize (*Zea mays*). (*a*) terminal male inflorescence (tassel), (*b*) axillary female inflorescence (cob).

from each ovary to above the top of the cob from which they protrude as a tight bundle of threads called the silk (Fig. 4.38). Pollen grains settle on the stigmatic surface of these threads, and on germination produce a pollen tube which continues growing downwards until it has reached the ovule where pollination takes place. Grain development now follows and the cobs mature inside the modified leaves forming the husk. Unlike in other grasses, there is thus no seed dispersal, which accounts for the fact that maize is not known as a wild plant.

Hybrid maize

Before maize was brought into intensive cultivation, there will have been a high degree of cross-pollination among the heterogeneous populations of

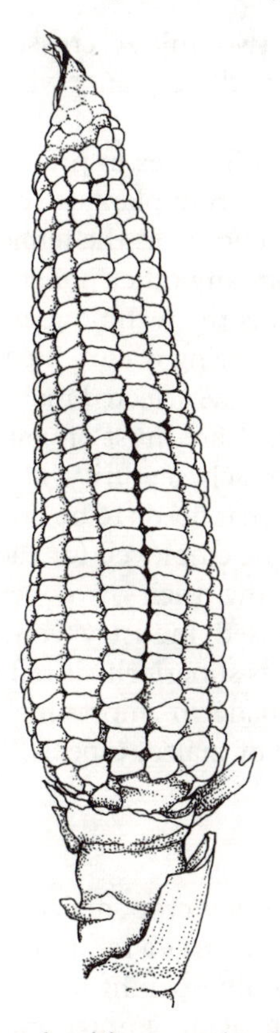

Fig. 4.39. Maize (*Zea mays*); cob with protecting leaf sheaths removed.

plants grown by man. With increasing selection for uniformity, self-pollination among plants of the same type will have occurred increasingly, although not to the same extent as in other grasses with hermaphrodite flowers. However, because of the separation of male and female inflorescences in maize, it is possible for man to control the degree of inbreeding or outbreeding quite simply, and this is the basis for the production of hybrid maize. If for example, the same line of maize (A) is self-pollinated for a number of generations, inbreeding depression results and the vigour of the progeny declines. Suppose that another line (B) had been similarly inbred and that after some generations of selfing A and B are now crossed. The resulting hybrid is normally superior to either parent owing to the expression of hybrid vigour or heterosis. Even better results are obtained by including two further inbred lines (C and D) which are eventually crossed, and this is followed by the production of a so-called double hybrid from the two initial crosses AB and CD. Superior production is obtained from these plants which are now widely used in modern agriculture.

There are few technical difficulties in producing hybrid maize because pollen can easily be collected from plants selected as the male parent and used to pollinate the plants chosen as the female parent whose stigma are subsequently protected from any other pollen. Easier still and applied to large-scale field production is the technique of growing breeding lines in alternating strips. When flowering occurs the tassels of the female parent plants are cut off so that only the selected male parents growing alongside can bring about pollination. Care must of course be taken to ensure that both pollen and stigmas reach maturity at the same time. There is, however, one further point that needs to be noted and this is to remember that new hybrid seed must be employed for each sowing and that it is not possible for farmers to sow their own seed. This has led to the establishment of a specialised industry continually producing hybrid maize, and not unnaturally its superiority is reflected in the cost of the seed. Great strides have been made in improving the productivity, quality, disease and pest resistance of maize, especially in the USA.

Uses of maize

Although predominantly a grain crop, maize has many other uses as well. In the first place it is often grown as a forage crop especially in areas where the climate is marginal for grain ripening but where advantage can be taken of the rapid growth of the plant. Forage maize is cut at an immature reproductive stage and is either fed to animals in the fresh state or after having been ensiled. Chopping and ensiling improve the feeding value of the material. Another use of maize which has grown considerably in importance over the years is as a vegetable, namely sweet corn. For this purpose the plant breeder has selected lines of maize in which the endosperm remains soft and sugary for some time, thus enabling the crop to be harvested and sold before starch formation has gone very far. It will be obvious that the quality of this vegetable will be at its best when it is cooked and consumed as soon as possible after harvest, and even more delicate is the flavour of baby corn, the immature cob of special cultivars.

Economically the most important product of maize is the grain which is a valuable source not only of starch but it also contains more oil than most other cereals. A typical analysis of maize grain would be 4–4.5% fat, 9.5–11% protein, 70–72% soluble carbohydrate and 11% moisture. Several different types of grain have been developed by plant breeders. Flint corn has a hard endosperm and, since there is thus little shrinkage at maturity, the grain remains rounded; it is also opalescent in appearance and is rich in protein. Dent corn has similar characteristics but here the inner part of the endosperm is soft and it shrinks as it dries out, giving the grain an indented shape. These two types of maize are by far the most important but there is also flour corn in which the endosperm is soft and floury, and this was the cereal grown originally by the American Indians. Maize grain is used for a great variety of purposes, mainly as an animal feed but also as a breakfast cereal, corn flour, a source of starch, syrup and glucose. In fact increasing quantities of sugar are now being obtained from maize, and this may be fermented to produce alcohol. Corn oil is rich in essential fatty acids and is used widely as a salad and cooking oil.

As a source of protein, maize is somewhat deficient in lysine. This does not matter very much if the grain is fed to sheep or cattle, because ruminants can synthesise this amino acid, but man and monogastric animals are unable to do so. Fortunately, it was discovered that the level of lysine was genetically controlled and that it could be increased through plant breeding by incorporating a gene called *opaque-2* in crossing programmes. Initially the high-lysine hybrids did not yield quite as well as expected, and the high moisture content of the grain contributed to crop losses in unfavourable weather conditions. Another gene, *floury-2*, also improved protein quality but also depressed yield. Similarly, genetic manipulation has led to improvements of maize as a source of oil, to increased content of amylose used in the manufacture of plastic films, and to increased carotenoid and vitamin A concentrations.

Finally, two other types of maize should be mentioned briefly. The first is waxy corn which produces a waxy endosperm which on milling gives a flour resembling tapioca, normally obtained from the root of cassava. The other is popcorn, varieties with small, pointed grains whose endosperm expands rapidly on heating, whereupon the grain bursts open turning the endosperm inside out. Suitably coloured and flavoured popcorn makes its appearance at fairs and similar events.

4.30 SORGHUM

The genus *Sorghum* is found in warm, dry climates, especially in Africa, India, Pakistan, China and the southern USA where its members are grown as important grain or forage crops. Because sorghums have been in cultivation for a long time and because interspecific hybrids are easily formed, the taxonomy of the genus is somewhat confused. Until recently the cultivated types were loosely grouped together in the species *Sorghum vulgare*, but a thorough revision of taxonomic relationships suggests that *S. bicolor* is the species to which the grain crops should belong. Very large numbers of types and local varieties are in existence, although five basic races have been recognised which can combine in pairs to form cultivated types. Sorghum is not quite as tall a plant as maize, but it resembles it in other respects by having few tillers, prop roots, and broad, waxy leaves with a well-developed midrib. The inflorescence is a compact panicle, highly branched except at the tip of the rachis. The spikelets occur in pairs, of which one is broad, sessile and grain-bearing, while the other is long and narrow, attached by a pedicel, and it contains male florets only. Each sessile spikelet has two florets, a sterile one represented by only a lemma, the other perfect containing a grain surrounded by a palea and a lemma bearing a short awn. The glumes are large, distinctly keeled, and thus useful in protecting the fertile floret. The caryopsis is 4–8 mm in diameter, rounded with a blunt point, but shape and size depend on variety, as does the colour which varies from off-white to yellow, brown, red and black.

The seed of grain sorghum, or dura as it is often called, contains no gluten, and hence is by itself not suitable for bread making. Normally for human consumption the grain is ground into a flour, mixed with water or fat and cooked to form a porridge or batter. Alternatively, the grain is fed to pigs or poultry, its starch may be used for a variety of purposes such as an adhesive or for sizing, or it may be fermented to produce alcohol. Sorghums are also grown as forage crops and may produce high yields of the order of 30 000 kg ha^{-1} from several cuts throughout the year. Best results are obtained from special forage types such as sorgo, sweet or sugar sorghum which is variously described as a variety of sorghum or as a separate species (*Sorghum saccharatum*). Sudan grass (*S. sudanense*), a tall and tufted tropical grass, is often used to produce hybrids with sorghum, as for example the productive forage plants Sudax or Sordan. One possible problem with these forages is that contained in the leaves there may be a cyanogenic glucoside, dhurrin, which when eaten by animals hydrolyses to form poisonous hydrogen cyanide. In ruminants, hydrogen cyanide is rapidly detoxified in the rumen and liver by reactions with sulphide or cystine, but the danger of toxicity remains if glucoside levels are high or sulphur intake is low. The problem can be avoided by the choice of safe cultivars, by not allowing stock to graze very immature growth, or by feeding as hay or silage.

Largely because of resistance to drought and through the release of highly productive grain and forage cultivars which make full use of their C_4 physiology, sorghums have in recent times increased in importance, especially in the United States. Infinitely less desirable is *Sorghum halepense*, Johnson grass, a noxious weed which occurs in subtropical to temperate regions in many parts of the world. It spreads readily by seed or through its rhizomes. Single plants could produce some 80 000 seeds each in a single season, and these can remain dormant for several years. The main problem caused by Johnson grass is its persistence and the danger of poisoning livestock through cyanogenic glucosides which occur in both green and dried plants, especially under adverse conditions such as drought, high temperature or frost. On the other hand, in parts of the USA, improved lines of this grass have been found to be persistent and productive enough to provide good forage under rotational grazing.

4.31 SUGAR CANE (*SACCHARUM OFFICINARUM*)

Sugar is a comparatively recent addition to the range of foods available to man. For very many centuries honey and secretions from certain trees provided sweetening materials, and it was only when sugar cane in the tropics, and later sugar beet in temperate regions, came into cultivation that commercial sugar production could begin. Several species of sugar cane are in existence, but the most widely grown of these is *Saccharum officinarum* which is commonly cultivated particularly in Cuba, the West Indies and Guyana, southern USA, Hawaii, Brazil, Mexico, India, Indonesia, Australia, and South Africa. This crop requires a warm, tropical climate, fertile soils and an abundant water supply. Most of the sugar occurs as sucrose in the solid internodes of the stems, and the leaves which contain mainly monosaccharides are of no commercial value.

Sugar cane produces little viable seed and is thus commonly propagated vegetatively by the use of short stem cuttings consisting of several nodes and internodes. These so-called setts, 20–25 cm long, produce fibrous roots on planting but these are gradually replaced by adventitious roots arising at the base of the new aerial shoots. These stems reach a height of 2–4 m, depending on variety and growing conditions. The internodes are thick, often coloured, and have at their base one or more rings of root initials, a narrow meristematic zone and a lateral bud. The long, pointed leaves have a papery ligule and an auricle. The edge of the leaf is often toothed and quite sharp. The inflorescence is a large, open panicle which is usually referred to as the tassel or arrow. Spikelets in pairs are borne on its branches, each containing two florets, one fertile, the other represented by only a lemma. Long, silky hairs below the spikelets give the whole panicle a feathery or fluffy appearance. Some seed is produced, but it is very small and remains viable for such a short period that in practice it is of interest only to the plant breeder.

It takes more than a year and up to 18 months from the time of planting to produce a crop of sugar cane. During this time the plants tiller freely and produce a dense population of tall stems, which contain the sucrose for which the crop is grown. The upper portion of the flowering stems and the leaves are usually removed before harvest because they also contain reducing sugar which is not required for sugar production, even though it is of value for fermentation to ethanol. However, in general it is commercially desirable to prevent flowering, and this is done either by choosing genotypes whose precise short-day requirements for flowering are not met in the photoperiodic conditions in which they are grown, or alternatively by interrupting the night by flashes of light which stops flower initiation in short-day plants. Such chemicals as diquat may also be used for the same purpose. After harvesting, the plant grows up again and several more so-called ratoon crops are taken, each needing about a year to mature. The main and traditional use of sugar cane is for various types of sugar but more recently, in answer to the liquid fuel demand, ethanol is being produced from it. The production of sugar involves the extraction of the juice which is boiled and concentrated to a thick syrup. Impurities are removed and the resulting raw sugar purified by successive washing and recrystallisation. One of the most important by-products is molasses, a dark brown, sticky liquid which contains some 50% sugar, about one third of it in the form of reducing sugar, which is widely used as an additive to stock foods and in the making of silage. Fermentation of molasses results in the formation of alcohol, and this may be distilled to yield rum. The fibrous residue left after sugar extraction, bagasse, is used as a source of fuel and for the manufacture of paper, fibre board, cardboard or as a raw material for the production of the aldehyde furfural and plastics. Increasingly, the by-products of sugar manufacture have gained in economic importance, and there is also a rapidly developing sucrochemical industry for which sugar serves as the basic raw material.

4.32 OTHER ANDROPOGONEAE

Several other members of the tribe Andopogoneae are of some botanical and economic importance. The first of these is Job's tears or adlay (*Coix lachryma-jobi*), which is grown as a grain crop in parts of Asia. This is a freely tillering annual grass, some 1–2 m tall, with large leaves bearing a short ligule but no auricles. The distinctive feature from which the plant derives its name is the occurrence of false fruits, round or ovoid bead-like structures in the inflorescence which contain a spikelet with a single female flower. The male spikelets extend above the bead but fall to the ground at maturity. Although grown mainly for its grain which is parched and ground as a source of hot beverage, the false fruits varying in colour from white to blue, pink, brown or black are articles of commerce for the production of necklaces and rosaries.

The second plant in this tribe worthy of mention is red-oat grass

(*Themeda triandra*) which is widely distributed in East Africa and other tropical areas of the same continent. This is a valuable fodder grass, although it recovers slowly from cutting or grazing. Its ecological importance hinges on the fact that its seeds bury themselves in the soil through the action of an hygroscopic awn (see also wild oats p. 39) and thus escape the effects of grass fires. Mature plants also survive fires and become the dominant species in many regions. *Themeda triandra* is common also in the natural grasslands of Western India, Pakistan, Burma and the Philippines, but owing to its slow establishment and poor seed yields it has so far not been used in sown pastures.

The genus *Cymbopogon* contains several species that are of economic significance because of the presence of aromatic substances in their leaves. Botanically they are distinctive through their panicle-type inflorescence which consists of a multiplicity of pairs of short racemes containing paired spikelets and subtended by a conspicuous spathe. Lemon grass (*Cymbopogon citratus*) is grown in the West Indies, parts of South America, equatorial Africa and southeast Asia for its essential oil. Lemongrass is valued for flavouring many dishes and as a source of citral and other aldehydes which are used commercially for the preparation of scents, notably that of violets, in the pharmaceutical industry as well as for the synthesis of vitamin A. Citral is also extracted from Malabar grass (*Cymbopogon flexuosus*) grown in India, and because of its greater solubility in alcohol is more easily incorporated in perfumes. Also occurring in Asia is *Cymbopogon nardus* which is cultivated in southern India and Sri Lanka for the citronella oil content of its leaves. A higher yielding but less hardy variety of this plant was developed by the Dutch in Indonesia where it gave rise to an important industry. Other countries, notably Guatemala, Brazil, Taiwan and some central African states, have since become producers of citronella oil. Although natural products are being increasingly replaced by synthetics, this oil continues to be used as a source of citronella and geraniol in the pharmaceutical industry and as a base for insect repellents and ointments. Another substance used in the manufacture of perfumes is khus-khus or vetiver oil, obtained from the rhizomes of *Vetiveria zizanioides*, also a member of the Andropogoneae.

Tribe Paniceae

This tribe contains some grain crops of local importance but also forage and pasture grasses which are widely distributed. In distinction to the tribe Andropogoneae, the glumes of this group of grasses are normally not hard and only one per spikelet may be present. The spikelets contain two florets of which the lower is either male or sterile.

4.33 *AXONOPUS*

Carpet grass or savanna grass (*Axonopus compressus*), native of the West Indies and Central America is a stoloniferous perennial which is usually propagated by dividing up the roots. It does not tolerate drought but grows well in tropical areas with well-distributed medium to heavy rainfall. Under these conditions it withstands repeated close grazing or mowing. Not only is it a reasonably nutritious species but it is also well adapted to growing in lawns or for soil conservation.

4.34 *CENCHRUS*

Of the 25 species in this genus, buffel grass (*Cenchrus ciliaris*) is the best known. Its appearance is best described by its alternative name, African foxtail, because the bristles surrounding each spikelet on the spike cause it to resemble meadow foxtail (*Alopecurus pratensis*) (see section 4.9). In fact, some other species belonging to this genus have such hard bristles as to cause contamination of sheep's wool. Buffel grass is a tufted or rhizomatous perennial with a strong and deep root system which makes the plant drought resistant and persistent. Swollen stem bases which store carbohydrate ensure rapid recovery after defoliation, drought or fire. Most of the growth occurs in the summer, and cold tolerance is only moderate. Tall cultivars growing up to 1.5 m have been released, but there are also medium-height lines which reach 1.0 m and short ones, only about 80 cm tall. Establishment from seed is quite easy, which makes buffel grass a suitable species for reseeding denuded grassland in warm, dry climates. The main areas over which it is distributed are in tropical and subtropical Africa, India and Pakistan, but it has also been successfully introduced into Australia and the southern United States.

4.35 *DIGITARIA*

Pangola grass (*Digitaria decumbens*) is one of the most important pasture grasses of the Caribbean and tropical America, although it is of South African origin. It has been successfully introduced as a constituent of improved pastures into many countries with tropical, subtropical and warm-temperate climates. Like several other tropical grasses it does not produce fertile seeds and hence has to be propagated vegetatively by stolons which are spread on the surface and disced into the soil. In fact, it is the vigorous growth of its stolons that makes this grass highly competitive and tolerant of heavy grazing. It grows well over a range of climatic and soil conditions except that it does not withstand waterlogging. A related species, African couch (*Digitaria scalarum*), a common weed of plantation crops, is difficult to eradicate owing to its extensive rhizomes. Another member of the genus is hungry rice or fundi (*Digitaria exidis*), which serves as a grain crop in parts of West Africa, especially under arid and infertile conditions.

4.36 *ECHINOCHLOA*

Japanese barnyard millet (*Echinochloa frumentacea*) is one of the quickest maturing of all cereals and may give a crop of grain some 45 days after sowing. It grows up to about 1 m in height and produces an elongated panicle, often tinged purple, on a distinctly angled stem. The spikelets contain two florets, of which the upper one produces a grain. The caryopsis is tightly enclosed by a whitish lemma and palea. This crop is considered a poor substitute for rice when the paddy crop fails, largely because of the speed with which it matures. It is grown as a cereal in the Far East but has also found a use as a fodder crop in the United States.

Barnyard millet (*Echinochloa crus-galli*) is a useful fodder plant especially adapted to poorly drained situations in the tropics, although in California and elsewhere it also features as a weed of rice crops. Another member of the genus is antelope grass (*E. pyrimidalis*), which is commonly found in swamps in tropical Africa but is being developed as a hay and pasture plant.

4.37 *MELINIS*

Molasses grass (*Melinis minutiflora*) is native to tropical Africa but is of interest as a pasture grass in other regions with a similar climate particularly as a pioneer on well-drained areas. It has the advantage that it can be propagated by seed. This grass is named after its strong smell and the viscous surface of its hairy leaves (Fig. 4.40). It is quite palatable, and stock take to it readily despite its odour, although it does not persist under heavy stocking.

4.38 *PANICUM*

This genus contains some widely distributed forage grasses as well as small millets grown as cereal crops. The forage grasses are represented by guinea grass or panic (*Panicum maximum*), a perennial growing up to 3 m in height, which is one of the most valuable sources of animal fodder in the tropics (Fig. 4.41). Although its panicles bear fertile florets, the seeds shed easily and are low in viability, and thus vegetative propagation is

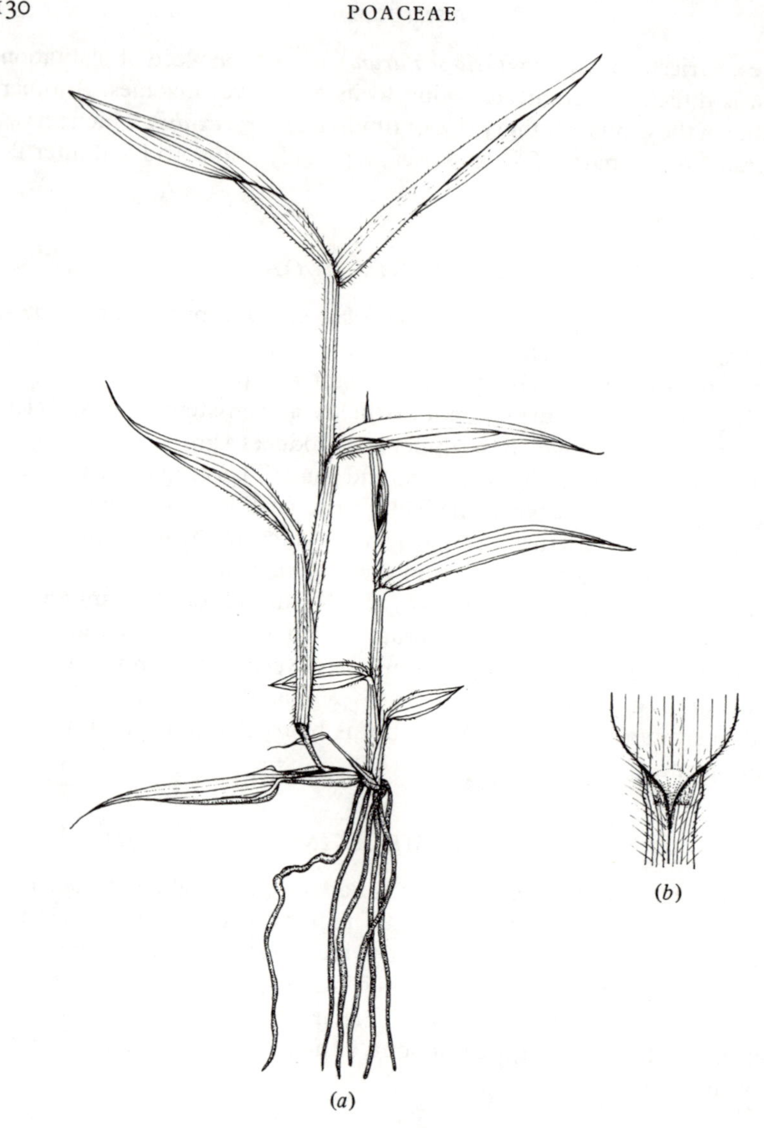

Fig. 4.40. Molasses grass (*Melinis minutiflora*). (*a*) plant, (*b*) leaf base.

predominant. The first cut can be taken some ten weeks after planting, and cutting can be repeated every six to eight weeks, producing very high yields per annum.

Common millet (*Panicum miliaceum*) is one of a group of cereals referred to as small millets which also includes members of other genera such as

Fig. 4.41. Guinea grass or panic (*Panicum maximum*). (*a*) plant (*b*) ligule and fringe of hairs.

Paspalum, *Setaria* and *Echinochloa*. Together they occupy several million hectares throughout the world, especially in India. Common millet, also known as proso, hog or Russian millet, has been in cultivation since the dawn of history, and was certainly well known to the Romans under the name of milium. Because of its low water requirement, lower than that of other cereals, nutritious grain, and its quick maturity, it has held its place as a crop in eastern Asia, Central Russia, the Middle East and elsewhere.

The plant stands up to 1 m in height, tillers moderately and produces drooping panicles containing spikelets with one fertile floret. The white, almost round caryopsis is tightly enclosed by a lemma and palea. Common millet is harvested as early as possible, because the grains are quickly lost through shattering. The hulled grain is cooked whole or ground into flour for preparation of chapati, porridge or gruel.

A close relative of common millet is *Panicum sumatrense*, little millet, which is a minor cereal crop restricted to parts of India. Its main characteristic of note is extreme hardiness, for it will grow and bear grain in very poor conditions, thus providing some sustenance when other crops fail.

4.39 *PASPALUM*

Kodo millet (*Paspalum scrobiculatum*) is a minor cereal, confined almost entirely to the southern states of India where it is valued for its drought resistance. It is an erect annual, some 70–80 cm high, with stiff leaves and an inflorescence consisting of two to eight spikelike racemes each bearing spikelets in two rows. The species is mostly cleistogamous, in that the florets rarely open and self-pollination is the rule. Seedlings of Kodo millet show some purple colour, but this pigmentation intensifies after ear emergence to the extent that crops can easily be identified from a distance. The caryopsis is tightly enclosed by a brown hardened lemma and palea which are difficult to remove. The grain is rather coarse but it is easily preserved and thus serves as an insurance against famine. Immature grains and husks are said to be unsafe for consumption possibly because of some fungus infection.

Several members of the genus *Paspalum* are useful forage grasses. Buffalo or sour grass (*P. conjugatum*) is a common species throughout tropical Africa and America, and Bahia grass (*P. notatum*) has its uses as a pasture plant and for soil conservation because of its deeply penetrating root system. Perhaps best known outside its native tropical South America is Dallis grass (*P. dilatatum*), also referred to simply as paspalum, which has been introduced as a pasture or forage species into many warm and humid regions of the world, including Queensland, New South Wales and northern New Zealand. In the vegetative state this grass may be recognised by its glabrous leaves bearing a few long hairs at the base and a conspicuous membranous ligule (Fig. 4.42). In New Zealand paspalum (*P. dilatatum*) has been shown to be a valuable component of a mixed grass/clover pasture, providing nearly 40% of annual production, particularly during the summer months. There are, however, indications that perennial ryegrass and paspalum are not necessarily compatible at all times, but the clover component is usually well maintained. Paspalum appears to be

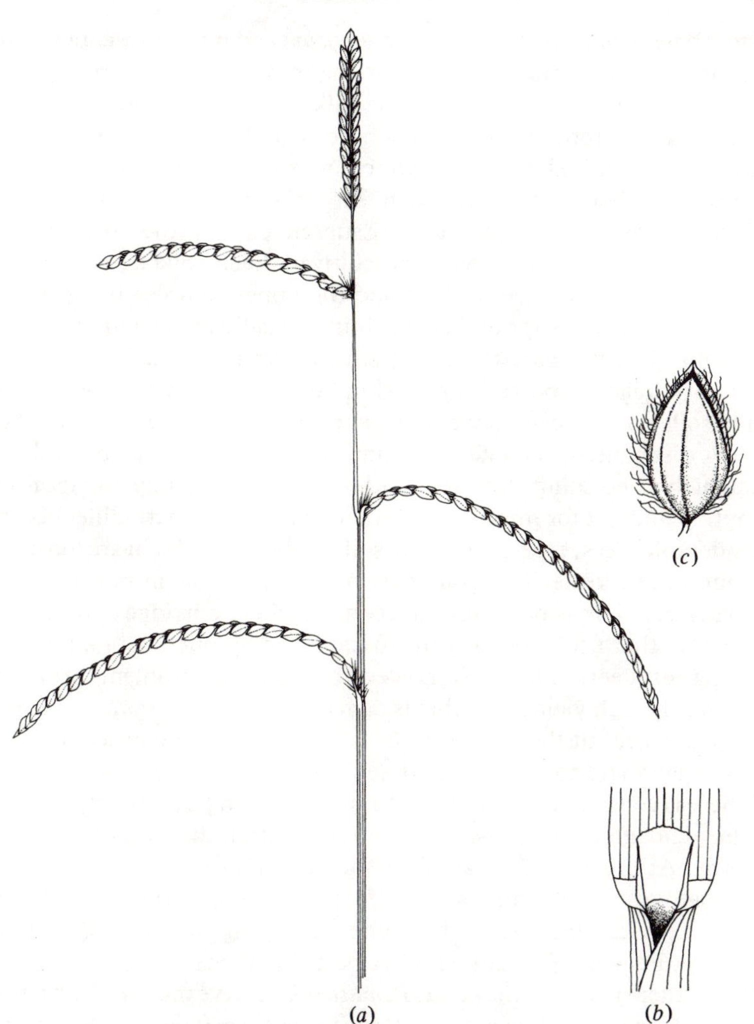

Fig. 4.42. Paspalum (*Paspalum dilatatum*). (*a*) inflorescence consisting of spike-like racemes, (*b*) ligule, (*c*) spikelet containing single floret.

better adapted to dairy swards than to tightly grazed sheep pastures, although it seems to be tolerant of heavy treading.

4.40 *PENNISETUM*

Bulrush, pearl or cat-tail millet (*Pennisetum glaucum*) is one of the most important grain crops in the drier parts of tropical Africa and in India. Its correct botanical name is *Pennisetum glaucum* and it is represented by four cultivated races of which *typhoides* is the most widely grown. It is a tall annual plant with solid stems and long, silky hairs on the nodes. The leaves are up to about 100 mm long with a narrow ligule and a fringe of hairs at their base. The inflorescence, which resembles a bulrush, gives the impression of being a spike but is in fact a highly contracted panicle. The spikelets usually occur in pairs and are subtended by 30 to 40 bristles which vary in length and colour with the cultivar. Each spikelet contains two florets of which the lower is male and the upper completely fertile. The caryopsis varies in shape and colour but is usually ovoid and off-white to dull light blue; it separates easily from the lemma and palea. One of the great advantages of bulrush or pearl millet is its ability to grow on poor, sandy soils and in relatively dry areas, provided that the rainfall is fairly evenly distributed. Unlike sorghum, it cannot withstand prolonged drought by becoming dormant. Although grown mainly for local consumption and not for international trade, bulrush or pearl millet occupies considerable areas, especially where soil conditions and rain are too poor to sustain other cereals. The grain may be cooked whole in much the same way as rice, or it may be ground and consumed as a porridge or unleavened bread. Another use, especially in Africa, is the production of malt for the brewing of beer. There is, however, a serious problem preventing consistently high yields, and that is damage by birds. Not only is the grain very palatable, but the inflorescence is conveniently constructed to allow birds to land and to cling to it while feeding.

There is considerable variation within cultivated pearl millet, and this is distinct enough to recognise separate races within the species *Pennisetum glaucum*. At present, there are four such races, of which *typhoides* is the principal pearl millet cultivated in Asia and the arid savannas of South Africa, Senegal and Egypt. The other races, *nigritarum*, *globosum*, and *leonis* feature as crop plants in other parts of Africa.

Two other species in the genus *Pennisetum* deserve mention. The first of these is elephant or Napier grass (*Pennisetum purpureum*), a tall perennial, native to high rainfall areas in tropical Africa but now widely distributed through introduction into many countries. Although it flowers profusely, the seed is difficult to collect, and hence propagation is normally by stem cuttings. If well maintained, especially by application of nitrogen, Napier grass is very productive and provides bulky forage from repeated cuts throughout the year. Grazing is quite possible but must be rotational and not too intensive. One of the by-products from this plant comes from the culms which are hard and strong enough to be used for fencing and as a building material. The second species of economic importance is Kikuyu grass (*P. clandestinum*), also a native of Africa where it occurs at high altitudes and in the presence of at least 1000 mm of rainfall near the equator. This grass has also been introduced into many other areas, such as the Andes, and extends as far afield as northern New Zealand where it shows some promise as a pasture grass. Propagation is one of the main problems, for it does not flower readily and the sparse culms that are produced are only a few centimetres long and bear few spikelets (Fig. 4.43). On a practical scale planting of stem cuttings is necessary, but the plant spreads quite rapidly through rhizomes and stolons, forming a dense sward with other grasses and clovers.

4.41 *SETARIA*

Golden timothy or setaria grass (*Setaria anceps*), probably better known to agriculturalists as *S. sphacelata*, is a tufted perennial with glabrous leaves and tightly folded sheaths. The inflorescence is a spike-like panicle bearing characteristic bristles below the spikelets. This variable species is widespread in Africa but its ability to grow well at relatively low temperatures of 20–25 °C and to withstand light frosts has allowed it to be introduced to southern Australia, USA and other non-tropical areas, provided that rainfall is over 750 mm annually. Establishment is by seed or by discing vegetative material into the soil. Setaria combines well with some legumes, it persists under grazing and provides palatable herbage for livestock. One problem related to this grass is its content of anhydrous oxalic acid which can be high enough in some cultivars to cause oxalate accumulation in the kidneys of cattle, leading to ill-thrift and even death.

Also well known and more widely distributed is *Setaria italica*, the foxtail millet or Italian, Hungarian, German or Siberian millet. This annual has been grown as a cereal since ancient times, probably one of the first crops to be domesticated in central and eastern Asia. Not only does it produce good grain for human consumption and for feeding to livestock and birds, but it is also used extensively in the USA and elsewhere as a fodder crop for hay and silage. The plant tillers freely and produces long leaves which are smooth above and slightly rough below. The inflorescence is a spike-like panicle, covered with soft, white hairs and bearing up to 12 spikelets at the end of its branches. Each spikelet is subtended by bristles and contains two florets, of which the lower one is sterile. The grain is tightly enclosed by a lemma and palea whose colour varies from pale yellow to orange, red, brown and black. Foxtail millet is widely grown in Japan and India extending into other parts of Asia, south-eastern Europe, and parts of North Africa and the USA.

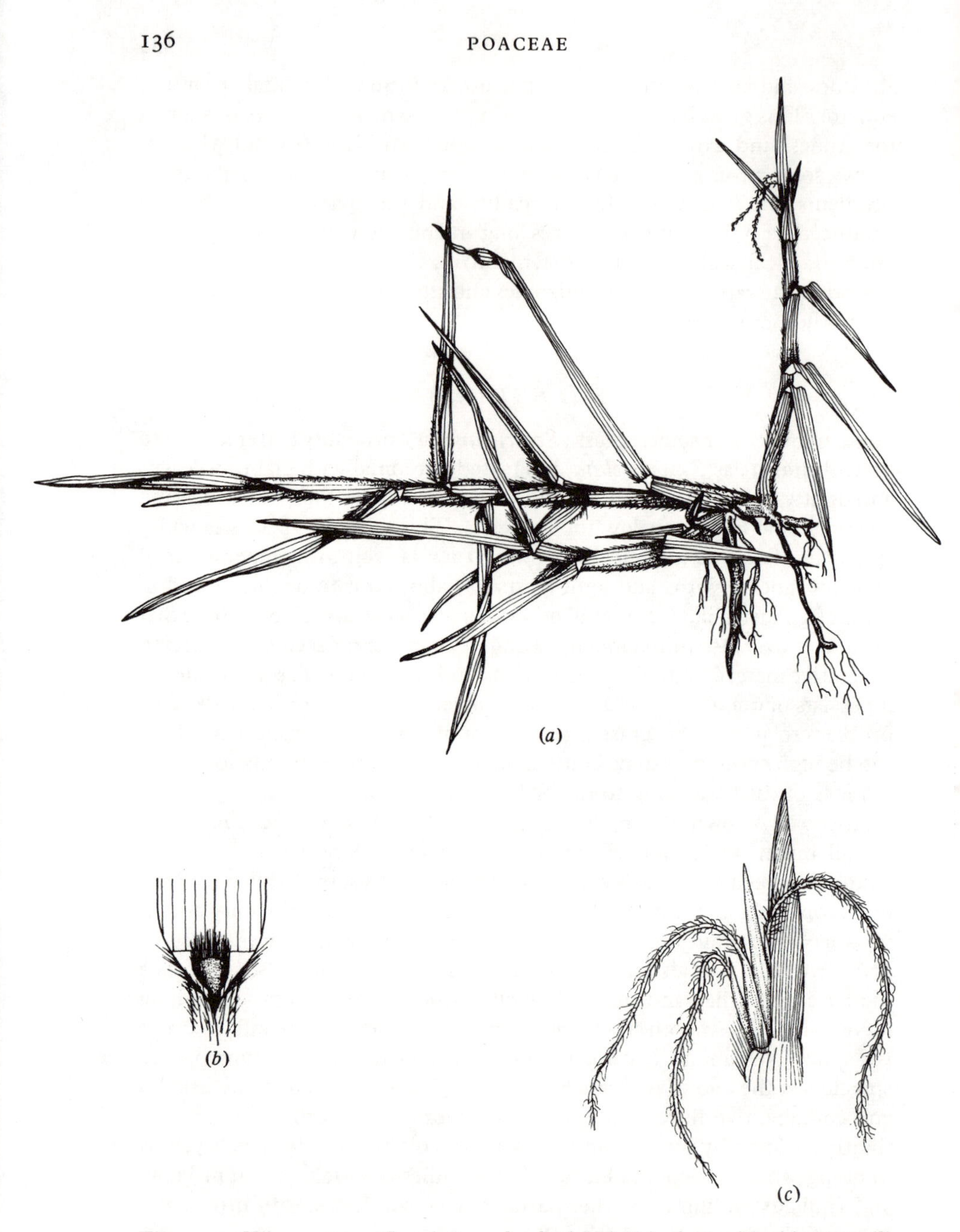

Fig. 4.43. Kikuyu grass (*Pennisetum clandestinum*). (*a*) plant with rhizomes, (*b*) ligule a fringe of hairs, (*c*) inflorescence consisting of two to four short-stalked spikelets with long stigmas showing.

Subfamily Chloridoideae

This group of grasses is adapted to tropical conditions, although its distribution is wider than that and includes North America and parts of Asia. Only three genera are selected for separate treatment.

4.42 *BOUTELOUA*

This is an American genus with about 40 species, distributed mainly in the south-west and mid-west of the United States, but extending into Mexico and the West Indies. Side-oats grama (*Bouteloua curtipendula*) is probably the best known species on account of its common occurrence on the American continent from mid-western USA to Uruguay and northern Argentina. It produces good grazing and is very useful in arid and semi-arid areas. Establishment from seed is readily achieved, and the plant tends to spread by means of short rhizomes.

4.43 *CHLORIS*

Rhodes grass (*Chloris gayana*) is native to southern and eastern Africa but is now widely grown in other parts of the world, notably New South Wales and Queensland where it is playing an important part in the development of sub-coastal scrub lands. It is an erect, tufted perennial, up to about 1.5 m in height, which spreads by means of stolons (Fig. 4.44). Leaf sheath and leaves are generally glabrous, the ligule is represented by a fringe of hairs but auricles are absent. The inflorescence is a digitate panicle consisting of about eight to twelve sessile spikes bunched together near the top of the rachis. The spikelets contain three to five florets, of which only the lowest is fertile. Rhodes grass is a summer growing species, best adapted to subtropical conditions with moderate annual rainfall. It is reasonably drought-resistant and can tolerate frosts as long as it is growing vigorously. Although versatile as regards soil conditions, it prefers reasonably high nutrient levels and tends to deteriorate if fertility is not maintained. It is superior to other grasses in tolerance to soil salinity, and it persists well after fire. As a pasture plant it withstands heavy grazing and it combines well with lucerne and other legumes. Nutritive value is high, as long as there is no preponderance of stemmy material, and this can be avoided by increasing grazing pressure. In addition, Rhodes grass is of value in erosion control because it covers bare ground very quickly.

Fig. 4.44. Rhodes grass (*Chloris gayana*). (*a*) plant with inflorescence, a digitate panicle consisting of sessile spikes, (*b*) ligule replaced by fringe of hairs, (*c*) spikelet.

4.44 *CYNODON*

Several species of agronomic importance belong to this genus, of which Bermuda grass (*Cynodon dactylon*) is the most widely distributed, ranging from Turkey to India and Pakistan, the USA, Australia and South Africa. This grass is a creeping perennial with stolons or rhizomes bearing flat, folded leaves which may be glabrous or hairy (Fig. 4.45). The inflorescence consists of a whorl of three to six spikes made up of green to purplish spikelets subtended by glumes of unequal size. Bermuda grass is a very variable plant, as shown not only by the many different names by which it is known locally, but also by the fact that some authorities recognise several botanical varieties within the same species. In the United States it used to be best known as a weed of arable land until in 1943 a cultivar suitable for grazing and forage production was released after hybridising a local with a South African variety. Coastal Bermuda grass, as it was called, soon became widely accepted throughout the central and southern parts of the American continent, the Philippines, and India, and it is said to have revolutionised the livestock industry in many places. Although the seeds may be sterile, propagation by pieces of stolon or rhizome is quite successful and, given time to establish and adequate levels of nitrogen, a good sward may be obtained. Mixtures with red clover or crimson clover (see section 11.2) have improved nutritive value, and very satisfactory liveweight gains may be obtained. Still better animal peformance is claimed for more recently bred cultivars of Bermuda grass which have greater digestibility and lead to greater intake.

Subfamily *Oryzoideae*

All genera belonging to this subfamily are usually included in a single tribe, of which rice (*Oryza sativa*) is the most important representative. Another genus is *Zizania*, a wild plant whose grains were collected by North American Indians as an important source of food. Grasses in this subfamily are either truly aquatic or they require ample water supply for their growth.

4.45 RICE (*ORYZA SATIVA*)

Rice is a cereal crop of outstanding economic importance, only second to wheat in terms of total world production but equally as essential as a staple food for a high proportion of mankind. Although rice can be cultivated in high latitudes up to about 50°, it grows best and is most widely distributed in the humid tropics and subtropics where it is often possible to obtain two

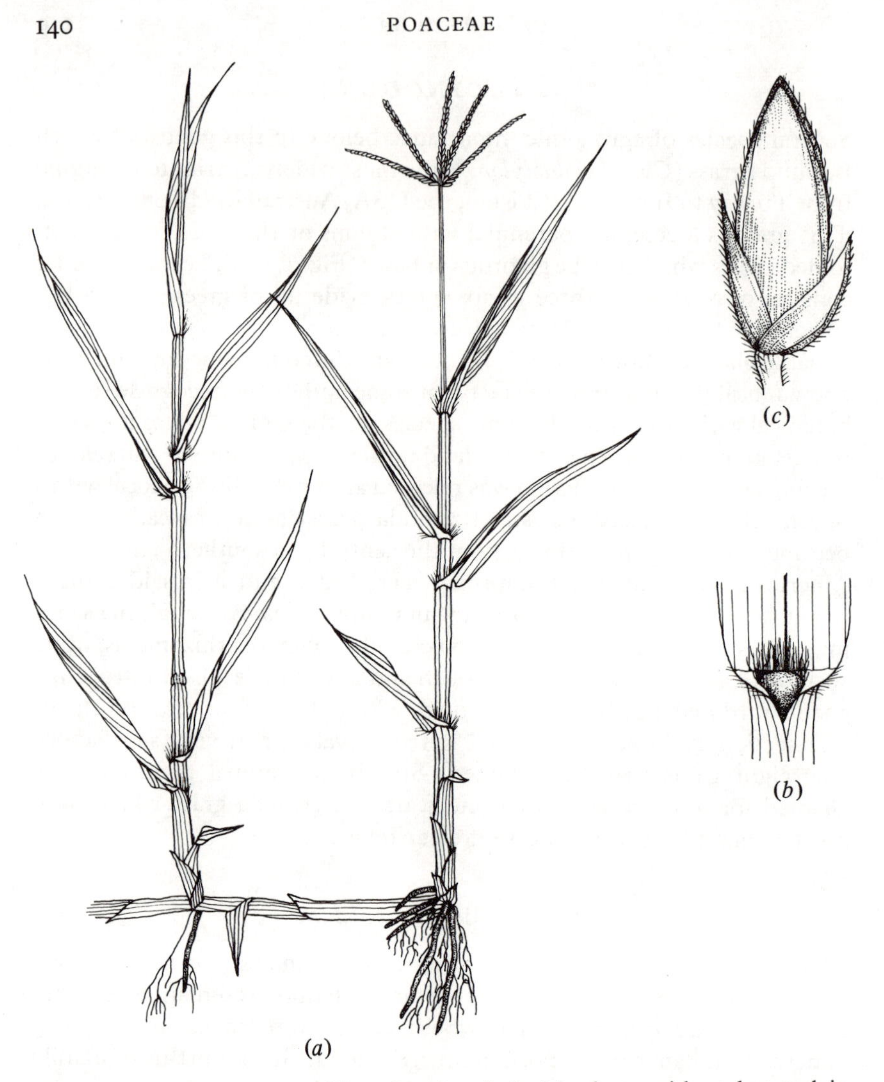

Fig. 4.45. Bermuda grass (*Cynodon dactylon*). (*a*) plants with stolon and inflorescence of sessile spikes, (*b*) ligule a fringe of hairs, (*c*) spikelet.

or more crops in the same year. China and India are the leading producers but, since rice is predominantly grown for local consumption in densely populated areas, not a great deal of it appears on the world market. Among the countries which do export rice to the temperate zone are the United States, Thailand and Australia.

Origin of rice

There are at least twenty different species of *Oryza*, of which Asian rice, *Oryza sativa*, and the less important African rice, *O. glaberrima*, are cultivated. Although these two are both diploid ($2n=24$), hybrids between them are highly sterile, but this does not rule out the possibility of a common progenitor before they became separated through continental drift. Domestication started much more recently but probably before the third millennium BC. *Oryza sativa* may have been derived from annuals which resembled the present-day weed species found in ditches and ponds adjacent to rice fields. The spikelets of these species shatter easily and the seeds are strongly dormant, characters that enable weeds to survive in climates with alternating hot, dry periods and monsoons but that had to be selected against during domestication.

The rice plant

Rice (*Oryza sativa*) is usually grown in paddy fields which are flooded for a large part of the growing period, an environment to which the plant is adapted by having large intercellular air spaces in its leaves and roots. The lamina is not wide, the ligule about 2 cm long, tending to split with age, and the auricles are very narrow and sickle-shaped (Fig. 4.46). The inflorescence is an open panicle bearing many one-flowered spikelets. At the base of each spikelet there occur two small glumes, although some authorities refer to them as lemmas of aborted spikelets, in which case a pair of ridges just below the lemma may be taken to represent the glumes. The boat-shaped lemma is covered with hairs and in some cultivars terminates in a short awn. The palea is also keeled and hairy. One unusual feature of the rice plant is the presence of six stamens, but in other respects the flower resembles that of other grasses in having two feathery stigmas and a single carpel. Two lodicules ensure the opening of the floret at anthesis, although only a small amount of cross-pollination occurs. The mature grain consists of the caryopsis tightly enclosed by the lemma and palea (the husk) with remnants of the glumes attached. Removal of the husk results in the production of so-called husked rice. The pericarp of the caryopsis is coloured, commonly brown, and since it is also strongly flavoured, the grain is usually milled to remove the pericarp, the aleurone layer and most of the embryo to produce white, polished rice. This serves as a source of almost unadulterated carbohydrates and thus requires the addition of vitamins, proteins and minerals from other foods to make up a balanced diet. However, it is possible to improve the vitamin content of

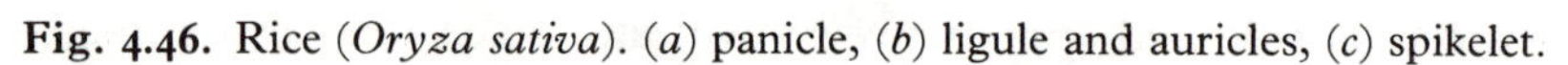

Fig. 4.46. Rice (*Oryza sativa*). (*a*) panicle, (*b*) ligule and auricles, (*c*) spikelet.

white rice by artificial enrichment, by parboiling before milling or by reducing the severity of the milling process.

Uses of rice

Rice is the staple diet of millions of people particularly in Asia but also in many other parts of the world. Although in large-scale production it is sown from the air into flooded fields, it is more commonly raised in seed beds and then transplanted to give reliable establishment. The fields are drained before harvest. Dryland or upland rice is drilled in moist soils or those capable of irrigation, but on the whole yields tend to be lower than those of paddy rice.

Because of its long history of cultivation in very many countries, rice is represented by thousands of local varieties which are adapted to local conditions and requirements. Broadly speaking, two major groups are recognised, the 'japonica' types which originally came from Japan and Korea, and the 'indica' types emanating from India, China and Indonesia. Varieties are also classified according to grain shape, whether long or short, and depending on the texture of the endosperm. Modern plant breeding, especially at the International Rice Research Institute (IRRI) in the Philippines has brought about rapid improvements in rice and some cultivars such as IR-8 have recorded yields of over 10 t ha^{-1} for a single crop. Since photosynthesis in the ears contributes only about 10% of grain yield, emphasis has been on short-strawed cultivars whose leaves are fully exposed to the light (see also p. 66).

FURTHER READING

Barnes, A.C. (1974). *Sugar cane*. Leonard Hill, London.
Baum, B.R. (1977). *Oats: wild and cultivated*. Canada Department of Agriculture, Research Branch, Ottawa.
Blackburn, F.H.B. (1984). *Sugar cane*. Longman, London.
Brown, L. (1979). *Grasses, an identification guide*. Houghton Mifflin, Boston.
Bushuk, W. (ed.) (1976). *Rye: production, chemistry and technology*. American Association of Cereal Chemists, St Paul.
Briggs, D.E. (1978). *Barley*. Chapman and Hall, London.
Chheda, H.R. and Crowder, L.V. (1978). *Tropical grassland husbandry*. Longman, London.
Datta, S.K. de (1981). *Principles and practices of rice production*. John Wiley, New York.
Doggett, H. (1988). *Sorghum*. Longman Scientific & Technical, published in association with the International Development Research Centre, Canada.

Evans, L.T. and Peacock, W.J. (1981). *Wheat science – today and tomorrow*. Cambridge University Press, Cambridge.
Findlay, W.M. (1956). *Oats, their cultivation and use from ancient times to the present day*. Oliver and Boyd, Edinburgh.
Gould, F.W. (1968). *Grass systematics*. McGraw-Hill, New York.
Grist, D.H. (1975). *Rice*. Longman, London.
Harlan, J.R. (1970). *Cynodon* species and their value for grazing and hay. *Herbage Abstracts* **40**, 233–8.
Heath, M.E., Metcalfe, D.S. and Barnes, R.F. (1973). *Forages. The science of grassland agriculture*. Ames, Iowa State University Press.
Holmes, W. (ed.) (1989). *Grass, its production and utilization* (2nd edn), Blackwell, Oxford.
Hubbard, C.E. (1984). *Grasses*. Penguin Books, Harmondsworth.
Hulse, J.H. and Spurgeon, D. (1974). Triticale. *Scientific American* **232**, 72–80.
Humphreys, L.R. (1978). *Tropical pastures and fodder crops*. Longman, London.
Johnson, V.A. and Schmidt, J.W. (1968). Hybrid wheat. *Advances in Agronomy* **20**, 199–233.
Jones, M.B. and Lazenby, A. (eds) (1988). *The grass crop*. Chapman and Hall, London.
Jugenheimer, R.W. (1976). *Corn. Improvements, seed production and uses*. John Wiley, New York.
Langer, R.H.M. (1979). *How grasses grow*. Edward Arnold, London.
Langer, R.H.M. (ed.) (1990). Pastures, their ecology and management. Oxford University Press, Auckland.
Lawes, D.A. and Thomas, H. (eds) (1986). *Proceedings of the second international oats conference*. Martinus Nijhoff Publishers, Dordrecht.
Leigh, J.H. and Noble, J.C. (ed.) (1972). *Plants for sheep in Australia*. Angus and Robertson, Sydney.
Lupton, F.G.H. (eds.) (1987). *Wheat breeding*. Chapman and Hall, London.
Mangelsdorf, P.C. (1974). *Corn, its origin, evolution and improvement*. Belnap Press of Harvard University.
Matz, S.A. (1969). *Cereal science*. Avi Publishing Co., Westport.
Miller, P.A. (1984). *Forage crops*. McGraw-Hill, New York.
National Research Council (1989). *Triticale: a promising addition to the world's cereal grains*, National Academy Press, Washington DC.
National Research Council (1988). *Quality – protein maize*. National Academy Press, Washington DC.
Oka, H.I. (1988). *Origin of cultivated rice*. Elsevier Science Publishers, Amsterdam.
Oram, R.N. (ed.) (1990). *Australian herbage plant cultivars*. (3rd edn). Australian Herbage Plant Registration Authority, Division of Plant Industry, Canberra.
Pearson, C.J. and Ison, R.L. (1987). *Agronomy of grassland systems*. Cambridge University Press, Cambridge.
Percival, J. (1921). *The wheat plant, a monograph*. Duckworth, London.
Peterson, R.F. (1968). *Wheat*. Leonard Hill, London.

Peterson, G.A. and Foster, A.E. (1973). Malting barley in the United States. *Advances in Agronomy* **25**, 527–78.

Price, D. Jones (1976). *Wild oats in world agriculture*. Agricultural Research Council, London.

Purseglove, J.W. (1972). *Tropical crops. Monocotyledons*. Longman, London.

Quisenberry, K.S. and Reitz, L.P. (eds.) (1967). *Wheat and wheat improvement*. American Society of Agronomy, Madison.

Rasmusson, D.C. (ed.) (1985). *Barley*. American Society of Agronomy, Madison.

Skovmand, B., Fox, P.N. and Willareal, R.L. (1984). Triticale in commercial agriculture: progress and promise. *Advances in Agronomy* **37**, 1–45.

Smith, D.F. (1972). *Hordeum species in grasslands*. Herbage Abstracts **42**, 213–23.

Spedding, C.R.W. and Diekmahns, E.C. (eds) (1972). *Grasses and legumes in British agriculture*. CAB, Farnham Royal.

Sprague, G.F. (ed.) (1970). *Corn and corn improvement*. American Society of Agronomy, Madison.

Tothill, J.C. and Hacker, J.B. (1973). *The grasses of southern Queensland*. University of Queensland Press, Brisbane.

Tsen, C.C. (1979). *Triticale: first man-made cereal*. American Association of Cereal Chemists, St Paul.

Wells, G.J. (1974). The biology of *Poa annua* and its significance in grassland. *Herbage Abstracts* **44**, 385–91.

Witcombe, J.R. and Beckerman, S.R. (1987). *Proceedings of the International Pearl Millet Workshop*. International Crops Research Institute for the Semi-Arid Tropics, Patancheru, India.

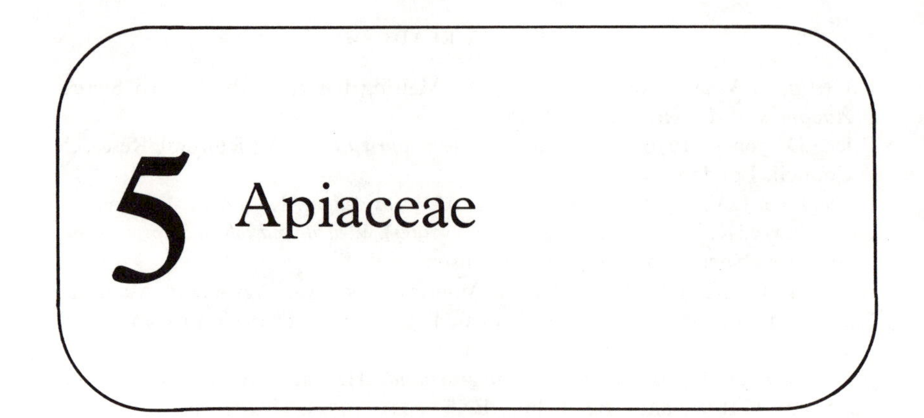

5 Apiaceae

This family is also sometimes known as the Umbelliferae after the distinctive umbel which its members bear at flowering. It comprises 275 genera with over 2800 species which are mainly confined to northern temperate regions but are sometimes found at high altitudes in the tropics. Besides three major vegetables – the carrot, celery and the parsnip – members of the family are utilised as herbs such as parsley and chervil or the seeds are used as spices such as aniseed, caraway, coriander, cumin and dill. The family also has classical connections, as the three kings who visited the infant Christ carried with them as a gift the perfume myrrh which is derived from *Myrrhis odorata*. On a less pleasant note, the Greek philosopher Socrates was executed by being given an infusion of hemlock (*Conium maculatum*), a species which is still a common weed in many temperate countries.

Members of the family are mostly biennial or perennial herbs. The stems are seldom woody but may be quite stout and are often hollow. The leaves are alternate and the petioles may be expanded or sheathing at the base. The leaves are compound and in most species are highly divided. As the former name of the family suggests, the inflorescence is an umbel (Fig. 5.2) which may be simple or compound, and in some species it may be subtended by bracts which form a whorl under the main or the branch umbel. The flowers are small and perfect (i.e. bisexual) and regular. The sepals are small and may be absent, there are five petals and five stamens which alternate with the petals and are inserted into a disc at the top of the ovary. The ovary which is inferior comprises two carpels and contains two locules. Each locule contains a single ovule which forms a single seed at maturity. The fruit consists of two dry ribbed or winged one-seeded indehiscent carpels or mericarps which separate at the base but remain attached at the top by a slender Y-shaped stalk. Structures on the carpel often contain longitudinal oil tubes which contain aromatic oil.

5.1 *APIUM*

This genus is easily recognised because plants belonging to it have an almost sessile umbel which is inserted on the stem opposite to the leaves. Only one species is of economic importance, celery or celeriac (*Apium graveolens*).

Celery and celeriac (*Apium graveolens*)

This species, which is consumed mainly as a salad vegetable, is usually grown for its greatly enlarged young petioles or swollen storage organs. The immature plant has a flat disc-like stem from which the petioles arise. When it becomes reproductive the stem which is solid and ribbed elongates and may reach a height of 1 m. The leaves are pinnate and have large well-developed petioles. There are usually three pairs of leaflets and a terminal one, and these are further subdivided into threes and are often coarsely toothed. Flowers are white, very small and are borne on small umbels among the leaves.

Celeriac sometimes called var. *rapaceum* is a member of the same species as celery but is not cultivated for its young petioles but for a swollen turnip-like structure which consists of stem, swollen hypocotyl and root. Celery, also referred to as var. *dulce*, is grown for its young swollen petioles.

Cultivation and uses

Celery and celeriac have similar cultural requirements and in the field are usually established from transplanted seedlings. They require a well drained fertile soil of pH 6–6.8. Planting rate is about 120 000 plants ha^{-1} and plants are usually spaced between 20 × 20 cm and 30 × 30 cm. Cultivated celery is frequently field-blanched by banking up soil around the plants, covering them with paper or excluding light from the petioles with boards. Blanching besides whitening the petioles is reputed to increase vitamin A content, improve flavour, and protect the crop from frost. The crop is harvested between 120 and 160 days after planting and this is done by pulling in celeriac or cutting off under the base of the plant in celery. Yields are 14–16 t ha^{-1} of celeriac and 38–63 t ha^{-1} of celery.

Celery has a distinctive odour and flavour and is usually utilised as a salad vegetable but may also be cooked. Celeriac, which is popular in Europe, is usually cooked prior to eating. The fruits ('seeds') of celery contain 2–3% volatile oil. The seeds may be used as a spice in their own right or the oil may be distilled off and used in the production of celery salt. In tropical climates of Asia, where the plant does not form large petioles, leaves and petioles are utilised as a herb.

5.2 *DAUCUS*

There are about 60 species in the genus *Daucus* but only one, *Daucus carota*, which exists in both wild and cultivated forms is of importance. The former is a minor weed while the latter is a popular root vegetable.

Carrot (*Daucus carota*)

The cultivated carrot was known to both the ancient Greeks and the Romans and had spread to the rest of Europe by the Middle Ages. It had become an extremely popular vegetable in England by the sixteenth century. In 1607 it was introduced into Virginia where it rapidly became popular with the Red Indians who spread the species throughout the American continent.

The cultivated carrot is an annual or biennial plant which usually has a very short stem until it becomes reproductive. It has a very large and well developed tap root. A single vascular cambium cuts off unlignified secondary xylem, secondary phloem and parenchyma cells. The roots formed may be long and tapering or of almost uniform diameter for their whole length. Finer non-thickened roots may penetrate the soil to a depth of 60 cm. The roots are bright orange in colour due to the presence of α- and β-carotene which are precursors of vitamin A. The leaves which are borne on long petioles with expanded bases, are compound and highly divided (Fig. 5.1). The flowers are small and white or yellowish and are borne on showy compound umbels.

Cultivation and uses

Carrots require well structured, freely draining soils with a pH between 5.5 and 6.8. As carrots are not easily transplanted they are usually sown directly into the field using precision seeders at the desired population density. For vegetable production the sowing rate is 1–1.5 kg ha^{-1} to give a final plant population of 800 000 plants while for baby carrots which are frozen or canned 2.5–5 kg ha^{-1} of seed are sown with the intention of establishing 3×10^6 carrots ha^{-1}.

As a result of plant breeding, varieties of carrots now exist that can be sown in all seasons of the year except the winter. Full-size carrots require 70–120 days to come to maturity, depending on time of sowing and variety, while baby carrots can be grown in 60–80 days. Yield varies depending on end use, varying from 10–15 ha^{-1} for baby carrots to 56 t ha^{-1} for main crop carrots.

Carrot roots are about 88% water, 1% protein, 0.5% fat, 10%

Fig. 5.1. Carrot (*Daucus carota*).

carbohydrate and 1% fibre. Carrots are eaten both raw and cooked. They are processed by canning, freezing and dehydration. They are also used as an important source of carotene concentrate.

5.3 *PASTINACA*

This genus which is distinguished by its yellow coloured petals contains one species of economic importance, the parsnip (*Pastinaca sativa*).

Parsnip (*Pastinaca sativa*)

The parsnip is thought to have originated in Eurasia and it can be found growing wild in most of Europe and the Caucasus. It was known to both the ancient Greeks and the Romans but its current thickened form was probably developed during the Middle Ages when it was commonly eaten with salt fish during Lent. Prior to the introduction of the potato the parsnip was a far more important vegetable in Europe than it is now.

The parsnip, like the carrot, develops a stout tap root and, although a biennial, it is usually grown as an annual. The root may be up to 10 cm across the crown and over 50 cm long. The flesh is either white or light brown in colour. When the stem elongates it is grooved, hollow and bulky reaching a height of about 150 cm. The leaves are pinnate with three to four pairs of sessile ovate to oblong toothed, lobed leaflets, up to 10 cm in length (Fig. 5.2). The flowers are greenish yellow and are borne in compound umbels.

Cultivation and uses

Like carrots, parsnips require a well-drained, well-structured soil. Optimum soil pH is 6.0–6.8. The crop is usually established by sowing with a precision seeder at 1.5–2 kg ha^{-1} to establish a plant population of 270 000 plants ha^{-1}. Depending on time of sowing, the crop matures in 90–120 days and yields of 26–40 t ha^{-1} of parsnip roots can be obtained. Parsnips are utilised almost entirely as a vegetable and are cooked by boiling, roasting or adding to stews.

5.4 OTHER APIACEAE

Herbs

A number of plants in the Apiaceae are important herbs. Perhaps the best known is parsley (*Petroselinum crispum*), grown for its extremely divided

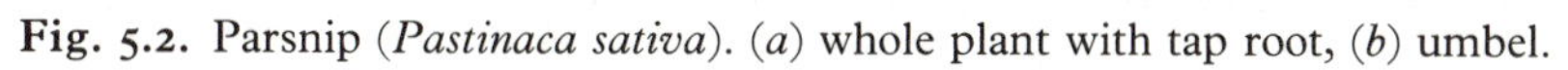
Fig. 5.2. Parsnip (*Pastinaca sativa*). (*a*) whole plant with tap root, (*b*) umbel.

compound leaves which are added to many dishes either during their cooking or as a garnish prior to serving. As with celery, there is also a form that develops a turnip-like root that is cooked as a vegetable.

Fennel (*Foeniculum vulgare*) is also used as a flavouring agent, both leaves and seeds (mericarps) being utilised. Florentine fennel has fleshy petiole bases which are blanched and used raw as a salad vegetable, the flavour being not unlike that of celery. Chervil (*Anthriscus cerefolium*) is not unlike parsley and is commonly used in Europe as a garnish and a flavouring. Angelica (*Angelica archangelica*) is not strictly a herb but another plant from the family which is grown for its stems which are crystallised and eaten either alone or added to cakes, icecreams and other sweet dishes for their distinctive green colour. This plant, which can grow to a height of 1.8 m, is also used in the flavouring of vermouth and various other liqueurs. In earlier times it was invested with an extremely mystical reputation in that it was claimed to be an antidote for poison and an agent reputed to prevent plague.

Spices

The presence of the aromatic oil tubes in the wall of the mericarp leads to a number of species being utilised as spices. Anise derived from *Pimpinella anisum* is utilised in curries, sweets, confectionery and in alcoholic beverages such as ouzo, raki and pernod. The essential oils distilled from the mericarps are used in medicines, perfumes and soaps. Caraway seed derived from *Carum carvi* is often sprinkled over the top of cakes for its distinctive flavour or incorporated in breads, biscuits, cheese, sausage seasoning and in pickles. It is also used to flavour the beverage Kümmel. The seed contains 3–8% oil which is used as a flavouring agent and in medicines. Coriander (*Coriandrum sativum*) is of Mediterranean origin but is now widely cultivated in India where it is utilised in the production of curry powders, chutneys and pickling spice. Coriander is also used in the flavouring of alcohol, particularly gin. The seed can only be utilised when ripe, for before maturity it is reputed to have an odour like bed bugs. In India the seed yield can be quite high reaching 1100–2000 kg ha^{-1}. Fresh coriander leaf is also frequently used as a garnish in Asian cooking in a similar way to parsley. Cumin derived from *Cuminum cymicum* is put to similar uses as caraway and is also a component of curry powder. Dill (*Anethum graveolens*) is used extensively in pickles particularly gherkin and cucumber pickles. It can be used either as a green plant or as seed. Dill water derived from the seed is supposed to cure flatulence in young babies. Its common name is even derived from the old Norse word *dilla*, which means to lull, presumably derived from its soothing effects on crying babies. Other uses of dill are similar to those for coriander and cumin. Finally, lovage which is derived from *Levisticum officinale* is utilised as its leaves or its powdered rhizomes which have the flavour of a mixture of spices, and in Germany it is known as *Maggiwurzel*. Its common name was derived from a former belief that it had magical properties in love affairs.

FURTHER READING

Atal, C.K. and Kapur, B.M. (eds.) (1982). *Cultivation and utilization of aromatic plants*. Council of Scientific and Industrial Research, Jammu-Tawi.

Heywood, V.H. (ed.) (1971). *The biology and chemistry of the Umbelliferae*. Academic Press, London.

Kowalchik, C. and Hylton, W.H. (1987). *Rodale's illustrated encyclopedia of herbs*. Rodale Press, Emmaus.

Little, B. (1986). *The complete book of herbs and spices*. Reed, Frenchs Forest.

Peirce, L.C. (1987). *Vegetables, characteristics, production and marketing*. John Wiley, New York.

Purseglove, J.W. (1981). *Spices*. Longman, London.

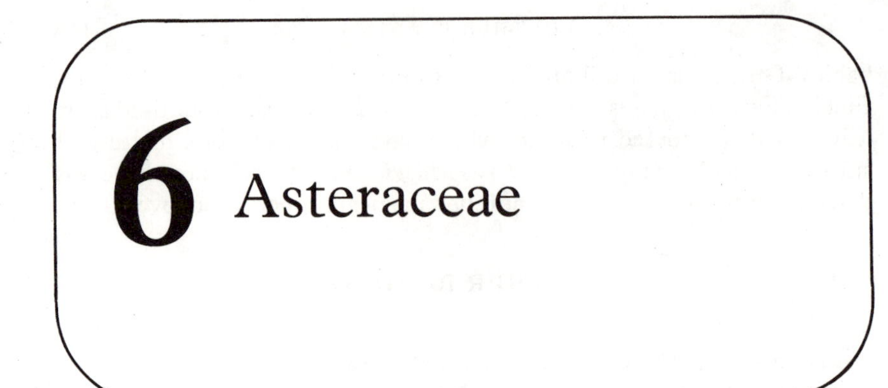

6 Asteraceae

The Asteraceae (Compositae) are one of the oldest groups of higher plants. They are also the largest family and include about one tenth of all species, over 20 000, which are distributed widely throughout the world. It is therefore somewhat surprising that the family should be best known for the weeds and wild plants it contains. Only a few genera have provided plants of economic value. The sunflower and safflower have been developed as oil seed crops, lettuce and endive as salad vegetables, pyrethrum as a source of insecticide, and a few other species because of their mineral or aromatic content. The family used to derive its name from the fact that the flowers do not occur singly but in composite inflorescences which take the form of a flower head or capitulum. Each capitulum consists of a receptacle surrounded by one or more whorls of bracts, jointly referred to as the involucre. Inserted on the receptacle are small flowers or florets, tightly packed together varying from a few to several hundreds according to the species. Each individual flower has an inferior ovary containing a single ovule. The single, usually forked style is surrounded by a tube formed by five stamens with their anthers joined. There are five petals which are united to form a corolla tube, but the sepals are either absent or represented by a ring of fine hairs, called the pappus, which play a part in seed dispersal. Two distinct types of flowers occur, either together on the same inflorescence or in other species only one at a time. Those with a corolla tube around the anthers and spreading out at the top into five lobes are referred to as tubular flowers or, since they occur tightly together in the middle of the inflorescence, as disc flowers (Fig. 6.1). In the other type the corolla tube is short and is surmounted by a long structure which appears to be one petal but has, in fact, small teeth at its end representing five petals. These are the ligulate or ray flowers. Pollination by insects is most common, because the stigma though close to the anthers is not receptive

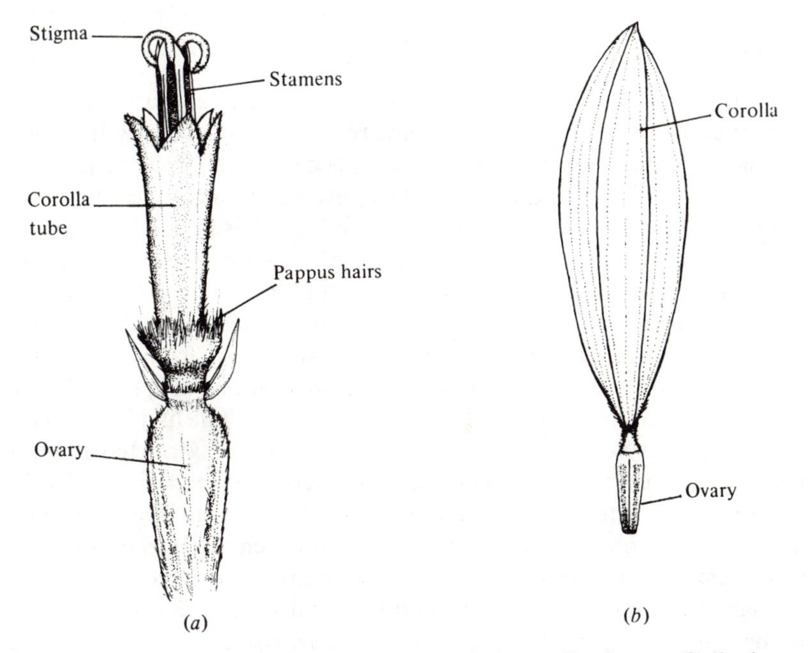

Fig. 6.1. Sunflower (*Helianthus annuus*). (*a*) tubular or disc flower. (*b*) ligulate or ray flower (sterile).

until it has emerged. The fruit containing one seed is termed an achene. In many species the seeds are dispersed by wind, with the hairs of the pappus acting like a parachute. The seeds contain little endosperm, and oil usually acts as the food reserve.

6.1 *HELIANTHUS*

The genus *Helianthus* contains some 70 species, among them two plants of economic importance, the sunflower and the Jerusalem artichoke or topinambour. Both are natives of the temperate regions of North America and served as food plants for the Indians. However, their present distribution is world-wide.

Sunflower (*Helianthus annuus*)

The sunflower derives its name from its large and striking capitulum with its dark disc flowers and golden-yellow ray flowers. Few other plants have inspired poets and painters to the same extent. For example, there can be few people who do not know one of van Gogh's most famous paintings.

Origin and early history

The sunflower is thought to have originated in the central highlands of Mexico and, once its food value became recognised by the early Indians, it may have become domesticated well over 2000 years ago. It lent itself to the semi-nomadic life of the Indian tribes because it had a relatively short growing season and produced an edible seed which was easily stored. The plant thus spread through human migration, particularly towards the immediate north where few other food plants could be grown in semi-desert conditions. By the time the Europeans arrived in the New World, the sunflower was cultivated in a broad belt stretching from the southern parts of Canada to Mexico, and several distinct types had appeared as the result of primitive selection and climatic differences. The first introduction from Mexico to Spain occurred in 1510, but initially it attracted attention in Europe as an ornamental plant, and early herbalists vied with each other to describe it. Its value as a crop plant appears to have remained unrecognised until the end of the eighteenth century when it reached Russia. Part of the reason for its ready acceptance may have been that the Russian Orthodox Church had promulgated a list of oily foods which could not be eaten on certain holy days, and not surprisingly the newly arrived sunflower had not been included. From then onwards systematic plant selection to increase the oil content took place and several improved types were isolated and hybridised. Russia has remained the largest producer of sunflower oil, but several Balkan states and Turkey also figure significantly in world trade. During the Spanish Civil War the sunflower became an important crop in Argentina because olive oil supplies from Spain had ceased. Other producing countries include South Africa, Tanzania, Uruguay, and Australia.

The sunflower plant

The stem of the sunflower is roughly hairy, growing to a height of 0.5 m to 3 or even 5 m depending on cultivar and cultural conditions, and bears large, alternate leaves on stout petioles. The amount of branching varies with plant density, soil fertility and other factors, but is uncommon in modern commercial cultivars. Until the time of anthesis the stems are heliotrophic and follow the direction of the sun. A very large terminal capitulum is produced, subtended by an involucre of several whorls of bracts (Fig. 6.2). Pollination is normally by insects, mainly honey bees. The fertile tubular flowers form a dark-brown disc in the centre, surrounded by 30 to 90 sterile bright yellow-to-orange ray flowers. The seed is a flattened, obovate structure, 1 cm or more in length, varying in

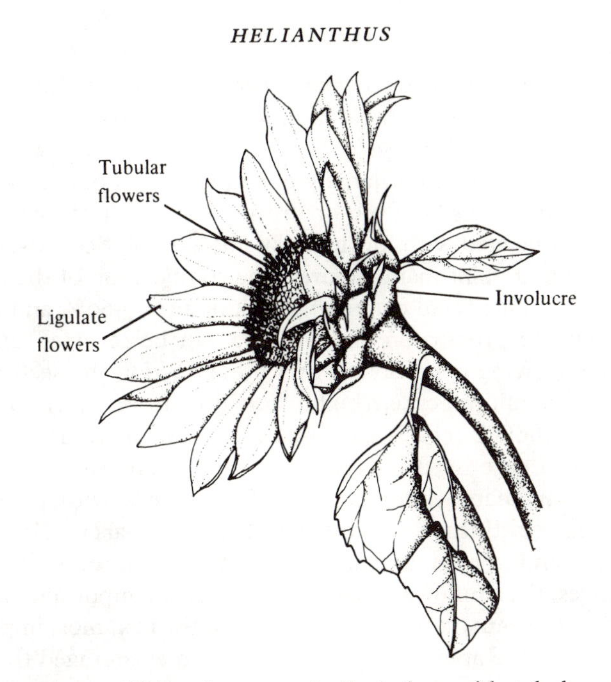

Fig. 6.2. Sunflower (*Helianthus annuus*). Capitulum with tubular and ligulate flowers, surrounded by involucre.

colour from black to dark-brown, or with dark-grey and white stripes in some cultivars. Their position in the centre of the capitulum and their palatability make the seeds most attractive and accessible to birds.

Cultivation and uses

Sunflowers require to be grown on good cropping soils but are adaptable to a range of temperate to cool-tropical climatic conditions. The seed is sown in wide rows to obtain a plant population of about 30 000 ha^{-1} in dry areas and about 60 000 ha^{-1} under irrigation. Optimum plant density is important, because large capitula from widely spaced plants are slow to dry and difficult to harvest, while small capitula from dense stands may produce very small seeds. The crop responds to fertilizers, but it competes poorly with weeds. It takes about four months for sunflower seeds to mature but harvesting has to be delayed until the fleshy receptacle has dried sufficiently. The seed should not exceed 10% in moisture content for safe storage but may be harvested at an earlier stage if artificial drying facilities are available. Shattering of seed and losses through birds are problems to be contended with. The crop is combine harvested or cut and threshed.

Good yields in excess of 2000 kg ha^{-1} can be obtained. Intensive plant breeding has raised the oil content of sunflower seed from less than 30% to over 40% and approaching 50%. The greater part of the world crop is crushed for the extraction of largely polyunsaturated oil, used predominantly as a table or cooking oil, or in the production of margarine and cooking fats associated with low levels of cholesterol. Second-grade oil is used for soap and paint manufacture. The composition of the oil is very variable, with linoleic acid commonly making up 55–60% and oleic acid 25–30%, but the proportion of these two can be reversed under high temperature growing conditions. After oil extraction the seed residue is processed into sunflower cake, a highly nutritious cattle feed rich in protein (40–46%). Sunflower seed is also used for birds but valued widely as a human food. Large seeds, roasted or salted, are eaten or used by confectioners in many different ways. Fried seeds with pepper are a delicacy in eastern Europe, and the flour is used in bakery. Because of its adaptability, and the great value of plant oils in present economic circumstances, the sunflower is likely to increase in importance as a world crop. After soya beans, cotton and peanuts it is the next most important oil crop in the world. Rapid development has been encouraged through the production of hybrid sunflower cultivars obtained, just as in maize (section 4.29), by crossing inbred lines which are superior in yield, oil content and suitability for mechanical harvesting.

Jerusalem artichoke (*Helianthus tuberosus*)

This is a minor crop grown to a limited extent in France, in Oregon and elsewhere as a tuber vegetable or as a stockfeed. How it acquired its name is not at all certain, although it is highly likely that the word Jerusalem does not refer to the ancient city but is considered by some to be a mispronunciation of *girasole*, the Italian for sunflower, or some other word. The plant occurs wild in the eastern United States, and early accounts relate that it was used by North American Indians as a food. It aroused considerable interest when first introduced to Europe, where it was called an artichoke because its taste resembled that of the globe artichoke (see section 6.6). The word topinambour by which the plant is also known derives from natives of a certain Brazilian tribe who happened to arrive in France at much the same time as the plant and were mistakenly associated with it. Jerusalem artichokes grow to a height of 1–3 m and in many ways resemble sunflowers. However, the capitula measure only 4–8 cm in diameter, about a quarter of the size of sunflower inflorescences. The edible portion consists of the tubers, storage organs some 10–20 cm long, produced below ground. Biochemically these consist predominantly of inulin, which yields laevulose on hydrolysis and because of this is said to be especially suitable for diabetics. Inulin also has other uses, among them as a raw material in the synthesis of artificial blood plasma. The tubers are cooked in as many different ways as potatoes, or eaten in soups or salads. An industrial use which may expand in the future is the fermentation of inulin to alcohol. It should be noted that, although Jerusalem artichokes grow well in a wide range of climates, the production of vigorous tubers depends on moderate temperatures and adequate water supply.

6.2 *CARTHAMUS*

In distinction to members of the genus *Helianthus* which have both ligulate and tubular flowers, the plants belonging to the genus *Carthamus* are like thistles in having only tubular florets. There are about 30 species but the only one of economic importance is the safflower, grown for the oil extracted from its seeds.

Safflower (*Carthamus tinctorius*)

Origin and early history

The safflower probably first came into domestication in the Middle East where related and inter-fertile wild species occurred. The Egyptians cultivated the plant as early as 1600 BC for the sake of its red or orange flowers which were, among other uses, sown onto strips of papyrus and then draped around the bodies of mummies. Probably as an extension of this custom it was discovered that the flowers contained a dye, a pigment named carthamin, giving a red shade in alkaline solutions, which became the chief product obtained from this plant. Together with indigo, a blue dye derived from species of *Indigofera* belonging to the Fabaceae safflower pigment became the most important colouring agent for fabrics and foods, and it was only when aniline dyes were discovered in the middle of the last century that both lost their pre-eminent position. In antiquity, safflower also enjoyed a reputation as a medicinal plant for the cure of constipation and to induce sweating. Use of the seed as a source of oil also started in antiquity but it appears to have been localised originally in Egypt and India until about 100 years ago when cultivation on a bigger scale began. As a result of isolation in several separate areas, a range of diverse types of plant have developed, from which modern plant breeders are able to select. The main producing countries are the United States, Mexico, Spain, Portugal and Australia. Safflowers are also grown widely in India.

The safflower plant

The safflower is a glabrous, highly branched annual plant growing to a height of about 0.5–1.5 m. The stem branches freely and bears dark-green, glossy leaves with spines at the margins and tip. Cultivars differ in the degree of spine cover, some being very spiny and others almost free of spines. The capitula, 2.5–4 cm in diameter, are fringed by an involucre of spiny bracts. Only tubular florets are present, bearing yellow to red-orange corolla tubes, five bright yellow, united anthers and a forked stigma. The achene is off-white or grey in colour and, in distinction to the sunflower seed, it is smaller, more angular and less flat. Both self- and cross-pollination occur. The safflower plant has a deep taproot which has been shown to penetrate two metres or more into the soil.

Cultivation and use

Safflower seed is sown in the spring and the crop matures within 120–200 days. It requires well-drained soils of high fertility and a near neutral pH, it can be grown successfully in saline soils, but it does not tolerate excess moisture because of the incidence of root rot. Excessive branching is not encouraged because primary stems produce the heaviest seed yield. Weed control is very important. The crop may be harvested when seed moisture levels have reached about 15%, although for storage 8% is a safe figure. The seed does not shatter easily. Combine harvesting is the usual method, but operators may need to be protected against skin and eye irritation caused by small bristles from the receptacle. These bristles also tend to block air inlets and radiators. Good yields without irrigation may reach 1500 kg ha^{-1} or more but irrigated crops may produce more than twice this amount. Safflower seed has an oil content of 36–45% but there are prospects of exceeding this range by appropriate plant selection. The oil extracted from the seed is very highly regarded for several purposes. It is valued by the food industry because its high unsaturated fatty acid content (about 78% linoleic, 13% oleic acids) make it a safe commodity for people threatened by high cholesterol levels in the blood. A high concentration of linoleic acid but virtual absence of linolenic and only a little saturated fatty acids make safflower oil most useful as the basis of fast drying paints that do not yellow with age, retain their colour and have good brushing characteristics. After removal of the oil, the seed is still valuable as a livestock feed with a crude protein content of some 43%. For this purpose the seed is either hulled to form decorticated meal, or left intact as whole pressed seed meal with a relatively high fibre content.

The flowers of the safflower plant contain two pigments, the yellow water-soluble carthamidine and the more important red-orange carthamine which is soluble only in alkaline solutions. Manufacture of the dye no longer occurs on any commercial scale and is confined to meet specific local requirements. Safflower florets were also used as a cheaper source of the pigment saffron which is obtained from the dried stigmas of the crocus.

6.3 *CHRYSANTHEMUM*

This genus contains about 100 species. Best known among these are probably the autumn flowering garden chrysanthemums which are derived from *C. indicum*, a native of China and Japan, and *C. morifolium* from China which were cultivated in the Far East for many centuries before reaching Europe some 200 years ago. When it comes to field crops, the genus is represented by pyrethrum, a plant grown for the insecticidal compounds, pyrethrins, which are obtained from it.

Pyrethrum (*Chrysanthemum cinerariifolium*)

Although there are two species with insecticidal properties in this genus, this account is restricted to the more important *Chrysanthemum cinerariifolium*. The other, *C. coccineum*, has lower insect toxicity and is not grown on a large commercial scale.

Origin and history

Pyrethrum is a European plant, native to Dalmatia on the Adriatic Coast of Yugoslavia. It was introduced to Japan at the end of the last century and grown there commercially as a source of insecticide. Interest in this crop also continued in several European countries, and trial plots were established in a number of localities, among them the Rothamsted Experimental Station in England. Seed from there and from Dalmatia was taken and planted in Kenya in 1929, where the crop soon became established as an important export commodity. Because the flowering season is very long and labour costs relatively low, Kenya, Tanzania and neighbouring African states have become the major producers of pyrethrins, with much smaller quantities coming from Ecuador, Papua New Guinea and Japan.

The pyrethrum plant

Pyrethrum is a perennial plant with an economic life of about two or three years. The stems are 30–60 cm high, covered with fine hairs, and bearing alternate, pinnately divided leaves on slender petioles. The capitula are 3–4 cm in diameter and consist of both ray and disc florets rather like a daisy. On the outside is a ring of 18 to 22 ray florets, each with a white corolla and a bilobed stigma but no anthers. The fertile disc florets are yellow in colour and conform in structure to the typical pattern of the Asteraceae. Cross-pollination by insects is required. The achenes are about 4 mm long, pale brown in colour and contain one seed.

Cultivation and uses

The crop is propagated by seed or by planting so-called splits obtained by dividing up plants, spaced about every 30 cm in rows 1 m apart. Good weed control is essential, and the crop requires adequate fertilizer applications and responds to irrigation. Harvesting of inflorescences begins some four months after establishment and picking continues at intervals during the flowering season which in favourable areas extends over many months. The flowers are picked by hand when fully open but before seeds have begun to form. Yields are very variable but range up to about 1000 kg ha^{-1} annually for three years, after which the stands are renewed. At the end of the picking season the crop is cut back to allow regeneration and further flower production. After harvesting the flowers are spread thinly on trays and dried to 8–10% moisture in hot air. The economic product extracted from the dried flowers, especially the ovary, consists of six compounds with insecticidal properties, collectively known as pyrethrins, of which pyrethrin I is the most toxic. Kenya pyrethrum contains at least 1.3% pyrethrins, often considerably more when grown at high altitude, with a high pyrethrin I content. The value of this material is based on its toxicity to a wide range of insects including fleas, flies, mosquitos and lice which appear to be unable to develop resistance against it. Pyrethrins achieve rapid knock-down and effective kill, produce no taints, have low persistence, and are effectively synergised. Above all, they are virtually non-toxic to man and mammals, and in this respect alone have enormous advantages over DDT or other chlorinated hydrocarbons and organophosphorus insecticides. Because it is biodegradable and is destroyed by light, pyrethrum is not persistent and can only be used as a contact insecticide. Despite this disadvantage, there is considerable potential for the expansion and improvement of this crop, including possibly hybridisation with related species, and development of less labour-intensive harvesting techniques.

6.4 *LACTUCA*

Some 100 species occur in this genus, most commonly in north temperate regions. They are all herbaceous, annual or perennial, capable of producing latex in their tissues. One species, *Lactuca virosa*, is used for the drug lactucarium obtained from its dried latex, but economic importance of the genus rests on two other species, *L. sativa*, the lettuce, and *L. indica*, a leafy vegetable grown in China, Japan and south-east Asia.

Lettuce (*Lactuca sativa*)

Although the origin of cultivated lettuce is uncertain, its cultivation goes back to at least 4500 BC when, according to paintings of what appear to be lettuce leaves on the walls of old tombs, it seems to have been grown in Egypt. Numerous references in the classical literature indicate that the Greeks and Romans knew and valued it, and that it quickly spread into other parts of Europe and Asia. The Spaniards and Portuguese took the plant to the New World, and it accompanied the early settlers to Australia and New Zealand, so that today the lettuce is distributed widely in all temperate countries. Its earliest use may have been as a source of oil from its seeds, and as a salad vegetable it consisted originally of bunches of loosely arranged leaves before hearted types were first developed in the middle of the sixteenth century.

The lettuce plant is an annual consisting of a rosette of glabrous leaves closely packed together to form a head. The internodes are very short, but on flowering an elongated, highly branched stem is produced bearing numerous capitula. The florets are all ligulate with a pale-yellow corolla, five stamens and a forked stigma. Self-pollination is the rule, and the achenes are 3–4 mm long, off-white, grey or brown in colour, ending in a beak terminated by two rows of pappus hairs. In an endeavour to provide a succession of supply throughout the year and also to cater for a variety of tastes, a great many different types of lettuce have been developed. Basically four groups are distinguished, recognised by some as botanical varieties in a formal sense. Butterhead lettuces form heads of rather soft, curly leaves rather like cabbages. Similar in shape, though often more conical, are the crisphead lettuces with somewhat crisper and more succulent leaves. Cos or romaine lettuce form fairly loose, oblong hearts composed of fairly rigid leaves with a well-developed midrib. Leaf lettuces consist of loose rosettes of leaves, often very curly and highly fringed. All types of lettuce are grown widely in home gardens, but their economic importance depends on intensive, large-scale production in glasshouses or in the open. Quickly maturing, tightly heading, non-bolting cultivars have been selected, suitable for mechanised handling, packaging and transportation. Apart from these characteristics, plant breeders have concentrated on resistance to leaf diseases and adaptability to local climatic conditions. Very many cultivars have been developed, among them Grand Rapids, famous for the discovery of dormancy control by phytochrome in its red or far-red light absorbing forms. The popularity of lettuce has increased rapidly in recent times as seen for example by the appearance of salad bars in many restaurants and the demand for a healthier diet. A growing preference for crisphead lettuce has led to thriving international trade and renewed activity by plant breeders to produce a range of acceptable cultivars with a long season of supply.

6.5 *CICHORIUM*

Two species in this genus contain plants of some economic importance, chicory and endive. Chicory (*Cichorium intybus*) is a perennial plant with a stout taproot and large, slightly hairy, serrated leaves with pointed lobes. The stems grow to a height of about 1–1.5 m and bear in the axils of their leaves clusters of small capitula composed entirely of ligulate flowers. The corolla is blue in colour, pollination is by insects, and the achene is about 3 mm long, grey-brown, angular and surmounted by a pappus of finely divided scales. Chicory is a common wild plant on chalky soils in England but has been cultivated since Roman times for a number of purposes. The leaves can serve as a vegetable for human consumption and, because of high mineral content and drought resistance, chicory is sometimes included as a constituent of pasture mixtures. Certain cultivars with deep and well-developed roots are grown as root crops in much the same way as sugar beet, and indeed there is growing interest in chicory as a source of fructose. The thick root stocks may also be replanted and, if grown in the absence of light, they produce a pale, oval head of delicately flavoured leaves called whitloof or Belgian endives. Traditionally chicory roots were also processed to form a substitute for coffee. Endive (*C. endivia*) is a very similar plant with glabrous leaves which are eaten raw or after blanching as a salad vegetable. It was cultivated by the ancient Egyptians for this purpose and has continued to be grown in several parts of Europe and USA where it is known as chicory or escarole.

6.6 OTHER CROP PLANTS

Several other plants belonging to the family Asteraceae deserve mention for their economic importance. Among these is guizotia or niger seed (*Guizotia abyssinica*) which, as the botanical name implies, is a native of central Africa and is now extensively cultivated in Ethiopia and India. This is an annual plant, 1–1.5 m in height, bearing yellow capitula in its leaf axils. There are both sterile ray and fertile disc flowers, and the achene is 5–10 mm long, angular and shiny-black. The seeds contain 38–50% oil and about 20% crude protein. Good quality oil is used for cooking, and inferior grades for manufacture of soaps and paint. The oil consists of 9–17% saturated fatty acids, 30–40% oleic and 48–55% linoleic acids. Niger seedcake, produced after oil extraction, is valued as a cattle feed. Although Niger does not yield well enough for large-scale commercial production, it is considered to be suitable for smallholders in areas where other oil crops do not thrive. Another plant of some importance in temperate countries is the globe artichoke (*Cynara scolymus*), which is a native of the Mediterranean area and the Canary Isles and is now grown predominantly in southern Europe and in California. The crop is cultivated for the sake of its fleshy capitula and the involucre bracts surrounding it which are blanched and served with a dressing. In appearance the globe artichoke is like a large, blue thistle. It should not be confused with the Jerusalem artichoke (see p. 158). Yarrow (*Achillea millefolium*) enjoyed a brief period of popularity when it was included in pasture mixtures as a mineral-rich constituent. More correctly it should be classed as a weed, common in pastures and on cultivated land which persists tenaciously through its extensive system of rhizomes and copious seed production. It regenerates readily from quite small rhizome fragments and is thus difficult to eradicate. This plant is easily recognised by its highly divided, feathery leaves and flat flower heads consisting of numerous capitula massed together (Fig. 6.3). The leaves and especially the rhizome contain bitter and slightly pungent compounds which may taint the milk if cows are allowed to eat excessive amounts of yarrow.

In addition to these cultivated plants, the Asteraceae contain a large number of species which are well known and important as weeds. Each country has its own list of these plants, depending on their spread and ease of control, but because of their common occurrence we should at least mention the genera containing the thistles (*Carduus*, *Cirsium*), groundsel and ragwort (*Senecio*), mayweeds and chamomiles (*Matricaria*, *Anthemis*), daisies (*Bellis*, *Chrysanthemum*), sowthistles (*Sonchus*), dandelion (*Taraxacum*), and the hawkweeds, hawksbeard and related plants (*Hieracium*, *Crepis*, *Leontodon*, *Hypochaeris*). Altogether, this is a large, important and widespread family.

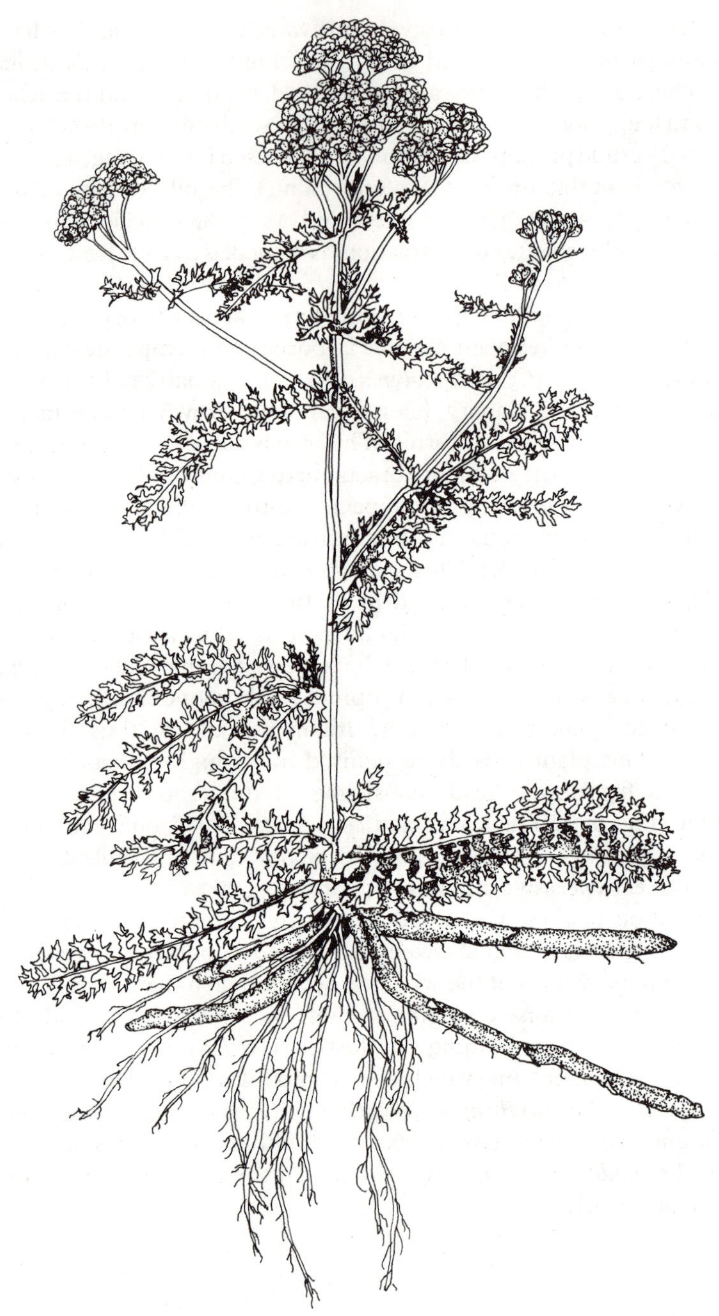

Fig. 6.3. Yarrow (*Achillea millefolium*) plant showing inflorescences, highly divided leaves and rhizomes.

FURTHER READING

Bassett, M.J. (ed.) (1986). *Breeding vegetable crops.* AVI Publishing Company, Westport.

Beech, D.F. (1969). Safflower. *Field Crop Abstracts,* **22**, 107–17.

Carter, J.F. (1978). *Sunflower science and technology.* American Society of Agronomy, Madison.

Casida, J.E. (ed.) (1973). *Pyrethrum. The natural insecticide.* Academic Press, New York.

Godin, V.J. and Spensley, P.C. (1971). *Crops and products digests. 1. Oils and oilseeds.* Tropical Products Institute, London.

Heiser, C.B. (1976). *The sunflower.* University of Oklahoma Press.

Heywood, V.H., Harborne, J.B. and Turner, B.L. (eds) (1977). *The biology and chemistry of the Compositae.* Academic Press, London.

Ryder, R.J. (1979). *Leafy salad vegetables.* AVI Publishing Company, Westport.

Weiss, E.A. (1971). *Castor, sesame and safflower.* Leonard Hill, London.

Weiss, E.A. (1983). *Oilseed crops.* Longman, London.

Whitehead, D.L. and Bowers, W.S. (eds) (1983). *Natural products for innovative pest management.* Pergamon Press, Oxford.

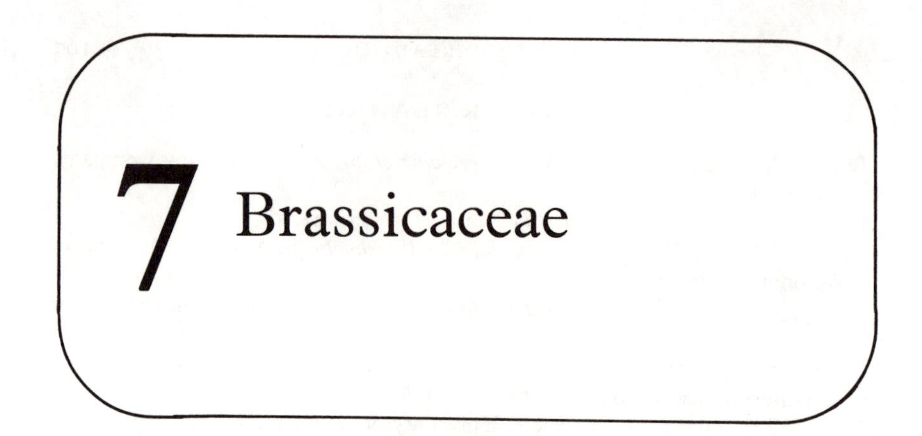

7 Brassicaceae

The Brassicaceae (Cruciferae) derive their old name from the shape of their flower which is said to be cruciform or resembling a cross when looked at from above. This family consists of some 300 genera and 3000 species, mainly herbaceous plants occurring predominantly in temperate regions. The genus of the greatest economic importance is *Brassica*, which will need to be described in some detail, but there are two or three other genera deserving at least a mention.

7.1 *BRASSICA*

There are about 40 species in this genus, most of them annual or biennial, several of them cultivated extensively in cool, temperate regions. Plants belonging to this genus usually have tap roots, and there is a tendency for storage organs of various kinds to be developed. The stems are erect, although internodes are often greatly foreshortened while the plant remains vegetative, the leaves are alternate, lobed, and glabrous or covered with simple hairs. The inflorescence is a raceme bearing numerous, conspicuous flowers, usually yellow in colour. There are four sepals, arranged as an inner and an outer pair, and four petals each consisting of a narrow, erect claw at the base with a broad, spreading limb above, together forming the configuration of a cross (Fig. 7.1). The six stamens are represented by a relatively short outer pair and four somewhat longer inner ones. The ovary is superior and consists of two united carpels divided by a membranous, false septum and a single, lobed stigma. Pollination is by insects, and there are nectaries at the base of the anther filaments. The fruit is a capsule, referred to as a siliqua when it is longer than it is broad. In other members of this family the fruit is broader than long, in which case it is described as a silicula. There are numerous ovules, arranged in rows,

Fig. 7.1. *Brassica* spp. Inflorescence and flowers.

which remain temporarily attached to the false septum when, on maturity, the valves of the fruit dry up and retract. The seeds have no endosperm and consist in the main of two cotyledons which act as reserve organs rich in oil. One of the distinctive biochemical characteristics of the genus *Brassica* is the presence of sulphur-containing glucosides, glucosinolates, which through the action of such enzymes as myrosinase break down to form strongly tasting substances. These sulphur-containing compounds, predominantly thiocyanates and isothionates, impart a bitter taste to wild plants in this genus protecting them from grazing animals, and in cultivated plants they are responsible for the typical volatile flavours with which they are associated. Blanching these vegetables with boiling water prior to freezing destroys the enzymes and thus alters the taste perceptibly. On cooking cabbage, dimethylsulphide is also released from cystine and this is largely the cause of the odour produced. The sulphur nutrition of the plant is said to have a bearing on the level of sulphur-containing substances in the tissue. These compounds are also not uniformly distributed. For example, the concentration of allyl isothianate has been shown to be higher in the heart of a cabbage than in the outside leaves. Another feature worth noting is that brassicas are prone to attack by insects

Table 7.1. *Economically important* Brassica *species*

Botanical name	Common name	Chromosome number (2*n*)
Brassica oleracea		
var. *acephala*	Kale, borecole	18
var. *botrytis*	Broccoli, cauliflower	
var. *capitata*	Cabbage	
var. *gongylodes (caulorapa)*	Kohlrabi	
var. *gemmifera*	Brussels sprouts	
var. *italica*	Sprouting broccoli	
Brassica napus		38
subsp. *oleifera*		
var. *biennis*	Winter rape	
var. *annua*	Summer rape	
var. *napobrassica*	Swede	
Brassica compestris	Field mustard	20
subsp. *oleifera*	Turnip rape	
subsp. *rapifera*	Turnip	
subsp. *chinensis*	Chinese mustard, pak-choi	
subsp. *pekinensis*	Chinese cabbage, pe-tsai	
Brassica juncea	Brown mustard, Indian mustard	36
Brassica nigra	Black mustard	16

such as aphids and caterpillars and plant diseases like club root, although plant breeders have made great strides in breeding resistant cultivars.

Table 7.1 shows the main economically important species in the genus *Brassica*. Since prolonged selection and cultivation have resulted in the differentiation of distinct types within each species, some authorities have created subspecific or at least varietal subdivisions. The danger of confusion is even greater when it comes to common names, especially as applied in different English-speaking countries.

The basic chromosome number in the genus is thought to be $n=5$, which would lead to the interpretation that *B. oleracea* ($n=9$) is an amphidiploid from a cross of two primitive species, followed by loss of a chromosome. *B. campestris* ($n=10$) would also appear to be of somewhat similar origin, and *B. napus* ($n=19$) probably is an amphidiploid hybrid of these two. *B. nigra* ($n=8$) is another basic species of the genus, and it is possible that together with *B. campestris* it gave rise to another amphidiploid, *B. juncea* ($n=18$). However, cytotaxonomic and serological evidence has been put forward by some workers to substantiate a somewhat different classification and explanation of relationships, and in particular the status of sub-specific categories is debatable.

Kales, cabbages, and related plants (*Brassica oleracea*)

This species contains a whole range of plant types that have become important vegetable and fodder crops in temperate zones. The plants belonging to this species have glabrous, blue-green glaucous leaves, and the inflorescence is a long raceme in which the unopened buds are above the open flowers (Fig. 7.1). The petals are pale yellow, occasionally white, and the six anther filaments are fairly similar in length. The fruit, 5–10 cm in length, is a rounded siliqua with a short beak. The seeds are grey-brown in colour and about 2 mm in diameter. Most of the types now grown are biennials. Although *B. oleracea* is native to the coasts of north-western Europe and the Mediterranean as the perennial wild cabbage, it is possible that the modern cultivated forms are not solely derived from it and that other plants may have played a part in their ancestry. Early cultivation appears to have taken place in the eastern Mediterranean, and the Greeks knew kales and what may have been kohlrabi (see below) as early as 600 BC. There is some dispute whether the plant described by Pliny a few 100 years later was cauliflower or kohlrabi, because it is thought that the former was not known before the Middle Ages. Sprouting broccoli, first mentioned in 1660, was originally known as Italian asparagus, and Brussels sprouts appeared in Belgium in the eighteenth century.

Kales (var. acephala)

Kales are biennial plants growing about 0.5–1.5 m in height, with leaves loosely arranged alternately up the stem. In growth habit they resemble the wild cabbage most closely. Those grown as a horticultural crop are relatively low-growing, bearing a rosette of leaves at the top of the stem. Most of the cultivars have crinkled leaves and are variously known as curly kale, cottager's kale or Russian kale. Because of their frost hardiness they are used as winter vegetables, several leaves or the whole top rosette being harvested at a time. Collards are grown in the United States as a leafy vegetable, a rosette of smooth leaves. All these horticultural kales are rich sources of vitamin C. Kales grown as fodder crops tend to be taller. Basically two types are available: 1000-headed kale and marrow-stem kale. The former has slender, woody stems that branch near the top and bear a rosette of slightly curly leaves, is fairly frost resistant and serves as a fodder crop during the winter. Marrow-stem kale, or Chou Moellier as it is known in some countries, is a similar plant, except that the stem is much thicker if the crop is grown at sufficiently low density to allow stem development. This enlarged stem, measuring 10 cm or more in diameter, has a large, succulent pith surrounded by a ring of vascular bundles. It can make up more than one half of the total crop by weight and with a protein content of about 12% compares quite favourably with the leaves. Both these kales are fed to animals *in situ* or cut and carted to feedlots. Although most valuable especially on dairy farms, their popularity depends on costs of production compared with other winter feeds, such as hay, silage or pasture.

Cabbages (var. capitata)

A cabbage could be described as a very large terminal bud with foreshortened stem and tightly packed, inrolled leaves (Fig. 7.2). Only if flowering is allowed to occur, is there any appreciable internode elongation terminating in the formation of an inflorescence. Plant breeders have produced a profusion of cabbage cultivars varying in colour, shape, leaf texture, keeping quality and other characteristics, with the aim of providing a continuous supply of vegetables throughout the year, especially in winter and early spring. Crops required in spring are sown and planted out in the autumn, and for this purpose rather loosely packed, conically shaped cabbages are used, apparently because these bolt less readily after the winter. Normally, flower initiation and stem elongation are induced by low winter temperature of the order of 4–7 °C for several weeks, provided the plant has passed the juvenile stage. Delayed response is an obvious advantage, enabling longer storage in the field before harvesting. Summer and autumn cabbages tend to be round, largely because they stack and transport easily. For winter consumption the savoy cabbage is preferred, round-headed cultivars with dark-green, crinkled and puckered leaves, which are frost-hardy and store successfully. Red cabbage cultivars, round-headed and rich in anthocyanin pigment are popular for pickling, although they make a pleasant vegetable if cooked slowly in water with a little vinegar, a few slices of apple and some peppercorns. In fact, cabbage is far too often overcooked and then develops undesirable sulphurous odours. Boiling water should be used and cooking time kept to a minimum. Alternatively cabbage is shredded raw and served as cole-slaw with other ingredients and dressing, or preserved as sauerkraut with salt and through the conversion of sugars in the leaf to lactic acid by anaerobic bacteria, causing the pH to fall below 4.0, a method practised in central Europe apparently since 200 BC. In addition to these culinary uses, cabbages can serve as a feed for cattle during the winter.

Fig. 7.2. Cabbage (*Brassica oleracea*).

Kohlrabi (var. gongylodes)

Superficially the kohlrabi resembles marrow-stem kale with the stem infinitely shorter and fatter. Soon after germination the stem above the cotyledons grows in thickness and little growth in length occurs. As a result a 'bulb' is formed consisting entirely of stem tissue, mainly a much enlarged pith surrounded by a ring of vascular bundles. The 'bulb' is round or conical in shape and arises from the basal, unthickened part of the stem. Towards its apex it bears petiolate leaves, below which the surface is marked by conspicuous leaf scars. The colour of the skin is green or green-purple, depending on cultivar. Kohlrabi is a well-known vegetable in central Europe, although not cultivated on a large scale, and both 'bulb' and the young leaves are eaten. Premature bolting is one of the problems in some cultivars.

Brussels sprouts (var. gemmifera)

Here we are dealing with *Brassica oleracea* plants in which the tall kale-like stem bears greatly enlarged axillary buds, and in fact the suggestion can be made that the original Brussels sprout may have been a mutant with inhibited terminal bud activity. The buds, occurring in the axil of petiolate leaves, grow to a large size once the leaves have dropped off, so that in course of time the outside of the stem becomes completely covered with tightly packed buds, or sprouts as they are commonly known. A well-grown sprout is about the size of a golf ball. It requires cool temperatures if it is to be firm and have its leaves tightly packed. Uniform sprout development is encouraged by topping the plants in late summer or autumn, and this facilitates the mechanical harvesting of the crop. Hand picking is slow, expensive and, in cold weather, unpleasant. Brussels sprouts are highly regarded as a late autumn to early winter vegetable and are capable of being preserved by freezing.

Cauliflowers (var. botrytis)

The white 'curd' for which this plant is grown as a vegetable is an enlarged terminal bud in an early flowering stage. The branches of the inflorescence are greatly swollen, with flower buds at an immature stage closely packed together (Fig. 7.3). The whole structure, enveloped by large leaves with a fleshy, white midrib, is borne on a foreshortened stem. Plant breeders have ensured a continuous succession of supply from summer cauliflowers which are planted out in the spring and harvested in summer to winter cauliflowers (broccoli) which are intended for consumption between autumn and early spring. In comparison with other types of *Brassica oleracea*, the cauliflower is not especially resistant to winter temperatures, and particularly the curd is easily damaged by frost if it is not protected by the surrounding leaves. It should also be noted that some cultivars behave as annuals because reproductive development resulting in curd formation occurs without exposure to low temperature when the plant has formed a certain minimum number of leaves. The optimum temperature in these early cultivars is about 17 °C, and at above 20 °C the curds do not form or are of poor quality. There is a gradual transition from annual to biennial types, with winter cauliflowers requiring prolonged exposure to cold. Sowing and planting times must be carefully adjusted to provide the

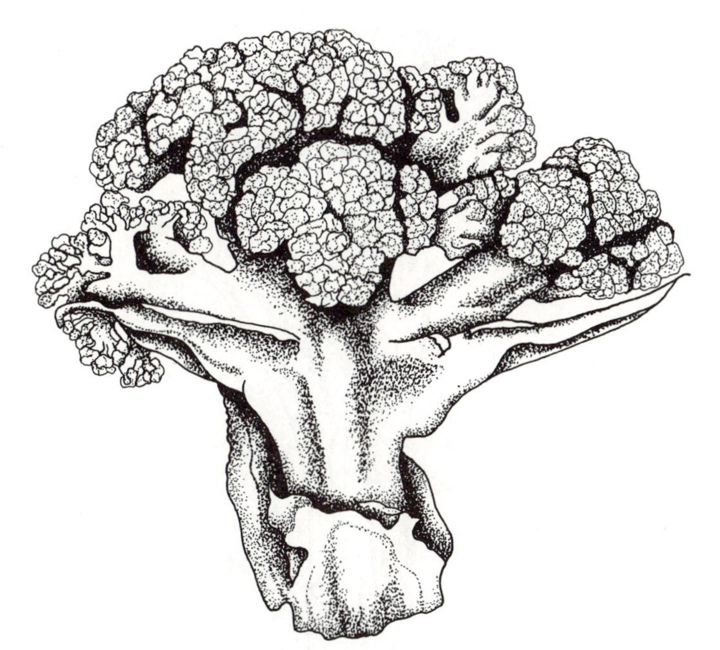

Fig. 7.3. Cauliflower (*Brassica oleracea*). Part of curd.

climatic conditions needed by each cultivar for successful head formation. If the plant is left to grow after the curd has formed, the inflorescence continues to develop and long branches with yellow flowers appear. Cauliflowers are very nutritious and readily preserved by freezing or drying.

Sprouting broccoli (var. italica)

This vegetable crop is also known as calabrese, denoting the fact that it was widely grown in Calabria in southern Italy. It was taken by migrants to the United States where it became established as a most popular vegetable, notably in California where over half the USA crop is grown. The name, broccoli, derived from brocco, the Italian word for branch, should be reserved for this plant and not extended to apply to winter cauliflowers, if confusion is to be avoided. Sprouting broccoli bears some resemblance to both Brussels sprouts and to cauliflower, in that immature inflorescences consisting of green to purple buds and thick flower stalks (Fig. 7.4) occur in the axil of leaves and at the apex of a tallish stem. Harvesting must be done before the buds open and the yellow petals become visible. Since the

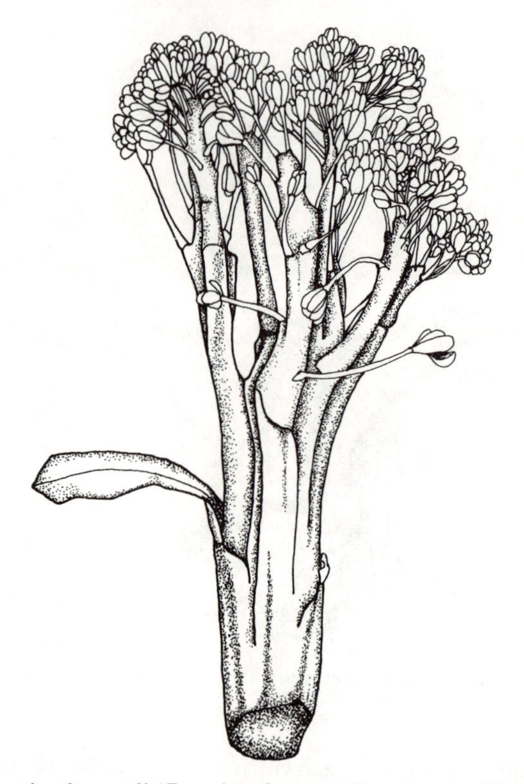

Fig. 7.4. Sprouting broccoli (*Brassica oleracea*). Immature axillary inflorescence.

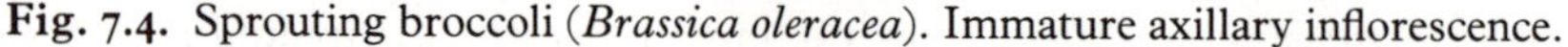

development of inflorescences in different axillary positions is not well synchronised, it follows that several pickings are necessary to obtain sprouts at their optimum stage. Appropriate plant breeding, possibly combined with topping of plants, should make mechanical harvesting possible. Sprouting broccoli is particularly suited to processing by freezing, which accounts for its popularity in the United States and elsewhere.

Swedes and rapes (*Brassica napus*)

This species is also called *Brassica napobrassica* by some authors. It has a diploid chromosome number of 38 and is thought to have arisen as an amphidiploid hybrid between *B. oleracea* and *B. campestris*, probably in the very early days of cultivation. The economically important plants to be considered are the swede and forage rape as sources of fodder and a vegetable, and the annual and biennial forms of oil-seed rape. Botanically they closely resemble *B. oleracea* apart from a few minor features. The leaves on young plants are slightly hairy, although later they are glabrous and glaucous. On the inflorescence, the unopened buds and flowers occur at the same level, petal colour is yellow to cream, the filaments of the outer stamens are slightly curved and the seeds are dark purple to black, some 2.2 mm in diameter. Leaves on the inflorescence tend to clasp the stem.

Swedes (var. napobrassica)

Swedes are grown as a fodder crop and to a minor extent as a winter vegetable for the sake of the swollen storage organ, colloquially referred to as the root or bulb. The plant has a normal tap root which only at its upper end becomes swollen and merges with the hypocotyl, the major component of the swede. The uppermost part, the so-called neck, is composed of a true stem, a conspicuous and distinct structure bearing very many leaf scars (Fig. 7.5). Anatomically, the swollen structure is composed predominantly of secondary xylem made up of unlignified xylem parenchyma cells. The secondary phloem outside the cambium is also thin-walled. Lignification occurs only after the plant has over-wintered, when reproductive development begins. Progressive hardening and withdrawal of food reserves then make the swede unpalatable. Until then the plant is regarded as a valuable fodder crop for sheep and cattle. Cultivars provide a spread of maturity dates from autumn to early spring, following sowing in early summer. The shape of the bulb is round, oval or tankard, and the outside skin is white below and coloured above. Depending on the amount of anthocyanin present, this colour ranges from a reddish purple to bronze and green, with the leaves displaying varying shades of blue-green. The flesh of the swede is mostly yellow or pale yellow, although there are also white-fleshed cultivars. Compared with other 'root' crops, swedes have a relatively high dry matter content (10–12%), many cultivars tend to be frost-hardy, they store well in the field, and are readily fed to livestock *in situ*. One of the more important considerations is resistance to clubroot, dry rot, aphids and virus disease. Swedes came into domestic and farming use only a few hundred years ago, the first record dating back to 1620, and introduction from Sweden to Britain is dated at 1775–80. In the United States this crop is known under the name of rutabaga or Swedish turnip.

Rape (subsp. oleifera)

The rape plant, in distinction to the swede, has a slender, often branched stem without any enlarged storage organ. The leaves are fairly small,

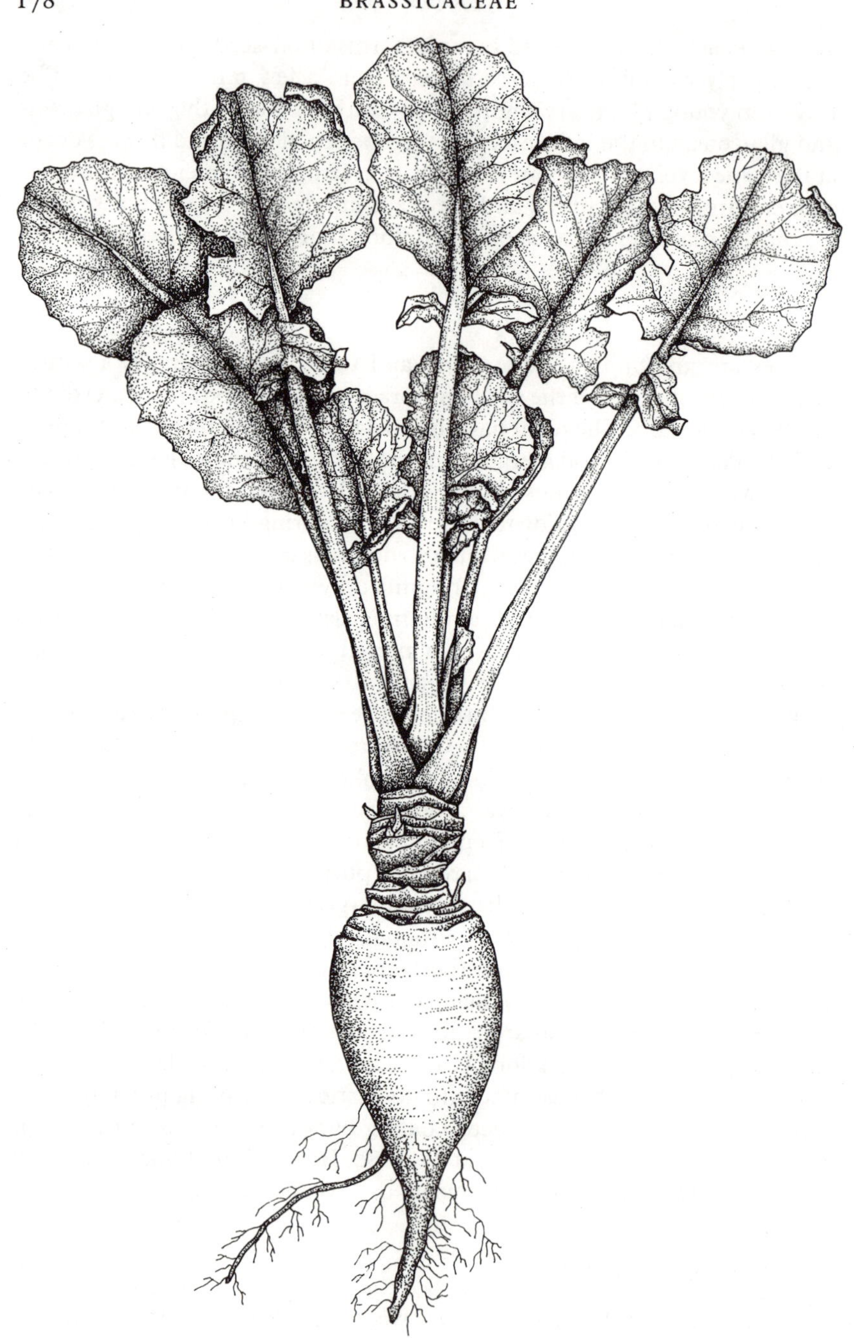

Fig. 7.5. Swede (*Brassica napus*).

Fig. 7.6. Rape (*Brassica napus*).

depending on cultivar (Fig. 7.6), the tap root is deeply penetrating and not swollen, and the height of the stem is up to about 140 cm. The crop grown for fodder is known as forage rape or if grown for seed as oil rape. Forage rapes are used mainly for sheep in northern Europe and New Zealand, where successive sowings in the spring commonly provide forage eight to twelve weeks later for weaned lambs and for ewes prior to mating. Basically two types of forage rape are distinguished, early giant rape and broad-leaved Essex. The former is represented by fairly tall cultivars with a distinct main stem and branches bearing large, drooping, pale-green leaves. Initial production is heavy and palatable, but there is little or no recovery growth following grazing. Because of its height and drooping leaves it has the tendency of keeping the sheep wet, and hence its use is restricted to relatively dry areas. Broad-leaved Essex rape has shorter and branching stems, with dark-green leaves forming a dense canopy. Though less productive or palatable than early giant rape, it recovers well from grazing and may be utilised several times. Plant breeders have attempted to combine the advantages of both types and also to incorporate aphid and clubroot resistance in recent cultivars. The range of variation does not appear to be great, and such plants as Siberian kale, hungry gap kale, or fillgap kale are very similar and best considered as forage rape. One possible problem with forage rapes is the occurrence of glucosides known to induce goitre in livestock, but it is possible to select for low concentrations of these substances.

Rapes selected for seed and oil production may be annual or biennial in habit. These are, strictly speaking, swede-like oil rapes in distinction to the turnip-rape (*Brassica campestris*), described in the following section, which is grown for the same purpose. Both species are closely related to one another (Table 7.1) and, although they can be distinguished by leaf characteristics, commercially their product has always been jointly referred to as rape oil. There is good evidence to show that the two species have been cultivated as oil crops for many centuries in countries where alternative sources of oil, such as the olive tree or the poppy, were not available. According to Sanskrit records the crop was known in India as early as 2000–1500 BC. Rape seed provided the most important source of lamp oil in northern Europe from the thirteenth century onwards until in more recent times it was replaced by animal oils and especially petroleum. Germany, Belgium and some Scandinavian countries grew this crop on a large scale, and in 1866 Denmark was reported to have had 15 500 ha devoted to it and Sweden to be producing 3000 t of seed. Declining importance since then has been replaced in more recent years by an upsurge in interest, and it is more than likely that present levels of production amounting to some 16 million tonnes throughout the world will be greatly exceeded. Rape seed is one of the few sources of plant oil that can be grown in temperate climates. Apart from northern European countries, the main producing areas are Canada, China, India, Australia, and Pakistan (*B. campestris*). Internationally the oil is often referred to as colza. When grown for oil, rape (*B. napus*) may be sown in the autumn or in the spring, since both annual and biennial cultivars are available. Autumn sowing, which should be early enough to obtain good crop establishment before the onset of winter, allows a longer period of growth and hence higher yields than sowing in the spring. The more productive biennial types are thus preferred unless winter conditions are too severe for them to be grown. Late spring or summer drought may also be a problem especially with annual cultivars. The crop is harvested when the seed still has a moisture content of between 12 and 20%, a relatively high figure, to avoid crushing or bruising. A safer storage moisture content is 6–8%. Seed shattering may be a serious problem, another reason for cutting before the crop is fully ripe. A number of different harvesting methods are available, including desiccation of the crop followed by combining. No single system should be adopted rigidly, but varied depending on climatic conditions in any particular season. Good average yields of winter rape may be as high as 2–3 t ha^{-1} in cool, temperate climates, but considerably less for summer rape and in warm, dry areas. The oil content of *B. napus* seed is 40–46%, of which normally 20–45% is erucic acid and 9–15% eicosenoic acid. The saturated fatty acid content is very low. For oil extraction the seeds are first crushed and then heated up to about 105 °C. Solvents are usually employed, and the crude oil is refined and bleached. The main use is for culinary purposes as salad or cooking oil or, if hydrogenated, in the manufacture of margarine. Less highly refined oils are used for lubricating, in the manufacture of soaps or for illumination. After the extraction of the oil, rapeseed meal is produced which is used as a protein supplement for livestock. Careful experimental work in recent years has shown that some constituents of rapeseed oil, notably the high erucic acid content, can have undesirable effects on the growth rate of animals and on heart and skeletal muscles. The remedy appears to be to control intake or, better still, to change over to cultivars selected for low or zero erucic acid, such as Canbra produced in Canada. This long-chain fatty acid is used commercially in the production of lubricants and plasticisers. If the trend to reduce erucic acid in rape oil continues, another source of this compound will need to be found. This may well be crambe (*Crambe abyssinica*), another annual member of the Brassicaceae which is under investigation in the United States and shows distinct promise as a new crop plant (see page 188). Another recent advance by plant breeders has been to develop so-called double-low cultivars of rape with a reduced erucic acid content but also containing less glucosinolates in the protein fraction which can have deleterious effects on monogastric animals.

Turnip and turnip-rape (*Brassica campestris*)

Plants belonging to this species resemble in general morphology other related groups in the genus *Brassica*, except that the leaves formed before flowering tend to be bright-green in colour as well as densely hairy. The leaves on the inflorescence are mostly glabrous, darker green, and the leaf bases project conspicuously so as to clasp the stem. In distinction to *B. napus*, the unopened flower buds are at a lower level than the flowers that have just opened. Individual flowers are also smaller, and the outer stamens have short and slightly curved filaments. There are hardly any differences in the structure of the siliqua, but the seeds of *B. campestris* are smaller as shown in Table 7.2. Seed colour also differs because turnip and turnip-rape seed is usually brighter and red-purple rather than purple-black. Some cultivars have yellow-coloured seed.

Table 7.2. *Seed weight (g per 1000 seeds) of* Brassica napus *and* B. campestris

Brassica napus:	
swede	3·5–4·0
winter rape	4·5–5·5
summer rape	3·5–4·5
Brassica campestris:	
turnip	2.0–2.5
winter turnip rape	3.0–4.0
summer turnip rape	2.0–3.0

Turnip (subsp. rapifera)

The turnip is grown for the sake of its storage organ, the hypocotyl and the swollen upper part of the root and lower part of the stem. In distinction to the swede, the stem portion bearing leaf scars is not drawn out into a neck-like structure, but instead it forms a contracted cone tapering towards the crown where the living leaves are inserted (Fig. 7.7). In other respects the turnip resembles the swede, including the virtual absence of lignified secondary xylem. These cells become woody only at the onset of flowering. However, when it comes to both shape and colour of the bulb, there is greater variability in the turnip, and this has been utilised by the plant breeder. Basically there are two types of turnip, those with white and those with yellow flesh. The white-fleshed turnips are low in dry matter content (about 8%). They grow quickly and mature early, they are generally not frost-resistant, and do not store well. Because they produce a useful crop within 10–16 weeks from sowing, they have a definite role in agriculture, especially as a sheep feed. Some types are also popular as a vegetable which deserves greater popularity than it appears to have. Others are so quick maturing that they may be used as stubble turnips or for grazing as part of a mixture including Italian ryegrass or other pasture species. The skin colour of the upper part of the bulb varies from white to green and light purple, depending on cultivar. Yellow-fleshed turnips are slightly superior in dry-matter content (about 9%), they mature more slowly and are thus sown earlier and utilised later. They also store better and in many ways they approach the swedes in agronomic characteristics. This type of turnip is strictly biennial in habit as opposed to the earliest white-fleshed cultivars which often run to seed in the year of sowing. Yellow-fleshed turnips, on the other hand, are mainly intended for feeding in late winter to early spring following sowing in the previous summer. Skin colours vary from

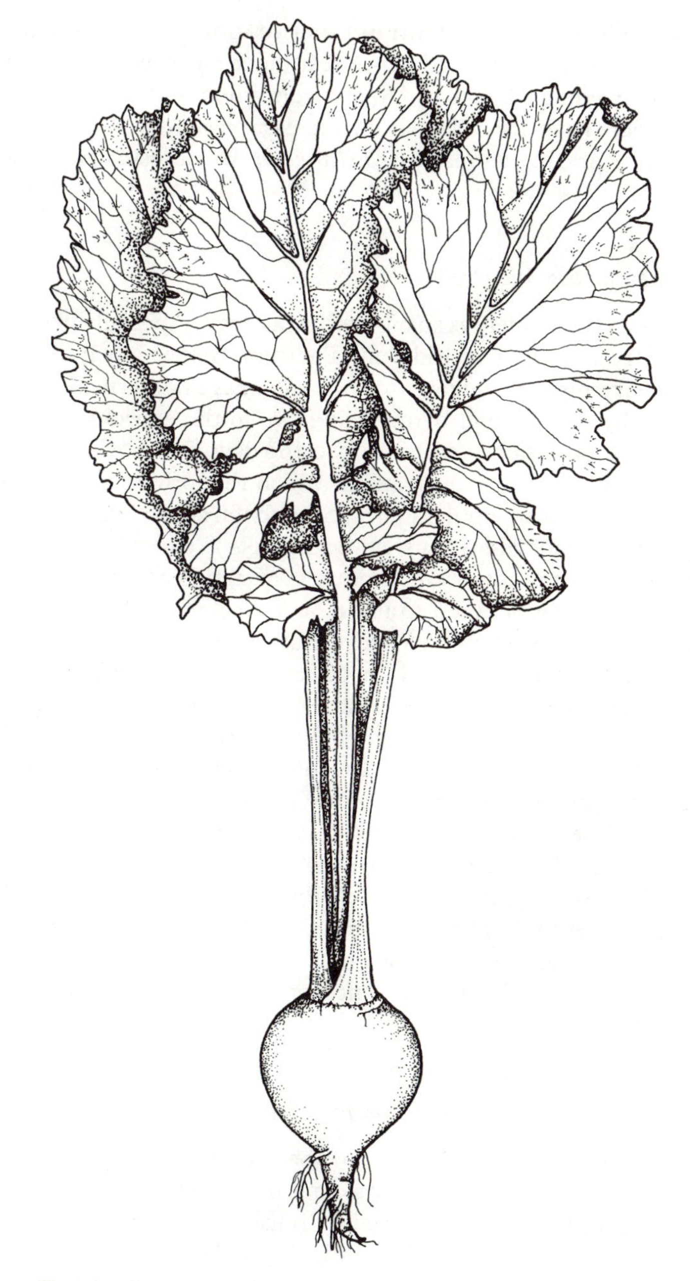

Fig. 7.7. Turnip (*Brassica campestris*).

green to purple. Both types of turnip are susceptible to clubroot and dry rot as well as to aphid attack, and hence resistance has been one of the most important breeding objectives. Apart from yield and date of maturity, plant breeders have also selected for increased leafiness so as to improve utilisation by grazing animals. *Brassica campestris* crosses easily with *B. napus*.

Turnip-rape (subsp. oleifera)

This is grown very widely as an oil-producing crop, especially in warm and relatively dry areas such as India and Pakistan. Botanically it differs from the swede-like rapes belonging to *Brassica napus* as described above, and agronomically it is distinct by maturing two to three weeks earlier and a seed crop may be obtained within 100 to 160 days from sowing. There are both annual and biennial cultivars, of which the former are most widely grown. They are most popular in dry, continental conditions on account of their quick maturity and short stem height (45–75 cm). Canada and India are probably the most important producers of summer turnip rape. The oil content of the seed is between 40 and 46%, the composition of the oil being similar to that of *B. napus* with a high unsaturated fatty acid content. Plant breeders have been successful in selecting for low erucic acid concentration, and they have also been able to reduce or eliminate glucosinolates or, as they were called previously, mustard oil glucosides or thioglucosides. These substances are split by the enzyme myrosinase to produce isocyanates which are not only pungent in taste and smell but also goitrogenic. Use of cultivars free of glucosinolates or inactivation of the enzyme by heat treatment would make rapeseed meal safe for animal consumption. As regards yield of oil, the greatest improvement to have taken place recently has been the selection of yellow seeded plants from India where annual rapeseed known as yellow sarson or colza has been cultivated for a very long time. Another widely grown type is toria or Indian rape.

Mustards (*Brassica* spp. and *Sinapis alba*)

The use of spices and condiments goes back thousands of years, probably to before 3000 BC when mustard was first mentioned in Sanskrit records. The Greeks and the Romans certainly relished mustard, and its use has continued unabated over the centuries, originally perhaps not so much to improve the flavour of meat but to mask the taste brought about by low keeping quality. The name mustard is derived from the sweet 'must' of old wine mixed with crushed seeds to obtain 'mustum ardens'. As a crop for export mustard is grown in Canada and in northern Europe, although for local consumption it is cultivated in many other countries. Botanically four species are involved each giving rise to a different type of mustard:

Brassica nigra	– black mustard
Brassica juncea	– brown mustard
Brassica carinata	– Ethiopian mustard
Sinapis alba	– white mustard

Brassica nigra is of importance not only as a crop plant but also because it played a significant part in the evolution of several species in the genus *Brassica*. It probably originated in or near Asia Minor but has spread through cultivation into every continent. Black mustard is an annual plant up to about 1 m tall with dark-green, somewhat hairy leaves. The bright-yellow flowers are small with spreading sepals and produce small, distinctly pitted seeds which are red-brown to black in colour. The base of the leaves does not clasp the stem of the inflorescence as in other *Brassica* species. The plant occurs wild in southern England. It used to be cultivated there and elsewhere in Europe for the sake of the pungent principle of its seed, an allyl thiocyanate produced through the action of the enzyme myrosin on the glucoside sinagrin. This was needed for the manufacture of mustard. However, black mustard was never grown on a big scale because its low yield and pungency made it unattractive as a fodder crop, and also because the seeds remain dormant for some years and tend to act as a reservoir for the invasion of other crops. Any further doubt about the place of this crop has been dispelled in recent times, when it was found to be unsuitable for mechanical harvesting. Brown mustard (*B. juncea*) has now taken its place.

Brassica juncea, brown mustard, originated in central Asia and has been cultivated for a very long time in India, where under the name of rai or Indian mustard it is still grown widely as an oil seed crop, together with other members of the genus *Brassica*. It is also grown for oil in China and Japan, but for use as a condiment its production centres are in Canada, USA, England and Denmark. It is an annual plant with leaves resembling those of the swede. The flowers are small and bright yellow in colour, and the seeds are small with a 1000-seed weight of about 2–3 g. Since the siliquas do not shatter easily, the crop is suitable for combine harvesting. Brown mustard is a self-fertile amphidiploid and does not cross readily with any other species.

Sinapis alba (white mustard) is the only crop species in the genus *Sinapis*, of which the common arable weed charlock (*Sinapis arvensis*) is probably a better known representative because of its widespread occurrence in Great Britain. Although white mustard resembles the other mustards in many ways, its chromosome number of $2n = 24$ isolates it from the genus *Brassica* in which it is sometimes included as *B. alba*. It is an annual with hairy, petiolated leaves. The siliqua is hairy and terminates in a beak which in wild forms contains a seed which cannot germinate until the surrounding tissue has rotted away. Cultivated types contain up to seven seeds in the siliqua itself. The seed is large and contains the glucoside sinalbin which on enzyme hydrolysis produces hot and pungent tasting compounds which are not as volatile as those obtained from black or brown mustards. *S. alba* is wind-pollinated. It has been greatly improved as a seed crop through the introduction of new cultivars, especially those bred in England, Sweden and Germany. Plant breeders have also striven to improve white mustard as a forage and green manure crop. Its leaf production is high enough for both purposes, and palatability is good provided the plant remains vegetative. Unlike in *Brassica nigra*, the seed does not remain dormant in the soil. White mustard in the seedling stage is also used as a salad vegetable, alone or together with cress (*Lepidium sativum*).

7.2 OTHER CROP PLANTS

Radish (*Raphanus sativus*)

Although known mainly as a popular salad vegetable, the larger types of plant belonging to the same species have been and still are important food crops in eastern Asia, especially in China, Korea, Japan and India. Black radishes were grown by the Egyptians as early as 2000 BC and by the Chinese by about 500 BC. As a human food radishes have many uses, for they can be eaten fresh or cooked or preserved by pickling or drying. Not only the 'roots' but also leaves and seed capsules are edible. Within the species there is great variation in size, colour and shape of the storage organ and other characters. Selection for annual or biennial growth habit has occurred. Most of the types have large, swollen storage organs, composed predominantly of hypocotyl and the upper part of the root. The flowers are white or mauve in colour with spreading sepals, and the siliqua is long and inflated bearing a distinct beak. In Europe, North America and Australasia the small red or red and white garden radish is grown very widely, even to the extent of producing out-of-season crops under glass. The somewhat larger, white variety is popular in southern Germany. Types without fleshy hypocotyl but with rapid leaf production have also been bred in western Europe, in the hope that fodder radish would be superior to rape and other cruciferous fodder crops. Intergeneric crosses of this genus with *Brassica* to give a new crop *Raphanobrassica* have opened up new promising possibilities.

Water cress (*Rorippa nasturtium – aquaticum*)

Water cress occurs as a wild plant and is also cultivated intensively as a crop. This is a diploid species ($2n = 32$) which used to be called *Nasturtium officiniale*. In addition there is also a tetraploid species and also a sterile triploid, referred to as brown cress. The wild plant has been known and valued at least since the first century AD, and its medicinal properties as an antiscorbutic were recognised from an early date. Systematic cultivation began in England and France at the beginning of last century, usually with the objective of providing large cities with a fresh salad vegetable. For its production the plant requires to be grown in water which is not acid and supplied with nitrates. Numerous dark-green leaves are produced on slender stems which root freely at the nodes. The flower is white and the slender siliqua contains small seeds in two rows. Propagation is by seed or vegetatively. As a crop, water cress requires considerable inputs of labour that cannot easily be mechanised.

Horse radish (*Armoracia rusticana*)

Horse radish is a perennial plant with a long, fleshy taproot that penetrates deeply into the soil. The flowers are white and the capsule is a silicula containing one to six seeds. Although the plant has previously been reported as being seed-sterile, present cultivars are predominantly fertile. Nevertheless, current commercial propagation does not occur from seed but vegetatively by planting secondary roots, about 30 cm long, obtained from the previous harvest. These sets are grown for one season only, the object being to produce annual crops of taproot. In home gardens the plant is usually treated as perennial and in fact, once established, it is difficult to eradicate. Horseradish is grown for the sake of the white flesh of its roots which is grated and used as a condiment, especially with beef. Horseradish sauce is for roast beef what mint is for roast lamb. The pungent principle is allyl isothiocyanate which occurs in the plant bound to glucose and sulphate ions as the glucoside sinigrin. When the root tissue is macerated the glucoside is broken down by the enzyme myrosinase, thus releasing the isothiocyanate. Commercial horseradish sauce consists of root tissue, acetic acid and salt. As a crop, horseradish is cultivated in areas with a temperate climate, especially in the United States where some 800 hectares are devoted to this crop producing about 6000 tonnes per annum.

Crambe (*Crambe abyssinica*)

Crambe is a genus closely related to *Brassica* containing some 30 species, of which *C. abyssinica* is the only one to be cultivated at present. This is a herbaceous, branched annual, about 1 m in height, which bears yellow flowers producing small brown seeds with an oil content of about 40%. Although adapted to a Mediterranean-type climate it tolerates many other conditions ranging from the sub-tropical to the cool temperate. Cultivation on a commercial scale began only some 60 years ago, but since then interest in crambe has grown considerably. Russian, Swedish and other strains have been evaluated in the USA and breeding programmes are in progress in many countries. One of the problems is the high erucic acid content of the seed oil which reduces its value for edible use and restricts it to industrial application especially lubrication. Because of the presence of glucosinolates, the meal remaining after oil extraction cannot readily be used as a stock feed, quite apart from the high fibre content of the original seed. Considerable modifications are thus required before crambe can become widely established as an oil crop.

Apart from the economically important plants described in this section and a few minor crops of local significance, the family Brassicaceae contains a large number of plants which are known as weeds. These species differ from country to country and it would be tedious to provide long lists. Apart from charlock (*Sinapis arvensis*), another one to deserve special mention because of its widespread occurrence is shepherd's purse (*Capsella bursa-pastoris*) which derives its name from the characteristic, almost heart-shaped silicula.

FURTHER READING

Appelqvist, L.-Å. and Ohlson, R. (1972). *Rapeseed; cultivation, composition, processing and utilization.* Elsevier, Amsterdam.

Bassett, M.J. (ed.) (1986). *Breeding vegetable crops.* AVI Publishing Company, Westport.

Bunting, E.S. (1969). Oil-seed crops in Britain. *Field Crop Abstracts*, **22**, 215–23.

Godin, V.J. and Spensley, P.C. (1971). *Crop and product digests. 1. Oils and oilseeds.* Tropical Products Institute, London.

Hinman, C.W. (1985). Potential new crops. *Scientific American*, **255**, 24–9.

McNaughton, I.H. and Thow, R.F. (1972). Swedes and turnips. *Field Crop Abstracts*, **25**, 1–12.

Matheson, E.M. (1976). *Vegetable oil seed crops in Australia.* Holt, Rinehart and Winston, Sydney.

Nieuwhof, M. (1969). *Cole crops; botany, cultivation and utilization.* Leonard Hill, London.

Scarisbrick, D.H. and Daniels, R.W. (eds) (1986). *Oilseed rape*. Collins, London.
Tsunoda, S., Hinata, K. and Gómez-Campo, C. (eds) (1980). *Brassica crops and allies*. Biology and breeding. Japan Scientific Societies Press, Tokyo.
Vaughan, J.G., MacLeod, A.J. and Jones, B.M.G. (1976). *The biology and chemistry of the Cruciferae*. Academic Press, London.
Ward, J.T., Basford, W.D., Hawkins, J.H. and Halliday, J.M. (1985). *Oilseed rape*. Farming Press, Ipswich.
Weiss, E.A. (1983). *Oilseed crops*. Longman, London.

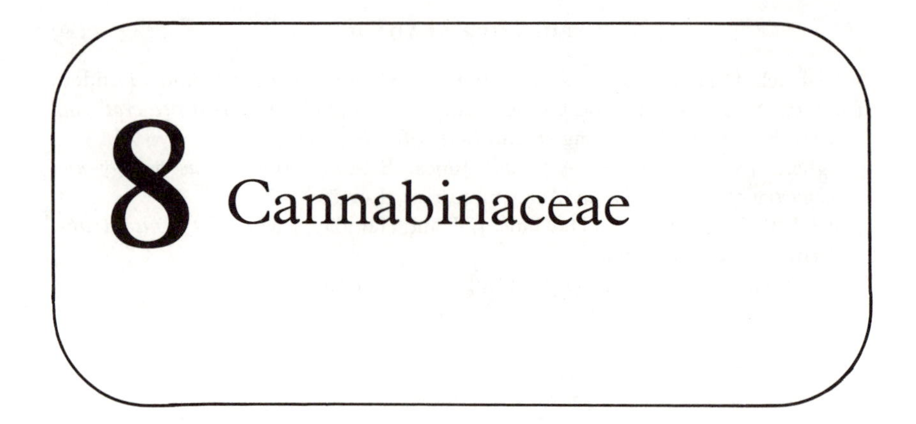

8 Cannabinaceae

This is a very small family comprising only two genera and three species. Of these, two are crop plants of economic importance, hop and hemp, while the third, the Japanese hop, is occasionally grown in gardens as a climbing ornamental. All have unisexual flowers and are dioecious.

8.1 HOP (*HUMULUS LUPULUS*)

The hop (*Humulus lupulus*) is a perennial plant with tall, climbing stems which need to be supported. The plant is grown for its female inflorescences which are used for the brewing of beer.

Origin and history of hop

There is some uncertainty concerning the origin of the hop but, since the Japanese hop and the related hemp both come from central Asia, it seems that *Humulus* is of Asian origin and that it spread to Europe and America from there. It appears that hops have been used in beer brewing for a very long time, probably for some 2000 years at least, according to an old Finnish saga, Kalevala. The crop was certainly well established in central Europe by the Middle Ages, especially what is now Czechoslovakia and Bavaria. Flemish planters in the sixteenth century brought it to England where it found ready acceptance in Kent. It was taken to North America by the early settlers and is now cultivated predominantly along the Pacific coast. Similarly, it was one of the crop plants transported to Australia and New Zealand when European settlement began there, and in fact was reported to grow in New South Wales as early as 1804. The main Japanese cultivar is derived from American or European material.

The hop plant

In the spring, annual stems arise from the perennial rootstock and climb in a clockwise direction up supporting strings to a length of 6–7 m. The stems are angular, covered with reflexed hairs, and they bear many large, three to five palmately lobed leaves which are rather variable in shape. In the summer, the male plants produce inflorescences, panicles with small flowers about 0.5 cm in diameter. Each flower consists of five perianth members with five stamens opposite each segment. About one male is planted for some 80 female plants, although the introduction of seedless hops has made the planting of male plants unnecessary. The female inflorescence, appearing at the same time, is a smaller axillary panicle, which gives rise to the strobilus or hop cone (Fig. 8.1). Each of these cones has a short axis which bears numerous pairs of stipular bracts. In the axil of these bracts there is a very short axis bearing four female flowers, each subtended by a bracteole. The individual flower consists of a perianth shaped like a cup, a single ovary and two large stigmas, about 4 mm long. When exerted these stigmas give the strobilus a distinct whiskery appearance. Pollination is by wind, following which the stigmas are shed and the stipular bracts and bracteoles enlarge rapidly. The strobilus now begins to look like a small pine cone, about 4 cm long and hanging down. The lower portion of the bracteoles, especially the perianth, and to a lesser extent the lower part of the stipular bracts, are covered with a golden-yellow dust-like secretion produced by lupulin glands. These are cup-shaped structures, one cell thick and covered by a cuticle. From pollination onwards the cells of the cup start producing a yellow resinous secretion, lupulin, which gradually accumulates, swelling out the lupulin gland into a spherical structure. As the resin solidifies the scales of the hop cone give the appearance of having been dusted with a yellow powder. Lupulin, a complex mixture of soft resins, is used in brewing, probably in the first instance because the resins have a bacteriostatic effect on Gram positive organisms and thus help to preserve the beer. The α-acid fraction of the soft resins is converted into isohumolones during brewing, imparting the characteristic bitter taste to the beer. This is now the main use of hops, since modern methods have made their preservative function redundant. Plant breeders have made great strides over the years, improving the level of lupulin production. This has been achieved in the main by hybridisation with American wild hops, although subsequent selection was necessary to overcome an undesirable effect on aroma that was not acceptable to the industry. Another important development has been the advent of seedless hops, because the presence of the seed, an achene, may be responsible for unwanted flavours and aroma. Diploid plants with a high degree of sterility

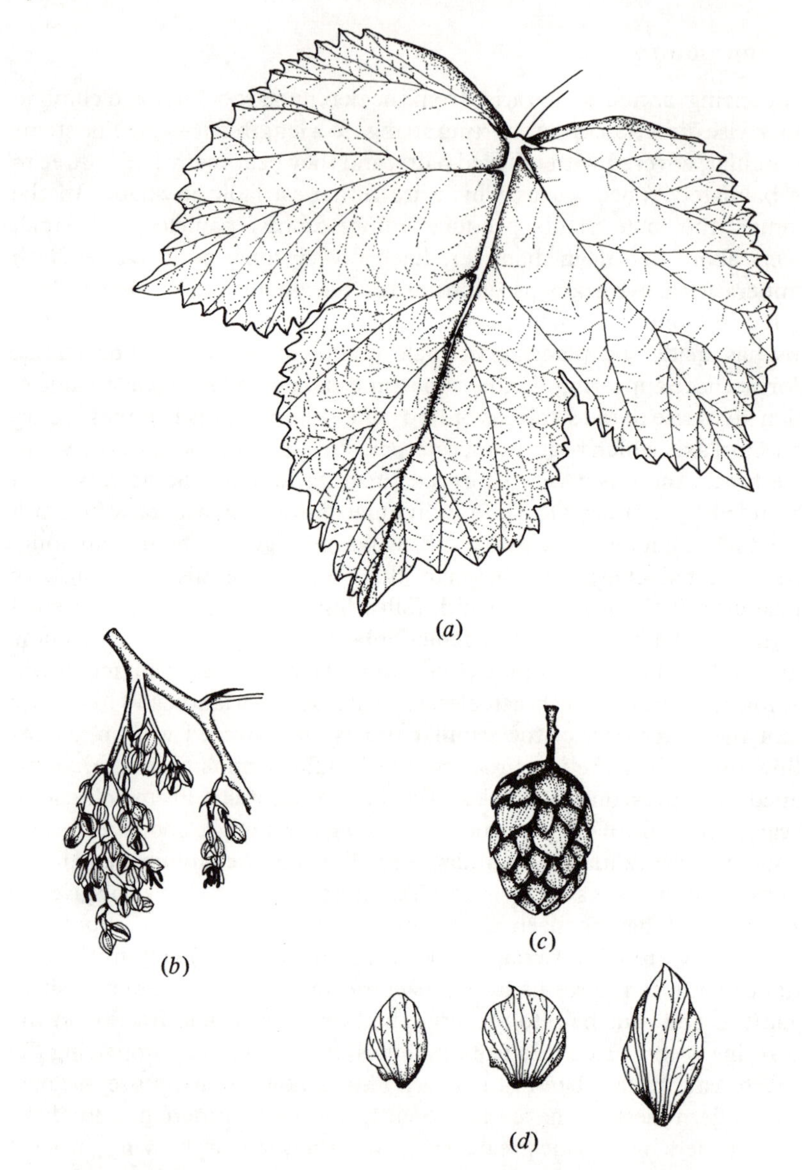

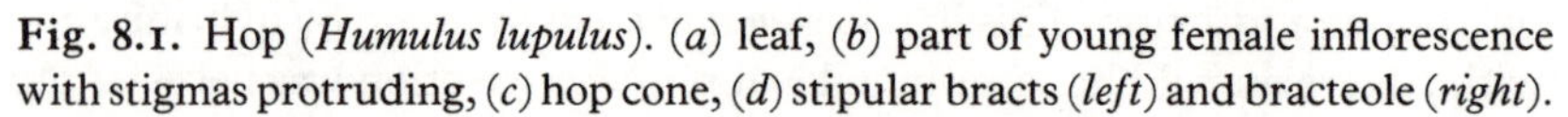

Fig. 8.1. Hop (*Humulus lupulus*). (*a*) leaf, (*b*) part of young female inflorescence with stigmas protruding, (*c*) hop cone, (*d*) stipular bracts (*left*) and bracteole (*right*).

are used, or alternatively tetraploidy is induced through colchicine treatment, and diploid male plants crossed with tetraploid females to produce triploids. These are grown successfully in some countries, although elsewhere they have turned out to be too late maturing and to produce more seeds than permissible in seedless hops.

Cultivation and use of hops

Hops are propagated vegetatively through the use of 10-cm long stem cuttings taken from the base of the plant. These are normally grown in nursery beds for a year before being planted in the permanent positions of the hop garden. Hops require a deep, well-drained, fertile soil and adequate shelter. Plants are usually spaced every 2–2.5 m in rows about the same distance apart. Four to five metres above, supported by poles, is a network of wires from which strings are suspended to each plant. About four healthy, vigorous stems or bines are selected in the spring and usually two trained clockwise around each string. Soon after training is completed, the plants receive a heavy dressing of NPK fertilizer, applied along the rows, and additional nitrogen may be given when the bines reach the top of the strings. A few weeks before harvest, it is customary in some areas to strip off the old leaves and laterals up to about one metre or so above soil level, partly to improve access but also to check the upward spread of red spiders. The crop is ready for harvesting in the autumn, when the hop cones are bright yellow-green and sticky, with the bracts and bracteoles still firmly attached. Until the middle of this century harvesting was done entirely by hand. The bines were cut down and the hop cones picked into bins. Labour for the harvest was recruited on a seasonal basis, and the annual exodus of whole families from the east end of London to the hop gardens in Kent used to be a well-known sociological phenomenon. However, increasingly the harvest has become mechanised, with machines stripping the hop cones off the bines. Freshly picked hops contain 65–85% moisture and must be dried to about 8–12% to allow them to be baled and stored. Drying is done in kilns, or oast houses as they are known in England, usually at fairly low temperatures to preserve the resins and aromatic constituents. After being cooled, the hops are allowed to cure slowly over the next 6–12 days or so during which the moisture content and biochemical constituents undergo slight change. Finally, the hops are compressed into bales and dispatched to the brewery.

8.2 HEMP (*CANNABIS SATIVA*)

Although this plant has acquired increasing notoriety as an illicit source of a drug, its economic importance rests on three products which may be obtained from it. The stem yields a fibre, the seed an oil, and the leaves and flowers produce resins that are used as narcotics. Different cultivars and methods of cultivation are used for each of these purposes.

Origin and history

Hemp is a temperate plant, most likely of Asian origin, which has been known by man since the dawn of civilisation. It is one of the oldest fibre crops in cultivation, and was grown by the inhabitants of northern China as long as some 4500 years ago. From there it appears to have spread to western Asia and into Egypt. Systematic cultivation in the Mediterranean area began at a very early stage, shown for example by a record of textiles made from hemp in Turkey around 700 or 800 BC. In Europe the crop became established in the first few centuries AD, especially in the areas where it is still widely grown as a fibre crop, in western USSR, Poland, and in Balkan and some Mediterranean countries. Hemp found its way to Chile in the middle of the sixteenth century and to New England and Virginia about 100 years later. Meanwhile it became naturalised in India as a drug plant.

The hemp plant

Hemp is an annual, herbaceous plant with a slender stem varying in height from about 1–4 m. The angular stems are slender. The degree of branching depends on the density of the crop, and in stands destined for fibre production is actively discouraged by a heavy seed rate. The leaves are palmately lobed, with seven to eleven pointed and serrated leaflets. As with other members of this family, hemp is naturally dioecious. The male inflorescence is an axillary or terminal panicle bearing flowers with five perianth members and five stamens. Female flowers occur in pairs, each of them surrounded by a bracteole covered with thick glandular hairs. There are five perianth members and two conspicuous but slender styles. The fruit is an achene. Hemp fibre, obtained from the stem of the plant, occurs in a ring of fibre bundles in the phloem parenchyma, in a similar position to those of linen flax (see section 13.1). The stem contains between 15 and 35 such bundles, each composed of ten to 40 individual fibre cells. Hemp consists of nearly 70% cellulose, about 16% hemicellulose and some lignin, making it a coarser fibre than linen flax. One of the problems of this crop grown for fibre is that from seed it produces a population of roughly one half male and female plants. Male plants flower and mature up to a month earlier than the females, and if left for harvesting until both are ready, they produce a harsh and brittle fibre. Although on a small scale and where the objective is seed production it is possible to pull male plants by hand, the long-term answer is to breed monoecious plants with both types of flowers developing at the same time. Some progress in this direction has been made.

Cultivation and uses of hemp

Hemp requires a deep, well-drained and fertile soil, an annual rainfall of at least 700 mm, a cool, temperate climate though an absence of frosts during early stages of growth. It is sown in spring, with seed rates varying from 60 to 100 k ha^{-1}. Given favourable conditions, it grows vigorously enough to suppress weeds. For fibre production the crop is ready for harvesting between flowering and seed ripening in female plants, some four to five months after sowing. After cutting, hemp is usually left to dry for some days, deseeded, and the straw retted by micro-organisms under the influence of dew and rain during the next few weeks. For this the straw is spread thinly and evenly, and it may have to be turned during the process. Alternatively, retting may take place by immersion in water in ponds or preferably temperature-controlled tanks. After drying the retted straw is subjected to a series of mechanical processes, the first being the breaking or crushing of the woody part of the stem, followed by scutching, a shaking, brushing and combing action. Hemp fibre is strong and durable, fairly coarse and not easily bleached as would be necessary for domestic use. The main outlet is for the production of rope, string, sacking, nets, and webbing as well as a raw material for the paper industry. Traditionally hemp was also used for the manufacture of sail cloth, and it was a consignment of *serge de Nimes* which was adapted by Levi Strauss to make the original denim jeans, although subsequently cotton was preferred. The USSR is the main producer of hemp fibre, but world production has generally declined through competition with plastics and other similar fibres such as sisal and Manila hemp.

Hemp seed contains 30 to 35% oil which, because of its drying properties, is suitable for the manufacture of paints and varnishes, and also for the production of soap. None of these end products of hemp are of more than minor importance on a world scale, nor is the use of the seed for caged birds and poultry. Sociologically far more important, if not notorious, are the narcotic properties of hemp which depend on the presence of resins produced by the glandular hairs of upper leaves and bracts of inflorescences. Resin production is most active in hot, dry climates, and in India branched cultivars of the crop are grown under licence as a source of ganja, a drug obtained from the female inflorescence and used medicinally as a sedative and hypnotic. In general, there appear to be wide differences both in amount and chemical composition of the drug produced by different types of hemp, and the names by which it is known are equally varied ranging from bhang, hashish, charas and kif to marijuana. It is commonly consumed by smoking, eating or by drinking as a tea. The primary effect is on the central nervous system producing dreams and hallucinations. Excessive heavy use is definitely harmful and may cause permanent damage and addiction to hard drugs. Cultivation, possession and use are strictly forbidden in most countries, although the debate continues whether modern society could not return to a greater degree of tolerance towards a substance which is probably less harmful than alcohol and tobacco.

FURTHER READING

Atal, C.K. and Kapur, B.M. (eds) (1982). *Cultivation and utilization of aromatic plants*. Regional Research Laboratory, Council of Scientific and Industrial Research, Jammu-Tawri.

Burgess, A.H. (1964). *Hops, botany, cultivation, utilization*. Leonard Hill, London.

Simpson, B.B. and Ogorzaly, M.C. (1986). *Economic botany. Plants in our world*. McGraw-Hill, New York.

Sundararaj, D.D. and Thulasidas, G. (1976). *Botany of field crops*. Macmillan, Delhi.

9 Chenopodiaceae

The Chenopodiaceae, not a large family, with only about 75 genera and 500 to 600 species, contains just one species of agricultural significance, the beets and associated plants (*Beta vulgaris*). Members of the family are distinctive for one or two reasons. Morphologically they differ from many other dicotyledonous families by having inconspicuous flowers without sepals and petals but with five perianth members. Physiologically the main point of interest is that many Chenopodiaceae are halophytes, plants that are tolerant of high salt concentrations in the soil and that in agricultural practice respond to applications of common salt. This ecological adaptation is usually associated with succulence in these plants and a fine, mealy covering of short, swollen hairs on the leaves. Also characteristic of these plants is their ability to accumulate considerable concentrations of nitrate and nitrite ions which can cause digestive disturbances in animals being fed on them.

9.1 BEET (*BETA VULGARIS*)

This is a variable species, sometimes divided into three subspecies, consisting of mainly biennial plants. They range from those with fleshy leaves with succulent midribs grown as leafy vegetables, such as the garden or silver beet, to plants in which the hypocotyl is swollen like the beetroot and those in which both hypocotyl and taproot are enlarged as in the fodder and sugar beet. Despite this wide variation, all these plants belong to the same species with a diploid chromosome number of 18.

Origin and early history

It is thought that most, if not all cultivated beets are derived from the subspecies *maritima* which occurs naturally near the sea in cool, temperate regions of Europe and parts of Asia, including the eastern Mediterranean where early domestication probably took place. Leaf beets, red and green chard, were known to the ancient Greeks, as shown in the works of such writers as Aristotle and Theophrastus. The Romans also used the plant for both human and livestock consumption, and from Italy the plant found its way to northern Europe. Beetroot in particular soon became well known, and some early Roman and later English and German recipes are on record. The sugar content of the beet first attracted attention in the middle of the eighteenth century in Prussia where the king granted a subsidy for experimental work including the development of processes for sugar extraction. Under a monopoly set up by royal decree, the first factory was established at Kunern in Silesia in 1801. The Silesian beet with a sugar content of about 6% was introduced into France and became the subject of further governmental protection when in 1811 Napoleon ordered some 28 000 ha to be planted and schools for the study of the plant to be established. Urgent action was needed to overcome the effects of the British naval blockade which prevented cane sugar from the West Indies being sent to France. A contemporary cartoon shows Napoleon squeezing a few drops of sugar beet juice into a cup of coffee. By 1840 some 5% of total world sugar production was derived from sugar beet, but by 1880 this had risen to over 50%, so determined were some countries in the temperate zone to become independent of cane sugar importation. However, the growth of the industry and the choice of raw material for sugar production have always been closely supervised and controlled by successive governments.

The beet plant

Germination in the beet is epigeal, which means that the cotyledons are carried above soil level by the elongating hypocotyl. A rosette of glabrous, dark-green glossy leaves soon develops, each with a conspicuous midrib and petiole (Fig. 9.1). Since the beet is a biennial, apart from some types with a predominantly annual habit, leaf production occurs throughout the season. Meanwhile the root and, in many types, also the hypocotyl undergo marked development. The primary root is diarch, giving rise to two lines of laterals, and normal secondary thickening takes place. However, as growth proceeds, there arises in the pericycle a ring of further cambium that forms xylem on its inside and phloem on the outside. This process is repeated several times over, so that the mature root and hypocotyl if cut transversely show characteristic, often pigmented rings. In the wild beet and forms cultivated for their leaves the xylem is heavily lignified and the root measures only a few centimetres across. In types selected as 'root' crops,

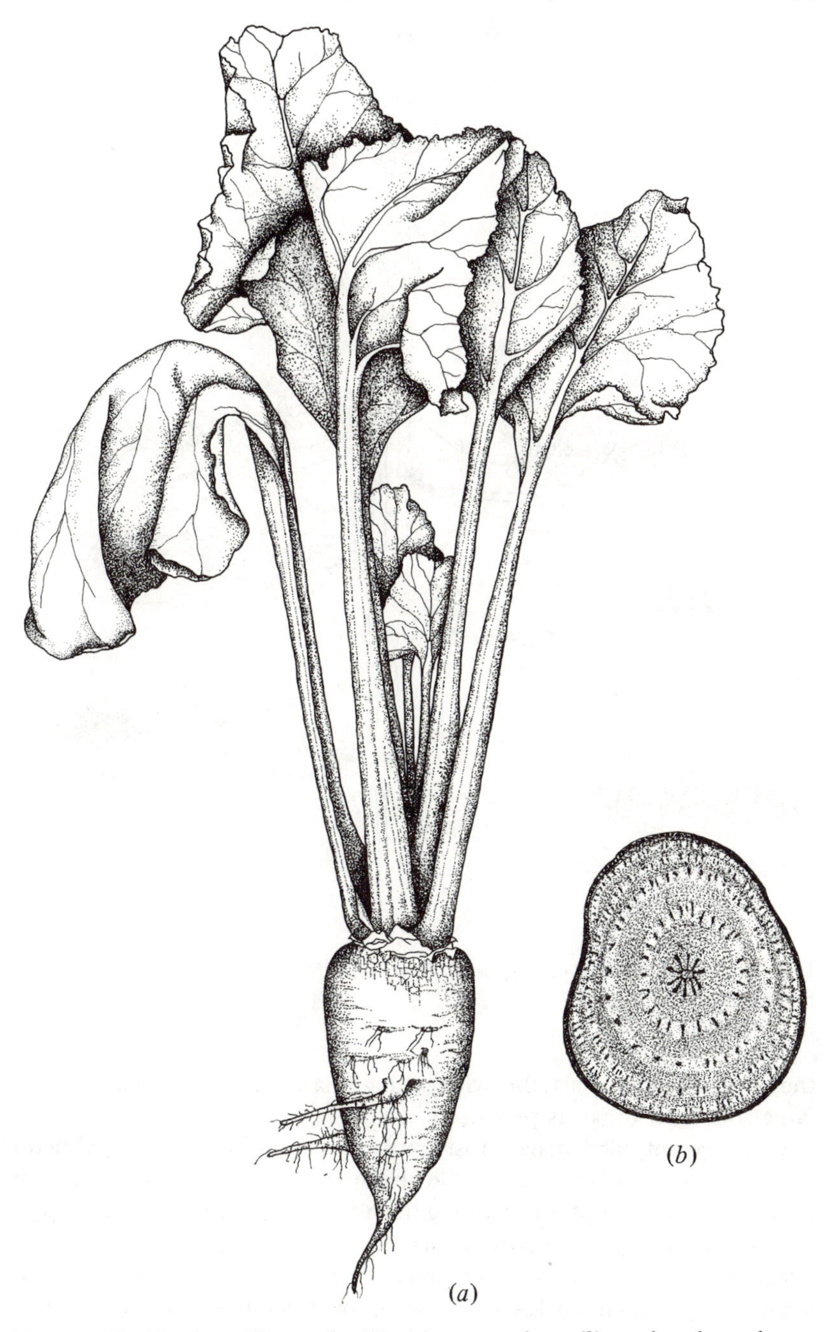

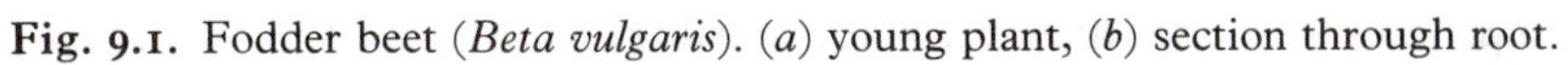
Fig. 9.1. Fodder beet (*Beta vulgaris*). (*a*) young plant, (*b*) section through root.

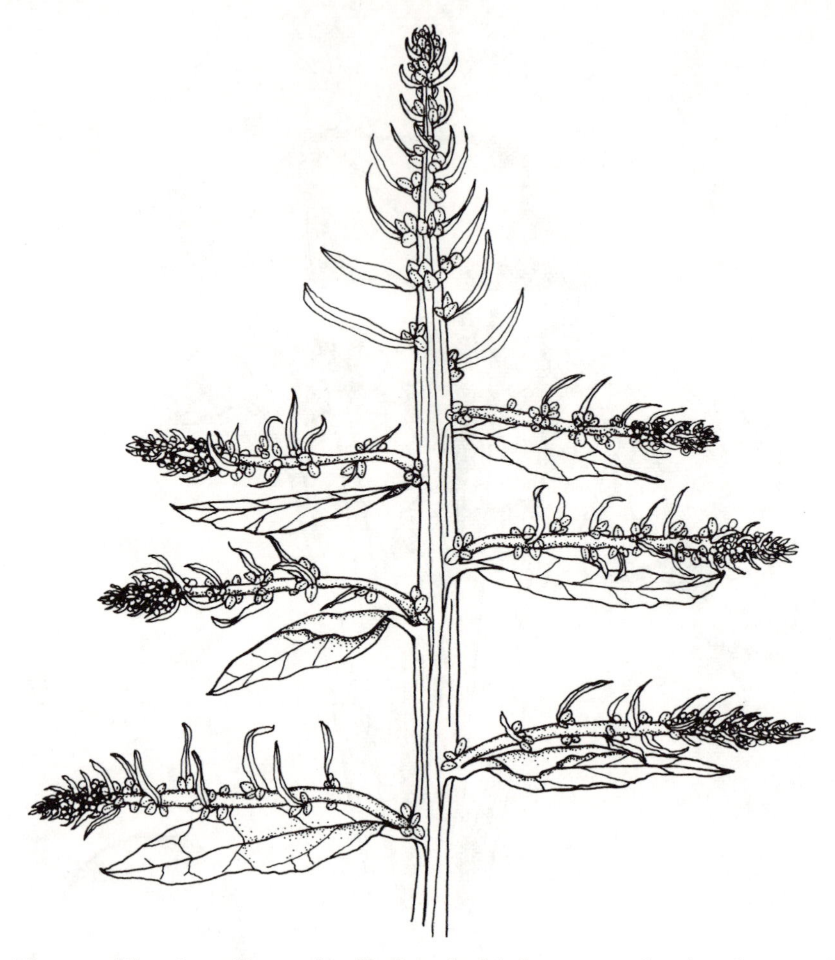

Fig. 9.2. Silver beet (*Beta vulgaris*). Branched inflorescence showing clusters of flowers in axil of bracts.

the rings are wider apart, the xylem remains largely parenchymatous, and a large succulent organ is produced.

Beets are not particularly frost-hardy and are usually harvested before the worst of the winter sets in. On the other hand, low temperatures are required for vernalisation, enabling the plant to respond to the lengthening days in the spring. Flowering is preceded by development of a tall stem which in the first year has remained short and insignificant. The tall, angled stem bears many leaves, the lower ones being large and attached by a petiole, the upper ones small and sessile (Fig. 9.2). Numerous branches

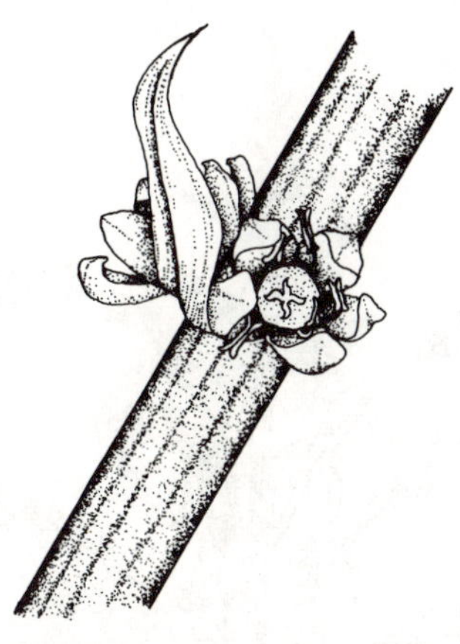

Fig. 9.3. Silver beet (*Beta vulgaris*). Close-up view of cluster of two flowers in axil of bract.

are produced and it is at the upper end of these and the main stem that inflorescences are formed. Each consists of a branching raceme on which, as on a spike, the flowers are sessile. Individual flowers are grouped in clusters of from two to seven in the axil of a small bract. Each flower has five green-yellow perianth members, five stamens opposite each segment and an inferior ovary usually with three styles (Fig. 9.3). Cross pollination by wind is normal. A single, black seed (about 2 mm in diameter) is formed but it becomes encased in the ovary as the perianth becomes extremely hard and woody. Not only that, but adjoining flowers in the same cluster stick together to form a hard, wrinkled particle, 3–5 mm in diameter, which serves as the agricultural seed. Several seedlings together may thus be produced by each 'multigerm' cluster, and to single them to one plant is an expensive and yet necessary operation. Basically two approaches have been adopted to overcome this problem: one mechanical, the other genetical. The first of these consists of breaking up the seed particles into fragments just large enough to contain one true seed. This can be achieved either by a rubbing action between a rubber belt and a rapidly turning emery wheel, which produces so-called rubbed seed, or by a chopping action which yields segmented seed. Neither method can necessarily be guaranteed to yield undamaged single true seeds, and germination percentage is usually depressed. A more reliable, though slower method is to breed 'monogerm' forms with only a single flower at each node of the inflorescence. Considerable success in this direction has been achieved, particularly in sugar beet. Since occurrence of a single flower is a characteristic of the inflorescence determined by the genetic make-up of the seed parent, it is

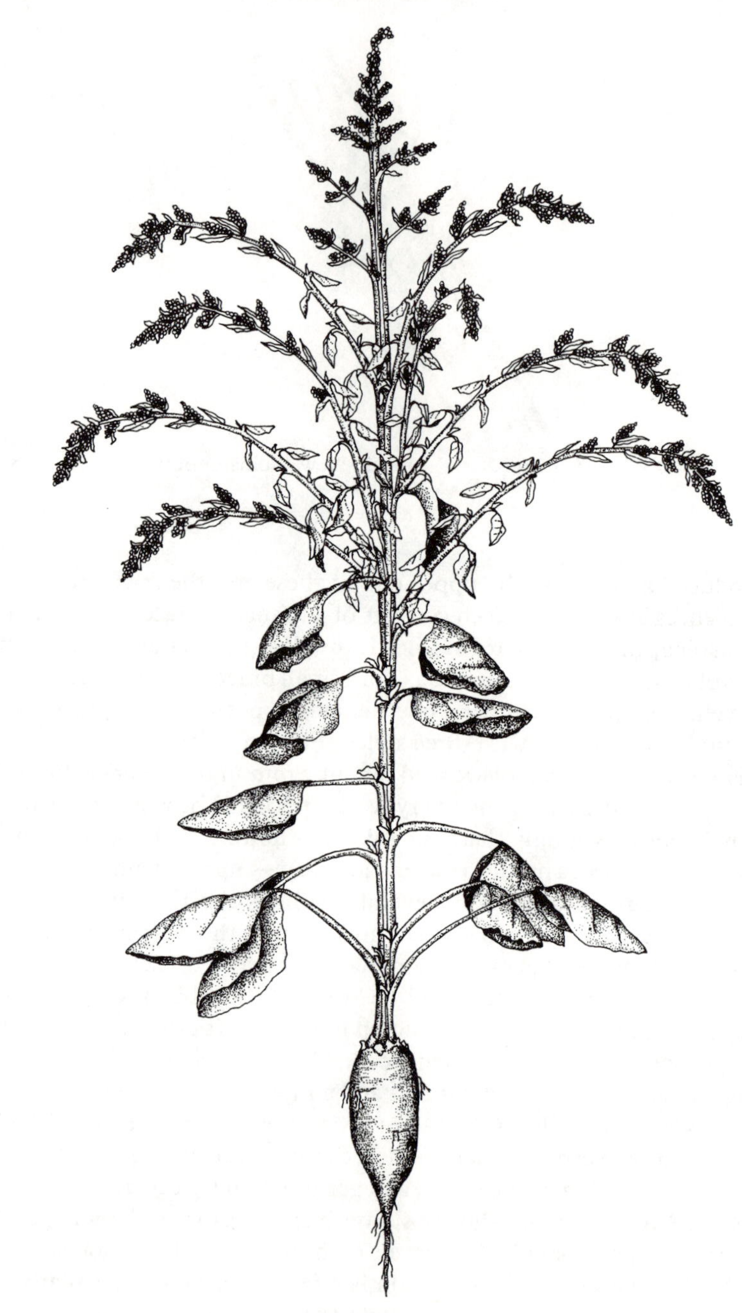

Fig. 9.4. Bolting fodder beet (*Beta vulgaris*).

possible in practice to cross a male-sterile monogerm with a multigerm plant as a source of pollen. Many monogerm cultivars are triploid. If this kind of seed is sown by a precision drill, there should be no need for any singling.

Another problem concerning flowering in the beets is the presence of so-called bolters, plants that produce an inflorescence and set seed in the year of sowing (Fig. 9.4). In leaf beets this stops further leaf production, and in root crops rapid lignification sets in to the detriment of quality. The plant breeder is doing his best to help to avoid this undesirable feature by selecting against the incidence of bolting, but there will always be some plants which require so little vernalisation as to be able to flower without passing through a winter. This problem has not been made any easier by the requirements of the agronomist who knows that high yields can usually be obtained only by sowing as early as possible in the spring. Since some genotypes respond to low temperature while still very small, it is quite possible that they will be stimulated to flower prematurely and bolt. This possibly should be borne in mind in cool districts. The positive lesson to be learnt from the reproductive physiology of the beet is that for seed production purposes sowing could be delayed until late summer, as long as the seedlings are old enough to be affected by the cold of the winter. Seedlings are either left to flower *in situ* or the plants, called stecklings, are transplanted in the spring.

Types of beet and their uses

Leaf beets

Beets that do not develop a swollen hypocotyl and are grown as leafy vegetables are thought to belong to the subspecies *cicla* of *Beta vulgaris*. Quite a range of types is in cultivation, ranging from spinach beet or perpetual spinach, grown for their laminae, to the silver or seakale beet or Swiss chard, plants with conspicuous petioles which are cooked and eaten with the true leaf. The petioles of the latter are usually white, although there are some selections in which they are red. By sowing in the spring it is possible to keep up a steady vegetable supply until flowering occurs after the following winter.

Beetroot

The usable part of the beetroot (Fig. 9.5(c)) is almost entirely hypocotyl, most of it above soil level, with the slender taproot penetrating more deeply. The shape is variable ranging from globe to long depending on

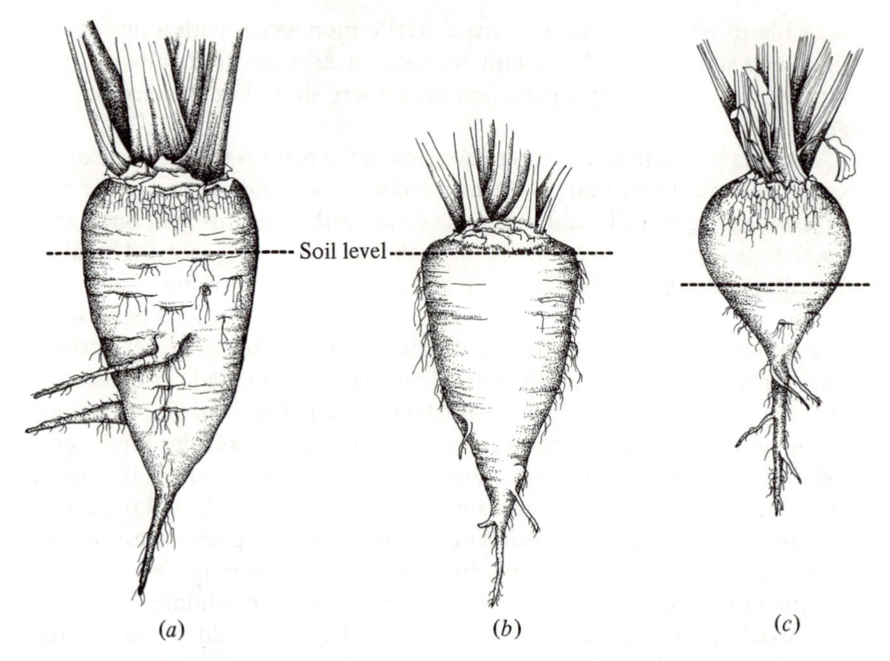

Fig. 9.5. Beets (*Beta vulgaris*). (*a*) fodder beet, (*b*) sugar beet, (*c*) beetroot.

cultivar. The most distinctive feature of the beetroot is its colour owing to the presence of the water-soluble red pigment betacyanin which differs from the more common anthocyanin by containing nitrogen in its molecule and by turning yellow when alkaline. Both pigments act as natural indicators and hence slices of beetroot in vinegar are a vivid, deep red showing up the typical rings of the species most clearly. The red colouration extends into the leaf veins, and in the laminae it mingles with chlorophyll to produce a greeny-bronze effect. The beetroot is a popular salad vegetable.

Sugar beet

Morphologically at the other end of the scale is the sugar beet (Fig. 9.5(b)) for here the taproot predominates and little more than the crown shows above soil level (Fig. 9.5). Years of plant breeding and efficient agriculture have resulted in high yields from cultivars rich in sugar. For example between 1920 and 1970 average root yield in the United Kingdom increased linearly to about 40 t ha^{-1}, and is capable of rising still further, while over the same period sugar yields rose from about 2.5 to 6.5 t ha^{-1}.

The sugar is deposited mainly in the phloem, and thus a large number of vascular rings, broad and rich in phloem, is taken as an indication of good quality. The highest sugar content is found near the widest part of the root decreasing above and below, and in fact the crown is removed at harvest when the beet are topped before being lifted. Although a sugar content of some 20% is possible, this figure is not consistently attained, except in certain seasons and when the accent is on sugar content rather than on yield. Three types of sugar beet are available varying in yield and sugar percentage. Cultivars with large 'roots' and relatively low sugar content tend to belong to the E (Ertrag) type, those with small roots rich in sugar to the Z (Zucker) type, with type N (Normal) being intermediate in both respects, although in practice closer to E than to Z.

Traditionally, plant breeders have concentrated on cultivars averaging 13–18% sucrose which were shown to produce the greatest total yield of sugar per unit area. More recently, however, the emphasis has shifted towards a higher sucrose content so as to reduce the number of beet requiring to be transported, handled and processed.

Sugar beet are sown as early in the spring as is possible without affecting the number of bolting plants unduly. Precision sowing is essential because the aim is to obtain a stand of about 80 000 plants ha^{-1} each weighing about 0.5–1.0 kg after topping. Control of plant numbers has been made easier through monogerm cultivars. Meticulous attention to weed control of the seedbed is highly important, because good establishment depends on freedom from competing weeds. Sugar beet responds readily to fertilizers, especially nitrogen and potassium (or sodium, reflecting its maritime origin), and an adequate level of soil pH is essential. The crop does not tolerate drought. Harvesting occurs in the autumn with the aid of special machines which first top the plants and then lift the roots and load them into a truck. The tops are valuable as feed to livestock either fresh or after conservation as silage. The roots are delivered to the factory where the beet are washed, sliced, and the sugar extracted by diffusion in a battery of large tanks filled with hot water. The raw juice is purified in a number of stages, decolourised and finally crystallised to produce sugar. The residual beet tissue, referred to as sugar beet pulp, has considerable feeding value for ruminants. It may be mixed with molasses, another by-product, and is stored in silos or dried before being fed to animals. Harvesting takes place over a period of a few weeks, but since factory production is of necessity spread over many months, it is necessary to store a high proportion of the crop. Apart from the expense involved, there are also risks of losses and chemical change. Fresh sugar beet juice consists almost entirely of sucrose, but with storage some of this is converted to monosaccharides which are not wanted for sugar production. However, if the raw juice is to be fermented to form ethanol, this problem is of no consequence. In fact, there is growing interest in sugar beet and fodder beet (see below) as possible sources of liquid fuel, perhaps initially by adding 10–15% ethanol to petrol without having to convert car engines. Rapid developments may be expected. For the present, sugar beet is grown extensively for sugar extraction, the main producers being the USSR, United States, Germany, Poland, France, Italy and other countries spreading from Europe through Iran to China and Japan.

Mangel

There is some doubt how exactly this crop obtained its name. The German word for leaf beets is, strictly speaking, Mangold, but it appears to have been used more loosely to describe other related plants as well. However, because of the value of the crop in times of need another similar word, Mangel (scarcity), came into use, or to give it its full name, Mangelwurzel. The mangel is a fodder crop, composed largely of hypocotyl, with only about one third of its length growing below soil level. Compared with the sugar beet it is considerably larger, probably weighing two to three times as much but also having only about half the dry matter content. The shape of the 'root' is variously described as globe, long or tankard, depending on cultivar, and the colour also varies accordingly from red to orange or yellow. The English-type mangels are generally high yielding but low in dry matter (8–13%) and easily lifted. Danish-type mangels have slightly more dry matter (13–15%) but are growing more deeply in the soil. Neither type is frost-hardy, and hence the 'roots' are pulled in the autumn and stored for feeding in winter and early spring. In any case, the high nitrate and nitrite content of the tissue precludes unlimited utilisation of the fresh material, because nitrites in particular cause animals to scour heavily with consequent loss of condition. During storage the levels of these compounds gradually decline.

Fodder beet

The relatively high water content and bulkiness of the mangel are disadvantages which can easily be overcome by growing fodder beet, hybrids between the sugar beet and the mangel. A whole range of cultivars has been produced varying in characteristics depending on the closeness of the cross to either parent. The objective is to maximise dry matter production per unit area without incurring any harvesting problems. Dry matter content increases and so does the depth of the 'root' in the soil (Fig.

9.5(a)), the more closely the fodder beet resembles the sugar beet, and in general 16–18% dry matter is considered optimal. Several excellent cultivars of this type have been bred in Scandinavia. Although, as the name implies, fodder beets have been bred for animal consumption, there is considerable interest in growing them as raw material for ethanol production. The sugar content of fodder beet, although inferior to that of sugar beet, is in the vicinity of 15–17%, high enough to warrant serious consideration in view of the superior yields of dry matter. This development would offset its declining use as a fodder crop, because cereals or grain legumes provide cheaper and more easily stored alternatives as animal feeds.

9.2 OTHER CHENOPODIACEAE

Chenopodium. Only three species in this genus are known to be cultivated, all of them only locally, of which *Chenopodium quinoa*, quinoa, is probably the best known. This crop is grown in the Andes for its seed which is of good nutritional quality, roughly on a par with wheat. Although very little breeding has been done, there has been selection over the years for increased seed size, reduced shattering and low dormancy. Quinoa flour is made into bread, biscuits or porridge, or it may be fermented into chicha. In Europe and wherever European agriculture has spread *Chenopodium album*, fathen or goosefoot, is a very common and widely distributed weed of arable land. This is an annual, bearing grey-green leaves with a mealy white covering, flowering profusely and producing an abundance of seeds (over 45 000 on large plants). A small proportion of the seeds are brown and are capable of germinating without delay, while the majority are black and remain dormant for short periods. They persist buried in the soil for several years, thus providing a continuous reservoir for new infestation. The leaves of this plant were at one time boiled and eaten like spinach, and in parts of Asia it is still used for this purpose. A related genus, *Atriplex*, is also a weed of arable and waste land, although occasionally grown as a leaf vegetable. In dry inland areas of Australia, the salt bush, represented by some 40 different species of this genus, provides valuable feed for sheep and cattle.

Spinacia. Spinach (*Spinacia oleracea*) is a leafy vegetable with dark-green, roughly triangular leaves. Two types are known, summer spinach with round and smooth 'seeds' and winter spinach which has rough or prickly 'seeds' owing to the persistence of projecting calyx teeth. The plant is smaller and lower growing than garden or silver beet but serves much the same purpose. It also differs botanically by having clusters of unisexual flowers, usually carried on separate but occasionally on the same plant. Although valuable in the garden, spinach suffers from the disadvantage of bolting and going to seed readily in warm summer weather.

FURTHER READING

Draycott, A.P. (1972). *Sugar-beet nutrition.* Applied Science Publishers, London.

Frick, G.W., Loomis, R.S. and Williams, W.A. (1975). Sugar beet. In *Crop Physiology, some case histories,* ed. L.T. Evans. Cambridge University Press.

Fisons (1972). *Sugar beet, a manual for farmers and advisors.* Fisons Agrochemical Division, Harston, Cambridge.

National Research Council (1989). *Lost crops of the Incas: little known plants of the Andes with promise for worldwide cultivation.* National Academy Press, Washington.

10 Cucurbitaceae

This family which comprises some 90 genera and 750 species is of wide geographic origin. Cultivated cucurbits are thought to have arisen in the Western Hemisphere, Africa, and Asia. Cucurbits are usually grown for their fruits which are large and are eaten fresh in salads or as a dessert. They are pickled and cooked and served as vegetables. Certain species commonly called pumpkins develop a very hard exocarp which allows them to be stored for considerable periods of time, thus providing an important source of winter vegetables in some countries. The hard exocarp can also be used in the production of gourds which are ornamented and used for storage and for carriage of liquids in many primitive cultures. One species, the loofah, is eaten in Asia but in Europe the fibrous vascular system of the fruit is retained and is put to a variety of uses ranging from a substitute for bath sponges to filters in industrial engines. In other cultures the seeds of cucurbits are dried and salted and are eaten in much the same way as peanuts or potato crisps. In the tropics the spreading tangled vines of *Momordica charantica*, besides providing food, are utilised as a cover crop in plantations.

General morphology

All cucurbits tend to be frost-tender although, as a result of the work of plant breeders, most cultivated species have now spread well into temperate environments in the summer time. Cultivated cucurbits are usually annuals which rapidly produce a dense vegetative cover. Because the plants have tendrils they are capable of climbing, but in general, except for glasshouse cucumbers, they are not provided with support and are allowed to grow over the surface of the soil. Stems are large, thick and rough and bear large simple leaves which may be highly dissected.

Tendrils arise in leaf axils as do further vegetative branches and flowers which are often solitary. In many cucurbits the flowers are unisexual, and the plants may be either monoecious or dioecious. The calyx is attached to the ovary and has five lobes. The flowers are actinomorphic, and there are five petals which may be fused or free. In male flowers there are three stamens, one of which is unilocular while the other two are bilocular. The ovary is inferior and contains numerous ovules. Inspection of the base of the flower allows female flowers to be distinguished from male flowers even before fertilization. Following fertilization the fruit which is formed is characteristic of the family in that it develops into a form of berry, a pepo, which has a hard exocarp or rind and a fleshy mesocarp which is the edible portion of the fruit. Cucurbit seeds are usually flattened and contain no endosperm.

10.1 *CITRULLUS*

The genus *Citrullus* differs from the other genera in this family in that its plants have deeply indented, pinnatifid leaves, in distinction to the entire or lobed leaves of the other genera.

Water melon (*Citrullus lanatus*)

In many parts of the world the consumption of water melon with its bright pink to red flesh, black seeds and variegated green and silver rind is associated with the hot days of late summer. Although modern bred water melons produce very large fruits up to 10 kg in weight, there are a number of hard-fleshed smaller and more primitive genotypes that are cultivated for their fruit which is used in the production of jam and pickles.

Water melons are best suited to hot dry climates where irrigation is available and they prefer sandy soils. Following germination which is epigeal the plant quickly produces an extensive shallow root system. Stems also elongate rapidly and are in effect runners although tendrils are produced in leaf axils. Stems are thin and hairy and can extend out to 5 m. Leaves are large (up to 12 × 20 cm) and, although simple, have from three to four pairs of pinnate lobes which are themselves further divided. The flowers are unisexual, and usually more male than female flowers are produced. They are solitary, borne in the leaf axils, yellow in colour and about 3 cm in diameter. Male flowers have from three to five stamens while female flowers have three staminodes. Following fertilization which is usually by insects the fruit grows rapidly. Although the fruit is usually pink or red it may also be green, white or yellow. As the name suggests, the fruit is mainly composed of water (94%) while the majority of the dry matter in water melon is carbohydrates, mainly reducing sugars and sucrose.

The melons are usually sown in groups, several seeds to a hill, and are thinned to one or two per mound, the mounds being about 3 m apart. Similarly, after the fruits form, they are also thinned to leave one or two on each mound, so as to give a more uniform fruit size. The melons are ready to harvest four to five months after sowing. Selection of ripe melons presents a problem. Among suggested methods are tapping the melon – a dull thud indicates maturity, picking when the tendrils begin to wither, and the appearance of the fruit such as its size and its external surface. A sure method which is not recommended because of the rapid spoilage that follows is to cut a small hole in the flesh. It has been suggested that the best method is to pick when the skin of the melon in contact with the soil changes colour from green to pale yellow, following the formation of the 'ground spot'. It is now possible to produce seedless water melons. This is achieved by crossing tetraploid female melons with normal diploid males. The resulting plants are triploids and form no viable seed. A good crop should produce about 1200 melons ha^{-1} of about 10 kg average weight. To maximise storage life the peduncle should be cut as far from the melon as possible when harvesting. To prevent bruising, harvested melons must be handled carefully. They cannot be stored for more than two to three weeks and may acquire taints after as little as a week in storage.

10.2 *CUCUMIS*

Although there are over 40 species in the genus, only two are of commercial significance. *Cucumis melo* from which a variety of different melons are obtained, and *C. sativus* the cucumber which is used as a salad vegetable and in pickles. There is also limited cultivation of *C. anguria*, the West Indian gherkin, for use as a vegetable and in pickles.

Musk melon, cantaloupe, or rock melon (*Cucumis melo*)

This species produces large sweet fruits, up to 2 kg in weight, which are known by a variety of common names throughout the world. In America they are called musk melons, in Europe cantaloupes and in Australia they are known as rock melons. One cultivar, called Honey Dew, is virtually known universally by its varietal name. The species is thought to have originated in Africa and appears to have reached Europe at about the time of the decline of the Roman Empire. Like most cucurbits it is frost tender and it grows best under hot dry conditions.

Presumably as a result of selection by humans, members of the species are extremely variable in appearance. The plants are annual vines which are covered in hairs. They form an extensive root system but this does not seem to penetrate far into the soil. Stems are ridged and tendrils arise from leaf axils. The leaves are large, up to 15 cm in diameter, and are borne on a long petiole, some 10 cm long. They may be orbicular, ovate or kidney shaped and may also be lobed. Sexual expression is complex, and plants may bear both male and female flowers on the same plant or may have hermaphrodite and male flowers on the same plant. Male flowers are borne in clusters while female and hermaphrodite flowers are solitary. Flower colour is yellow. Following pollination which is usually by insects but can be done by hand, fruits are formed. Within the species based on selection there is considerable variation in size and shape of the fruit, texture and colour of the skin, and the colour of the flesh. The rind may be smooth or rough and netted, and can be green, yellow, pink or orange. The central cavity of the fruit is filled with numerous seeds which are flattened and usually white to buff in colour.

In large-scale production the crop is now usually sown direct into the field with 1.2–2 m between the rows. Each plant is allowed to bear about four melons, and a good crop of melons should yield about 9 t ha^{-1} three to four months after planting. As with water melons, there are problems in establishing the optimum time to pick by observation of the crop. Fruit is at its optimum when sugar content is at a maximum, and this can only be established by sampling the ripening crop. As maturity nears in some cultivars, an abscission crack develops at about the junction of the peduncle with the stem. When this crack has spread completely around the stem the melon is at peak sugar content and is said to be at 'full slip'. For shipping, melons must be picked slightly before this stage. The Honey Dew melon presents an additional harvesting problem, because by the time the abscission crack forms the melon is already over-ripe. Other methods of assessment have to be used, and fruits are often treated with ethylene to promote even ripening after premature harvest.

Because of the ease with which the fruit can be bruised, harvesting is usually done by hand, usually every two to three days or daily in hot weather. Because heat reduces the quality of the picked crop and melons are normally chilled for shipment, harvesting is best carried out in early morning. Although the fruits of *C. melo* may not be of particularly high nutritional quality, being mainly water and carbohydrate, they are now a popular delicacy and are transported considerable distances from growing areas to markets in the United States, Europe and Australia.

Cucumber (*Cucumis sativus*)

Although the cultivated cucumber is used mainly as a salad vegetable, immature fruits are processed to produce gherkins, dill pickles and pickled cucumbers. The crop is believed to have originated in India and the species

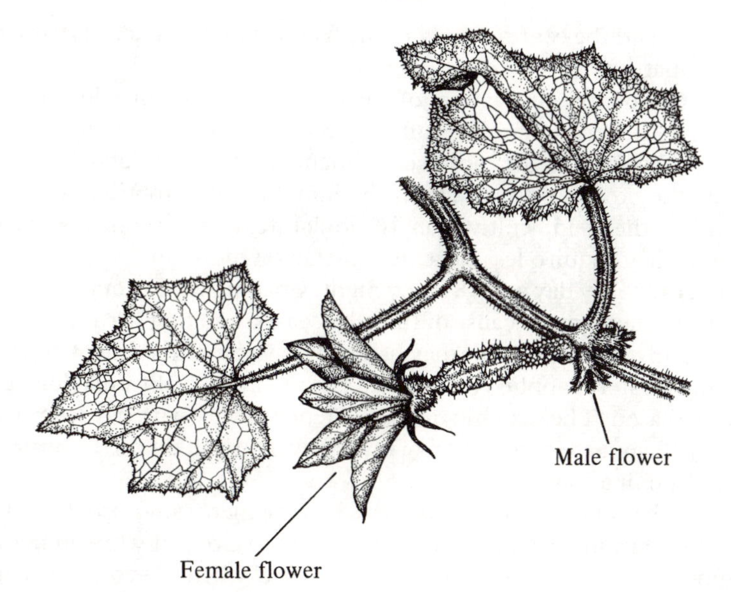

Fig. 10.1. Cucumber (*Cucumis sativus*).

is morphologically and on the basis of chromosome number ($n=7$) quite different from the other members of the same genus derived from Africa ($n=12$). It was known in ancient Egypt and is now cultivated in most parts of the world either as a summer annual crop or in glasshouses.

Cucumbers are trailing or climbing monoecious vines which are covered in stiff bristle-like hairs (Fig. 10.1). The root system, although extensive, tends to be shallow. Cucumber stems are four-angled and the tendrils which arise in the leaf axils are unbranched. The leaves are entire and nearly triangular in shape and are up to 20 cm across on petioles up to 15 cm long. Cucumber flowers are yellow, and more male than female flowers are produced. Male flowers are borne in axillary clusters while female flowers are solitary and borne on a short stout peduncle. Cucumber fruits vary considerably in size, shape and colour, although at maturity the flesh of the cucumber is always pale green to white. They may be almost spherical, as in apple cucumbers, or extremely long and cylindrical as in English glasshouse varieties. Skin colour at maturity is also variable and ranges from white through yellow to dark green. A feature of young cucumber fruits is the presence on the skins of warts and spines which tend to grow out as the crop matures. Some cultivars develop fruit parthenocarpically and contain no seed although most cucumbers are produced as a result of pollination and the central cavity of the fruit contains numerous flat white seeds at maturity.

The method of cultivation of cucumbers varies with location. In temperate countries they are often grown in heated glasshouses over supports. The cost of the capital equipment and the extra labour required to grow the crop is offset by the premium obtained from its sale out of season. In the field, cultivation is similar to that for other cucurbits although they require less heat than melons. Optimum temperature for growth is 30 °C by day and 18 °C by night. Cucumbers do not grow well in humid environments because of fungal diseases, and they are intolerant of waterlogging. In spite of this they have a high water requirement and up to 380 mm of water is applied to irrigated cucumbers in California during the growing season. The sex ratio of flowers tends to vary with day length, and long days favour the production of male flowers. Fruit set can be improved by hand-pollination.

Unlike the other cucurbits discussed so far there is no problem about picking of cucumber fruit as they are usually harvested while immature. Cucumbers for fresh consumption are picked when 13–20 cm in length although some glasshouse cultivars are up to 45 cm. For pickling, the fruit is picked when very young, and machines have been developed to harvest the crop, top quality fruits for gherkin production being 22×70 mm. Harvesting of cucumbers starts about two months after sowing and is continued every two to four days. In the United States yields of cucumbers are from 4.5 to 6.8 t ha^{-1}. Cucumber fruits when harvested show chilling injury if stored at temperatures below 10 °C and thus cannot be stored for long periods.

10.3 *CUCURBITA*

There are four cultivated species in the genus *Cucurbita* which are grown for their fruits. These fruits are usually eaten in human diets as a cooked vegetable, but in some societies leaves, flowers and seeds are also consumed. In three of the species there is considerable confusion in the use of common names and, because of their similar appearance and cultural requirements, the three will be considered together. The species are *Cucurbita maxima*, *C. moschata* and *C. mixta*, all three of American origin and sharing the common names of pumpkin and winter squash. *Cucurbita pepo* from which the vegetable marrow is derived also produces cultivars that are commonly known as summer squash, winter squash, pumpkins and ornamental gourds. Like the other three species, it is considered to be of American origin, possibly Mexico. It would appear that *C. pepo* is more tolerant of cool climates than the other members of the genus.

Pumpkin or winter squash (*Cucurbita maxima*, *C. moschata* **and** *C. mixta*)

All three species are annuals with long rambling vines which may root at the nodes. Leaves are large and usually entire, branched tendrils arise in the leaf axils, and the flowers are yellow in colour and unisexual. *Cucurbita moschata* and *C. mixta* which are very similar are tolerant of warm climates. At maturity the fruits of *C. maxima* may be either soft- or hard-shelled but the shell of the other two species is always soft. The distinction between squash and pumpkin appears to be one of fruit texture, in that fine-grained fruits with a mild flavour are described as squash, while coarse-grained more strongly flavoured fruits are called pumpkins. Fruit colour ranges from pale yellow to deep orange. Besides their use as a vegetable, the fruits are used as pulp in the production of jams and as a feed for livestock. Dried seeds are salted and sold as a confection in some countries.

Pumpkins are ready to harvest about four months after sowing and are usually harvested after all the foliage has died away. To prolong storage life they can be cured by heating at 27–30 °C for ten days at a relative humidity of 80%. Following this they should be stored at low humidity at 10–15 °C in single layers for maximum storage life. If carefully harvested and stored they can be kept for up to six months.

Marrow (*Cucurbita pepo*)

The species is extremely variable, both in the appearance of the whole plant which can vary from long scrambling vines to determinate bush types, and in the form of the fruit. Common names applied to members of the species include vegetable marrow, summer squash, winter squash, pumpkin and ornamental gourd. Like the three previous species it is of American origin and archaeological evidence dating it back to 7000 BC has been found in Mexico. The species was evidently grown widely in northern Mexico and the south-west of the United States prior to the arrival of Columbus.

The cultural requirements of the species are similar to those of other cucurbits in that they are intolerant of warm, humid conditions that favour growth of fungal pathogens and they require a well-drained fertile soil. The species is more cold tolerant than *C. moschata* and *C. mixta* and, although of tropical origin, it is now grown in the summer at considerable distances from the equator. Immature fruits of the marrow types are harvested and sold as courgettes or zucchini when about 10–30 cm in length and can be ready to harvest within eight weeks of sowing. Cultivars which are harvested at maturity take longer to grow, and harvesting is usually begun three to four months after sowing. As is suggested by the diversity of common names, the fruits, like the plants, are extremely variable in size, shape and colour, flavour and keeping quality.

FURTHER READING

Anon. (1980). *Cucumbers*. Grower, London.

Bulwalda, J.G. (1986). *Melons: physiology and culture*. MAF, Pukekohe.

Jones, C. (1981). *Growing tomatoes: with a note on cucumbers, peppers and aubergines*. Penguin, Harmondsworth.

National Research Council (1989). *Lost crops of the Incas: little known plants of the Andes with promise for worldwide cultivation*. National Academy Press, Washington.

Peirce, L.C. (1987). *Vegetables: characteristics, production and marketing*. John Wiley, New York.

Whitaker, T.W. and Davies, G.N. (1962). *Cucurbits*. Leonard Hill, London.

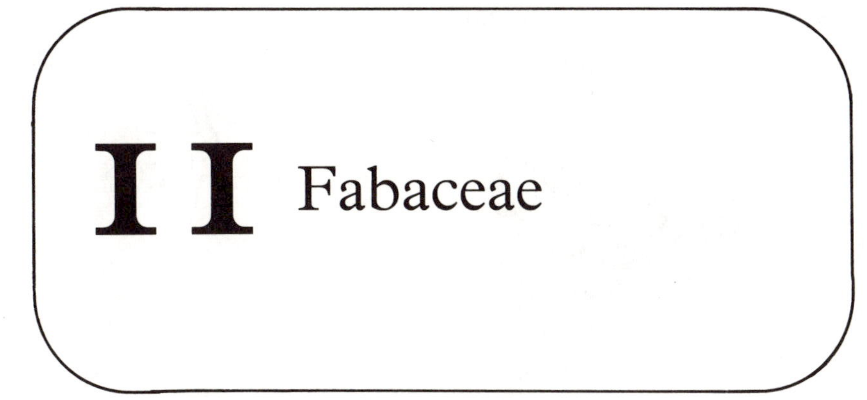

11 Fabaceae

After the Poaceae, the Fabaceae commonly known as legumes are the most important family of cultivated crop plants in both the tropical and temperate world. They also form a very large family comprising some 700 genera and 18 000 species, and in fact taxonomists have argued that, rather than being a single family, the legumes should be considered as an order divided into three families or that, at the very least, subfamilies should be recognised. Since we are not concerned here with taxonomic subtleties, we will for the sake of simplicity follow the second, and more universally accepted alternative. Of the three subfamilies in the Fabaceae two are of little interest in temperate regions and do not include plants of major economic significance. The first of these is the Mimosoideae, of which the wattles in Australia are a good example. In these plants the flowers are regular and tend to be massed together in large numbers to form conspicuous heads, with the stamens often forming the most visible feature. In the Caesalpinioideae the flowers are zygomorphic, that is their shape is irregular, and in particular the position of the uppermost petal differs from that in the other groups of legumes. This subfamily is also of tropical distribution and of little economic significance. The largest subfamily, and by far the most important in terms of agricultural production is the Faboideae (also called Papilionoideae or Lotoideae) which consists of nearly 500 genera containing 12 000 species. Their distribution is worldwide and the plants range from large trees to small annual herbs.

11.1 GENERAL CHARACTERISTICS

Flowers of the Faboideae are irregular or zygomorphic, with only one plane of symmetry. There are five joined sepals, but it is the five petals that

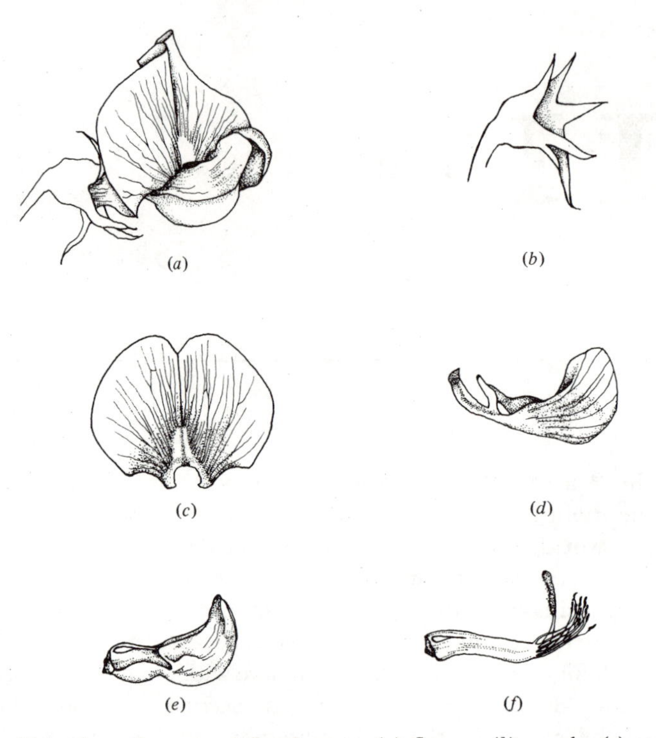

Fig. 11.1. Faboideae flower and floral parts. (*a*) flower, (*b*) sepals, (*c*) standard petal, (*d*) wing petals, (*e*) keel petals, (*f*) staminal column and stigma.

are the most distinctive features (Fig. 11.1). The large, posterior petal, called the standard, is very prominent. The two lateral petals, the wings, are broad above but tend to taper towards their base. Inside these are two small anterior petals, which are loosely joined together and because of their boat-shaped appearance are termed the keel. Within the two keel petals lie ten stamens which form a tube surrounding the ovary. Either all ten stamens are joined (monadelphous flower) or nine make up the tube and the tenth is free (diadelphous flower). A single carpel usually containing numerous ovules lies within the staminal tube, with the style emerging and curving upwards between the anthers. The structure of the flower together with the presence of nectaries are adaptations for pollination by insects, mainly bees, although in some species only bumble bees can reach the nectar at the base of the stamen tube. For pollination, the stigma and the anthers need to be exposed, and this occurs quite easily as the insect alights on the wing petals and at the same time depresses the keel. But in some plants, such as lucerne, the keel petals interlock tightly and, as they are depressed, often with some difficulty, the stamens and stigma curve upwards explosively. The anthers start shedding pollen before the flower is opened, but the stigmatic surface is not receptive until it has been ruptured. This occurs as it is released from the keel petals or, as the process is usually referred to, when the flower is tripped. In the vast majority of species this requires the action of an insect and thus cross-pollination is ensured, but there are some in which tripping occurs spontaneously with the result that self-pollination takes place. The fruit is a pod or a legume, containing a variable number of ovules. In some species at maturity, the pod opens instantly along the sutures of the carpel, ejecting the seeds some considerable distance from the parent plant. The leaves in this family are usually compound and petiolate, and the stipules range in size from being conspicuous to small hairs. The number of leaflets is variable, and the terminal leaflet is often modified to form a tendril. At the base of the petiole and leaflets there occurs a modified tissue called a pulvinus, which is very sensitive to changes in turgor. As a result of the reaction of these pulvini, both leaves and leaflets move, often on a diurnal basis as in the so-called sleep movements. Some species, like *Mimosa pudica* are famous for their sensitivity and quick response to touch which brings about changes in the pulvinus.

Nitrogen fixation

Irrespective of subfamily a feature commonly shared by nearly all legumes is the ability to fix atmospheric nitrogen in their roots in a symbiotic association with bacteria of the genus *Rhizobium*. The bacteria, which are either present in the soil or are applied by farmers to the seed at sowing, invade the roots of the plant via the root hairs. Once established within the plant there is a proliferation of both plant cells and bacteria to form organised structures attached to the roots, called nodules. Active nodules are usually red or pink in colour due to the presence of the chemical leghaemoglobin which has a similar chemical composition to the haemoglobin found in blood. It has been claimed that under favourable conditions the legume–rhizobium association can fix up to 600 kg ha^{-1} of nitrogen per year. This is equivalent to over a tonne of urea. A more normal figure would probably be from 150 to 250 kg ha^{-1} nitrogen fixed. In countries such as Australia and New Zealand little nitrogen fertilizer is used in either pasture systems or in cropping, as plants rely on nitrogen fixed by legumes for their growth.

To obtain good nitrogen fixation, however, a number of conditions must be fulfilled. Firstly the legume cannot fix nitrogen unless it is associated with the right rhizobial strain. Many legumes are not particular, or promiscuous, in their rhizobial requirements, other legume species are extremely strain-specific and will only fix nitrogen when inoculated with one selected strain of bacteria. Soil conditions can have a considerable effect on bacterial population and efficiency, some rhizobia cannot survive for long in soils in the absence of the host plant, while others can survive for many years. High soil acidity kills some rhizobia while others are destroyed by basic soils. In recent years there has been a considerable increase in knowledge of the legume – rhizobium symbiosis and the factors that control it. In many countries it is possible to purchase commercially prepared cultures of *Rhizobium* with which to treat seeds or to obtain seed which has already been treated and is ready for sowing.

Hard seeds

Another feature which commonly occurs in legumes irrespective of subfamily is hard seeds. Legume seeds are usually without endosperm, comprising a small embryo surrounded by two large cotyledons, the whole being enclosed in a thick testa. On the outside of the seed, the hilum is prominent, a scar denoting where the seed was attached, and close to it is a small pore, the micropyle. In many species the testa is impervious to water and, even if conditions are favourable to germination, the seed will not germinate. If a portion of the seed coat is removed the seed will germinate without any further treatment. Seed hardness is therefore a physical constraint on germination rather than physiological. For a long time it was thought that hard seeds were the result of a waxy layer on the surface of the seed, but more recently it has been claimed that they will germinate following damage to the strophiolar region of the seed. In nature hard seeds probably break down due to either the alternation of temperature in the soil or as a result of the action of weak soil acids. For legumes that grow in harsh environments, hard seeds provide insurance for the survival of the species because not all seeds germinate in one year. The utilisation of subterranean clover leys in Mediterranean environments depends upon the carry-over of seed in the soil from season to season (see p. 230). Similarly for weed species the presence of hard seeds adds to the chances of survival. For example, in some parts of the North Island of New Zealand it has been estimated that there are up to 80 000 seeds of gorse (*Ulex europaeus*) per square metre of soil surface. On the other hand in crop species hard seeds are a nuisance. To obtain good cover either very high sowing rates have to be used or the seeds must be scarified before they are sown. There are a number of methods of seed scarification which usually utilise heat, mechanical abrasion, or sulphuric acid to improve the germination of hard-seeded commercial legumes.

Because of their ability to fix atmospheric nitrogen through their association with rhizobia, legumes are of great value in agriculture, both in arable and grassland farming. Many species, particularly those belonging to the genus *Trifolium*, grow in association with grasses and, even though they rarely dominate, they form a most important constituent of a pasture. The herbage produced by these plants usually has a higher protein content and greater digestibility than grass alone. Some herbaceous species are equally valuable, even though like lucerne (*Medicago* spp.) they grow best by themselves, and there are also shrubs which provide useful feed for browsing animals. We shall describe several of these pasture or forage legumes. Also as a consequence of their nitrogen economy, the Fabaceae tend to have protein rather than starch as their main storage compound, and hence the seeds have considerable nutritional value. Legume seeds are frequently used as a protein source in the formulation of livestock rations, and particular attention is given to a correct balance of amino acids. In countries such as India legume seeds provide the majority of the protein that is required for human nutrition to supplement the carbohydrates of cereals. These seeds are often referred to as pulses. However, since some legumes like the soya bean (*Glycine max*) and the peanut (*Arachis hypogea*) are grown not so much for their protein as their oil, it is probably best to use the term grain legume to describe all members of the Fabaceae predominantly cultivated for their seed.

Pasture or Forage Legumes

11.2 *TRIFOLIUM*

The genus *Trifolium* derives its name from its characteristic compound leaf which is trifoliate. About ten species are of agricultural importance, although the total number probably exceeds 250. Clovers are annual or perennial in habit. Many of them have weak stems or are prostrate, thus making them highly persistent under grazing. Although trifoliate, the first seedling leaf is simple. The midrib of each leaflet terminates at the leaf margin and does not project as it does in the genus *Medicago* (see p. 235) with which some of the true clovers could be confused. At the base of the petiole may be found a pair of bracts, called stipules, which vary in size, shape and colour according to the species. These structures are useful in plant identification. Stipules in the genus *Trifolium* have entire margins, whereas in *Medicago* they are serrated. The flowers of clover are small and occur crowded together in racemes, the number varying from a few as in subterranean clover to many as in red or white clover. The straight pods contain one to eight seeds according to species. Clovers are commonly components of natural grasslands in the temperate zone or they are sown alone or in mixtures with other species for pasture of varying duration. The eastern Mediterranean area is the centre of diversity of the clovers, and there is evidence that they have been in cultivation in southern Europe for many centuries. Their fertility restoring role was recognised at a very early stage, long before anything was known of *Rhizobium* and nitrogen fixation from the atmosphere. Clovers are now an integral part of agriculture and an essential component of pastoral systems throughout the temperate world.

White clover (*Trifolium repens*)

Pride of place for its outstanding importance in grassland farming goes to white clover. This species is especially well adapted to grazing because of its prostrate growth habit and branching network of stolons. Persistence is also ensured through the ready formation of adventitious roots at the nodes of these creeping stems. The plant is entirely glabrous. The margin of the leaflets is distinctly serrated, and in most types the leaflet is further distinguished by a light-coloured, almost white V – or horseshoe-shaped leaf mark (Fig. 11.2). The stipules, on the other hand, are inconspicuous, pointed structures which become almost papery in appearance with age. The inflorescence is borne on a peduncle which arises in the axil of a leaf on the stolon, the first one appearing at the seventh to tenth node from the base. The number of flowers is variable but rarely exceeds 100, except in particularly well-grown specimens. The corolla is white, sometimes with a pink tinge, but after pollination it withers to a brown colour, and the flowers turn downwards. There are no leaves subtending the inflorescence. Pollination is carried out predominantly by bees, and white clover is a common source of honey. There are five to six seeds per pod, each seed being heart-shaped, bright golden-yellow with a touch of yellow-brown. As with most clover seed, the colour becomes progressively darker with age, a good indication of whether the sample is fresh. The weight of 1000 seeds is about 0.5 g.

There are three types of white clover which vary in size and growth habit. At the extreme end of perenniality and persistence is wild white clover, a prostrate, closely stoloniferous plant with an abundance of adventitious roots. Leaves, stolons and inflorescences are relatively small. Plants of this type are commonly found in permanent grasslands, and some like the Kentish wild white clover have been used by plant breeders in the production of cultivars suitable for close grazing. At the other end of the scale is the Ladino type of white clover, derived from the Mediterranean. Plants belonging to this group have a more upright growth habit and are larger in all respects. The system of stolons is less well developed and

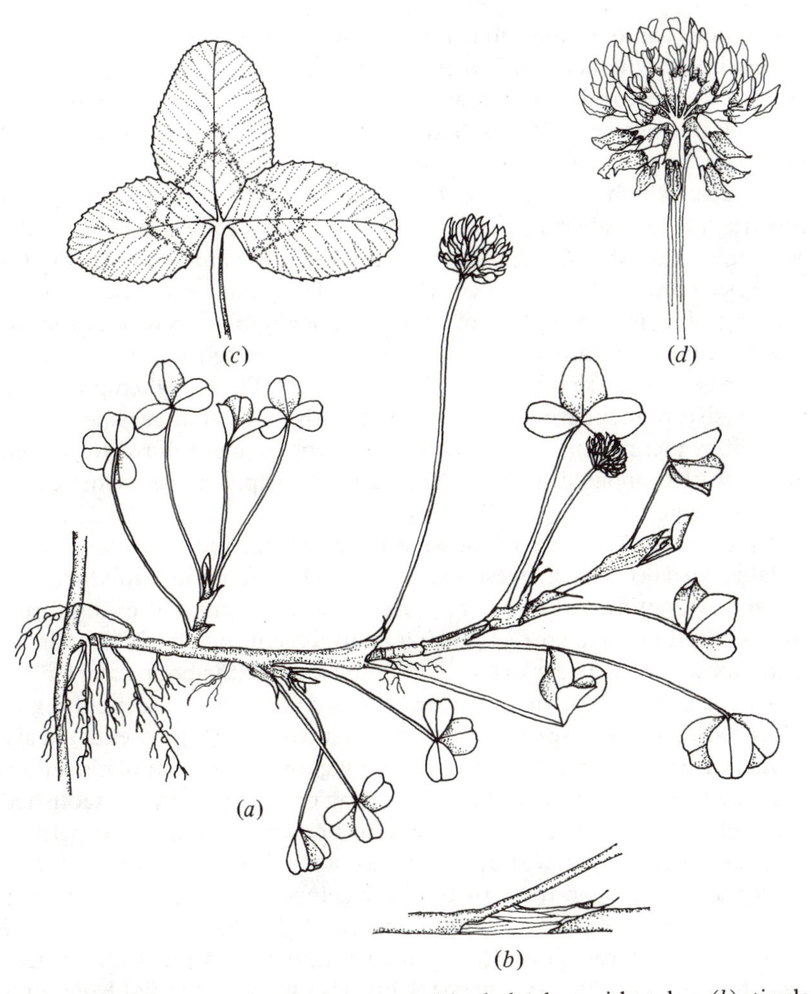

Fig. 11.2. White clover (*Trifolium repens*). (*a*) whole plant with stolon, (*b*) stipule, (*c*) trifoliate leaf, (*d*) inflorescence.

rooting at the nodes is less profuse. These plants are highly productive with ample moisture supply but they do not tolerate close and continuous grazing. This type of white clover has also featured widely in plant breeding, especially in the United States. The third group of white clover is intermediate in all respects. Flowering occurs earlier and productivity is greater than in wild white clover, while persistence is superior to the Ladino type. European and New Zealand cultivars intended for intensive grazing conditions tend to belong to this third group. In addition to these distinctions based on size and habit, white clovers also differ biochemically depending on whether or not they are cyanogenic. Many genotypes possess cyanogenic glucosides which are hydrolysed through the action of the enzyme linamarase to form hydrogen cyanide, the presence of both glucoside and enzyme depending on one dominant gene. In other genotypes, which are said to be acyanogenic, one or both of these constituents are absent. Selection for different gene frequencies has occurred under the influence of temperature and natural predators. For example, Ladino clovers are acyanogenic. It should be stressed that the amounts of hydrogen cyanide produced are negligible in relation to animal health but, on the other hand, they are large enough to allow different genotypes to be distinguished in the laboratory. The so-called picric acid test consists of incubating crushed leaves with strips of filter paper soaked in sodium picrate, in the presence of toluene or chloroform to prevent decay. With cyanogenic plants the colour of the paper changes from yellow to red-orange.

White clover is adapted to a wide range of climatic conditions and is suitable for all types of livestock. It is usually grown in a mixture with grasses. An optimum balance between grass and legume is ensured by a suitable seed rate and correct grazing management. A typical, very simple seed mixture would be 3 kg ha^{-1} of white clover with about 20 kg ha^{-1} of perennial ryegrass, sown in spring or early autumn but, depending on climate and the purpose of the pasture, other clovers and grasses could also be included. Early grass growth is more vigorous than that of clover and thus needs to be controlled through cutting or grazing. Once established, white clover grows very vigorously and becomes strongly competitive, especially in spring, early summer and autumn. However, peak productivity depends largely on the growth of the companion grass, because white clover suffers easily from competition for light. By the same token it recovers quickly from grazing, not only because its prostrate habit protects the sites necessary for its regrowth, but also because the light regime is suddenly improved. Active photosynthesis and sugar production are equally important to sustain growth and to provide the *Rhizobium* organisms with energy substrate. The value of white clover as a pasture constituent is at least twofold. It is very palatable in all seasons, rich in protein and minerals and thus highly nutritious, but at the same time it adds nitrogen to the system at a substantial rate, estimated as being between 150 and 250 kg ha^{-1} annually under favourable conditions. Both these roles depend on the vigour of white clover being maintained by correct management and fertilizer supply. White clover requires phosphates, sulphur, potash, lime and, among trace elements, molybdenum in particular. On the other hand, it is easily suppressed by applications of nitrogen which encourage the grasses to become strongly competitive. Flowering takes place in the summer and, if a seed crop is required, grazing should cease when the first flower buds appear. Until then a mixed grass/clover stand needs to be kept closely grazed to encourage good light penetration. Seed is usually taken in the second productive season. The crop is cut when the majority of inflorescences are brown and seed can be rubbed out by hand. The crop may either be wind-rowed into swaths and left for some days or desiccated with a herbicide, mown and then threshed with a combine harvester. A good seed yield in New Zealand is about 750 kg ha^{-1}, although national averages are considerably lower.

White clover is a tetraploid ($2n = 32$), and manipulation of chromosome numbers with the aid of colchicine as in diploids would be inappropriate. Plant breeders thus have to resort to more conventional hybridisation and selection. Apart from improving seasonal productivity and other agronomic characters, one of the aims in recent years has been to increase petiole length to enable the leaves to capture full light. In Europe there is some interest in rhizobial strains capable of functioning in the presence of artificial nitrogen, although the pendulum may swing the other way, as fertilizers increase in cost and because of the energy required for their production. Another interesting move has been to use Mediterranean ecotypes for crossing with the aim of improving winter growth where this is possible climatically without prejudicing resistance to cold.

Red clover (*Trifolium pratense*)

Red clover is widely distributed throughout temperate to sub-arctic regions. Its soil-improving properties have been recognised for several centuries, and it was this clover which from about 1730 was included in 'Turnip' Townsend's Norfolk four-course rotation as a fertility restoring crop. Red clover is a much branched semi-erect perennial with a taproot (Fig. 11.3). The aerial parts are hairy in varying degrees, and the upper surface of the leaves generally have fewer hairs than the underside. The leaflets are large, broad, dark-green in colour and their margin is entire. There is a distinct leaf mark. The stipules are conspicuous, often marked with purple-coloured veins, broad but narrowing abruptly in a long, sharp point with a tuft of hair. Inflorescences arise in the axils of leaves, but also at the tip of each shoot thus preventing further leaf production. Each inflorescence is globular, about 2–2.5 cm across, and contains some 27 to 30 flowers. It is subtended by two almost sessile leaves which hold it like a cup or a pair of hands. The corolla is pink to dull or purple-red in colour. Red clover is largely self-sterile and requires to be pollinated by insects. Bumble bees are mainly responsible, for the nectaries are deep-seated at

Fig. 11.3. Red clover (*Trifolium pratense*). (*a*) whole plant, (*b*) stipule, (*c*) trifoliate leaf, (*d*) inflorescence.

the bottom of a long tube formed by the base of the petals. The necessity to provide suitable pollinators was not always understood in the past when red clover was introduced to new areas. The seed is oval in shape, about 1.5–2 mm long, with a prominent projection like a Roman nose where the radicle is situated. The colour of the testa is yellow with shades of purple at one end. The pods, containing a single seed, are short, open transversely and terminate in a long point.

Red clover is represented by a range of types varying in the time of flowering. Early cultivars have a low photoperiodic requirement, while conversely late cultivars cannot flower until the long days of summer have arrived. In general, three types are recognised: wild red, broad red and late-flowering red clovers. The first of these is a relatively small and prostrate plant which persists naturally in temperate grasslands, although its productivity is low. Broad red clover represents a very different type and in fact was not derived from the wild plant but was introduced from Holland to England in the seventeenth century as a fertility builder in crop rotations. It starts growth early in the spring and forms a lax rosette bearing relatively bright-green, broad leaves. Flowering occurs early, the peduncles are hollow, and the first inflorescence appears between the fifth and thirteenth nodes. If cut for hay at first flowering, the plant recovers quickly through the presence of actively growing, basal buds. A second cut or further grazing are thus possible, but survival into a second or third year is less successful. Late-flowering red clover is also known as Montgomery red clover, because it has been cultivated for a long time in Wales and provided breeding material for many cultivars. This type of clover forms dense, semi-prostrate rosettes bearing grey-green to blue-green leaves. The stems are solid, and the first inflorescence appears between the tenth and 26th node on the stem. In distinction to broad red clovers, there is only one wave of shoot elongation, but at the end of the season enough dormant buds remain to provide good growth in the following spring, thus ensuring the plant persists.

Red clover establishes easily in most areas but, as with other clovers, it requires to be inoculated with the appropriate rhizobial strain before the seed is sown. Perhaps the best use of red clover is in special-purpose mixtures or by itself as a hay or silage crop, and here the choice lies between the broad red type giving two cuts for a year or two only, and late-flowering cultivars providing a single good yield for a number of years. However, both types have their uses under grazing. Broad red cultivars are sometimes included in pasture mixtures to provide forage in the initial period until other species take over. Late-flowering clovers are capable of making a valuable contribution to mixed swards for several years, provided rotational management is practised, preferably using cattle. A good level of production from mid-spring to the autumn can be expected.

Red clover is a diploid ($2n = 16$). A great many cultivars have been bred, varying in productivity at different times, ability to recover from cutting or grazing, and other agronomic characteristics. Resistance to clover rot (*Sclerotinia trifoliorum*) and eelworm (*Ditylenchus dipsaci*) is an important consideration. Plant breeders have also used colchicine to produce tetraploid cultivars which are capable of high annual yields. Seed size is greatly increased with a 1000 seed weight of 3–3.5 g as compared with 1.8–2.2 g in diploids. It is thus necessary to adjust sowing rates. Some tetraploid lines suffer from reduced seed fertility.

Alsike clover (*Trifolium hybridum*)

Alsike clover is not a hybrid, as its botanical name may suggest, but a species in its own right. It is named after a village near Uppsala in Sweden where it was first noted. In growth habit it resembles red clover, but in other respects, notably the absence of hairs and the serrated leaf margin, it is similar to white clover. In contrast to them both, alsike has no leaf mark, and another distinguishing feature is the continuously tapering stipules with greenish veins and long points without tufts of hair. The inflorescences occur in axillary positions but, unlike in red clover, there is no terminal inflorescence and thus the stems continue to grow and become semi-prostrate. There are also no subtending leaves below the flower head. The corolla is white or pale pink, and the flowers shrivel and turn down when mature. The seeds are about 1 mm in length, heart-shaped with a prominent radicle position, and light- to dark-green in colour turning dark-brown as they age. Alsike clover comes from the cool-temperate or sub-Arctic regions of Scandinavia and is thus better adapted than either red or white clover to wet soils and cold climates. It grows well on fairly acid soils and tolerates more alkalinity than most clovers. It is also quite resistant to clover rot. Although its productivity may not compare favourably with red or white clover under fertile conditions, alsike scores when it comes to difficult habitats. It has, for example, been sown successfully from the air in the tussock hill country of the South Island of New Zealand.

Crimson clover (*Trifolium incarnatum*)

Crimson clover, also known as Italian clover, was first cultivated in Southern Europe and spread from there northwards and to the United States. It is an annual, erect species with extremely hairy stems and obovate, hairy leaves. The stipules are broad and blunt. The inflorescence is a terminal raceme, cylindrical in shape and about 5 cm long bearing deep crimson flowers. The calyx is covered with stiff hairs and these are sufficiently dense to cause digestive problems in animals. For this reason crimson clover needs to be grazed before the flowers are fully emerged, or alternatively it should be used for hay or silage. Crimson clover, alone or with Italian ryegrass, used to be grown as a winter annual to provide early spring forage for sheep. However, since it is not very hardy and since it lacks the ability to recover adequately after grazing, it has to a large extent been replaced by more productive species, although it continues to occupy an important position in the United States.

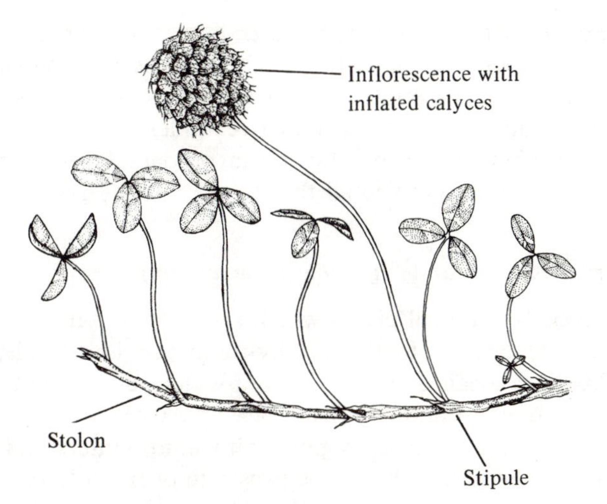

Fig. 11.4. Strawberry clover (*Trifolium fragiferum*).

Strawberry clover (*Trifolium fragiferum*)

This clover has a growth habit like white clover, except that it is slightly less prostrate nor quite as freely rooting at the nodes. It can be distinguished in the vegetative state by the presence of some hairs on the petioles and veins on the undersurface of the leaflets and by the stipules which are larger than those of white clover, encircle the stem and come to a long point (Fig. 11.4). The veins of the leaflet curve towards the end and meet the margin at right angles, whereas in white clover the angle is less than 90°. Any confusion between the two species is removed when flowering occurs, because the calyx of strawberry clover swells after fertilization to enclose the ripening seed and becomes slightly pink in colour and rather papery in texture. As a result, the whole inflorescence assumes the appearance of a dried-up raspberry, and does not really look like a strawberry, after which the species is called. The corolla is white to pink, and below the inflorescence is a collar of bracts. The seeds are heart-shaped, 2 mm across, light-brown in colour and speckled with dark-brown flecks.

Strawberry clover is of Mediterranean origin, and one of the best-known cultivars, Palestine, was bred from an ecotype found in the vicinity of the Dead Sea. Adaptation to this environment explains two characteristics of this clover. Not only is it superior to white clover in autumn and winter growth, but it is also extremely tolerant of moderately high salt concentrations. In South Australia it thrives equally on sandy soils with a pH of 5.5 and on alkaline peats and clays with a pH of 9.0. It grows well on poorly drained land with a high water table, prolonged immersion in moving water does it no harm, but it cannot withstand stagnant water for lengthy periods. Its prostrate habit enables strawberry clover to tolerate close grazing and to remain productive under heavy stocking. These attributes together with its broad spectrum of adaptability suggest that this clover may not have received all the attention it deserves.

Subterranean clover (*Trifolium subterraneum*)

There can be few annual plants which are as persistent as subterranean clover, or better adapted to difficult environments. This species originated in the Mediterranean area but is now widely distributed throughout southern Europe, around the Black Sea, in Australia and parts of New Zealand. It is best suited to regions with mild winters and warm, dry summers. Subterranean clover is a prostrate plant with long and highly branched creeping stems or runners which, however, do not root at the nodes (Fig. 11.5). Stems, petioles and leaves are generally densely hairy. The leaflets are heart-shaped, and in many cultivars they are marked by irregular red fleckings. A leaf mark is present, and the stipule is broad with a short point and distinct veins which are coloured pink to purple in some selections. The inflorescences are small axillary racemes bearing three to six fertile flowers, each about 1.25 cm long, with a corolla which may be white or cream, often with a faint pink blush. The calyx is hairy and may have one or two red rings around it, but also present in the inflorescence are a number of upper sterile flowers which consist of only a calyx with stiff teeth. Subterranean clover is self-pollinated, and after fertilization the peduncle bends over and grows towards the soil. Before long the inflorescence is pushing into the soil and held there by the reflexed calyces of the sterile flowers which form a network of barbs on the outside. The whole structure constitutes a burr with the developing seeds naturally buried. The seeds which occur singly in small pods are about 3 mm across, larger than in any other clover, ovoid to globular and purple-black in colour with a pronounced hilum. Flowering occurs in late spring and, when the seeds have set in summer, the plants dry off and die. Subterranean clover as a plant is not especially drought-resistant but, by burying its own seed just before the dry season begins, it manages to ensure its own continuance. Since a proportion of the seeds are hard, it may persist almost indefinitely, as a reservoir of seeds builds up in the soil. Germination occurs in the following autumn, and the life cycle is repeated.

Subterranean clover is an invaluable species in areas with severe summer drought where perennial plants are unable to survive or remain productive. Its large seeds allow it to germinate rapidly and to produce abundant

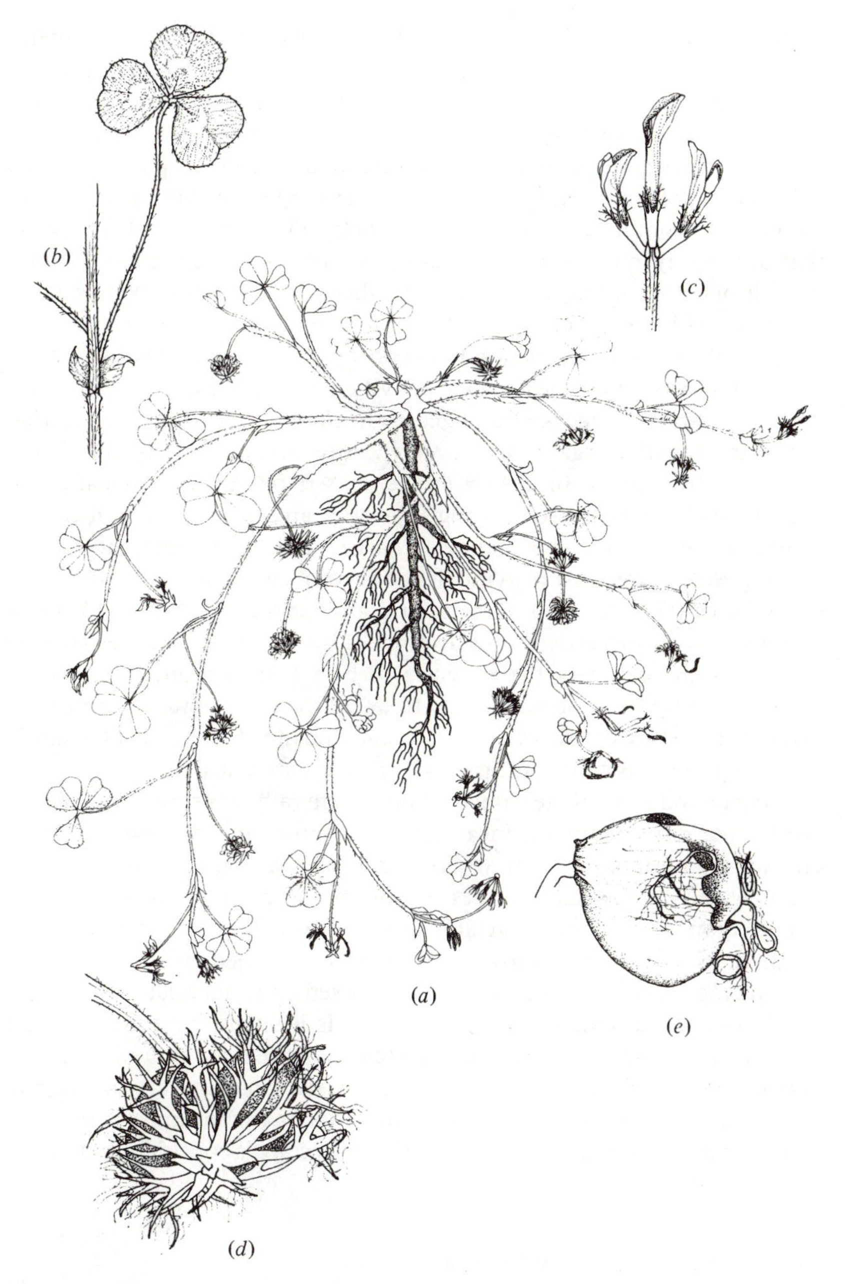

Fig. 11.5. Subterranean clover (*Trifolium subterraneum*). (*a*) whole plant, (*b*) trifoliate leaf and stipules, (*c*) fertile flowers, (*d*) mature inflorescence with calyces of sterile flowers, (*e*) fruit after removal of calyces.

herbage in late winter to early spring, depending on the local environment. It is best suited to a Mediterranean-type climate with predominantly winter rainfall. A growing season of five to nine months is required with at least 500–600 mm of rain, followed by a hot, dry summer. By selecting for different degrees of vernalisation requirement Australian plant breeders have succeeded in producing cultivars with a range of flowering dates to fit the climatic conditions of different districts. They have also bred plants that are highly nutritious and, thanks to their prostrate growth habit, capable of withstanding close grazing by sheep. There is, however a serious problem with some cultivars of subterranean clover which contain appreciable concentrations of phyto-oestrogens, substances with a chemical composition and physiological activity generally similar to female sex hormones. In sheep marked changes in the normal structure and function of the reproductive organs may occur, the most serious consequence being a serious drop in lambing percentage caused by reduced ovulation or lambing difficulties in ewes grazing subterranean clover pastures at critical times. Wether lambs may show enlarged teat and mammary gland development as well as a high incidence of renal and urinary calculi. Dairy and beef cattle suffer no infertility and are generally not affected. Plant breeders have succeeded in selecting cultivars with low oestrogenic activity. Agronomically the measure to be taken to reduce the risk of interference by these substances is to avoid grazing subterranean clover dominant pastures with ewes before and during the period of mating. Other legumes, notably lucerne (see p. 235) may also contain phyto-oestrogens and thus create similar, though generally less serious problems. In all other respects subterranean clover is a most valuable plant wherever climatic conditions are too severe for perennial legumes. Because its persistence depends on a succession of annual crops, it requires to be managed in such a way that abundant flowering and seed setting are ensured. New introductions, either as part of a seed mixture or when oversowing existing grassland, should be based on an adequate sowing rate and it should be remembered that the seeds individually are larger and heavier than those of other clovers. In selecting the most suitable cultivar, care should be taken to make sure that flowering occurs early enough to form and mature adequate seed for germination in the autumn. In Western Australia about 600 kg of seed produced per hectare is considered adequate for survival.

Yellow suckling clover (*Trifolium dubium*)

Yellow suckling clover is another annual legume but its economic value is not very great. It occurs commonly as a wild plant in temperate grasslands, where it persists through self-seeding, but its inclusion in pasture seed mixtures cannot be recommended. Suckling clover has rather weak, semi-prostrate stems bearing small, grey-green leaves in which the central leaflet is found on a longer stalk than the other two. The stipules are broad at the base but taper to a sharp point. The axillary inflorescences contain 12 or more flowers, each with a yellow corolla which withers but persists after pollination, as the flower turns down. The seeds are oval, about 1.5 mm long, pinkish-yellow to light brown in colour. Suckling clover seed does not uncommonly contaminate samples of white clover seed from which it cannot be readily separated. This greatly reduces the value of the white clover seed, because yellow suckling clover is much less productive and persistent. It may contribute something to herbage production, especially on dry, light soils, and it may also add some nitrogen to the system, but in both respects it is decidedly inferior to sown legume species. Yellow suckling clover is often confused with black medick (*Medicago lupulina*), a wild plant occurring in similar situations (see p. 240). Another similar plant is hop clover (*Trifolium campestre*) which has slightly larger inflorescences containing 35 to 60 flowers.

Rose clover (*Trifolium hirtum*)

Rose clover is also of Mediterranean origin and adapted to climates typical of that region. It is a semi-erect annual with densely hairy stems, petioles and leaves. The leaflets tend to have a crescent-shaped leaf mark, often with a dark line well above the centre. The stipules are narrow and terminate in a fine hair or bristle tip. The inflorescence is globular and contains 20 to 40 flowers with hairy calyces. The petals are light to deep pink, hence the name of the species. The seeds range in colour from cream to buff, and are broadly ovoid, about 2 mm in length. Rose clover serves a similar purpose to that of subterranean clover by producing most herbage in the spring, following germination in the autumn and slow growth in the winter. It is claimed to be of particular value under conditions of low soil fertility and rainfall. Several cultivars have been developed in Western Australia from material collected in the eastern Mediterranean area.

Berseem or Egyptian clover (*Trifolium alexandrinum*)

This clover probably originated in Asia Minor and spread from there to eastern Mediterranean countries where it has assumed considerable importance. It is also grown in Sicily and southern Italy, in India, Pakistan, and in North and South America. Berseem clover is an annual plant with oblong trifoliate leaves, lanceolate stipules, ovate inflorescences with creamy white flowers, and a deep root system. It is a vigorous species thriving in tropical or subtropical winters and warm temperate climates with at least 250 mm of rainfall but without any severe frosts. Like several other clovers, it can tolerate fairly high salt concentrations and a soil pH on the alkaline side of neutrality. If irrigated or in response to rainfall it recovers quickly from defoliation, so that under good growing conditions it may be cut every 35 to 45 days during a season lasting eight to ten months. It produces palatable and nutritious forage which is fed green or conserved as hay or silage. Because of their high moisture content, the stems are difficult to dry, whereas the leaves drop off easily during hay making. Berseem is grown mainly as a winter annual, generally under irrigation to take full advantage of its vigour and rapid recovery after cutting or grazing.

Caucasian clover (*Trifolium ambiguum*)

This clover, also known as Kura, is native to the alpine grasslands of the Caucasus and Armenia, but has been developed successfully for use in the mountain areas of other countries such as Australia and New Zealand. It is a perennial species capable of spreading through a system of rhizomes which produce new plants along their course. The stems are prostrate, some 20 cm in length and only sparsely hairy. The leaflets are glabrous, oval to elliptical, with strongly developed veins, a serrated margin, and a pale-green V-shaped leaf mark. The stipules are pale, membranous and have distinct veins. The inflorescence is round to slightly oval, 2.5–3.5 cm across, bearing white flowers which become pink as they mature. The seeds range in colour from dull yellow to red-brown. Caucasian clover is well provided with nectaries and for this reason has attracted attention in the United States as a honey plant. However, its main use is as a persistent pasture plant in alpine or hill-country environments or for soil conservation above the tree line. It tolerates waterlogged conditions, it is superior to white clover at low pH and, although it responds to fertilizer application, it grows well at low levels of soil phosphate. Unlike those clovers of economic importance that originated in the Mediterranean area, *Trifolium ambiguum* is adapted to a cool, continental-type climate and thus remains dormant during the winter. By the same token it is slow to begin growth in the spring. The basic chromosome number is $2n = 16$, although not only diploid but also tetraploid and hexaploid lines occur naturally, and even octoploids can be produced by using colchicine. The most promising cultivars so far have been $4n$ or $6n$.

Other clovers

Several other clover species have been developed as pasture plants, especially annuals with characteristics similar to those of subterranean clover. Cupped clover (*Trifolium cherleri*) is one of these species. It originated in the eastern Mediterranean area and has created some interest in Western Australia where several cultivars have been bred. This clover derives its name from an involucre formed by the inflated stipules of the uppermost pair of leaves just below the inflorescence. It is a prostrate annual with hairy leaves which are serrated along the margin. Cupped clover grows well in the winter and is adapted to a Mediterranean-type climate with a rainfall of 330–450 mm, as, for example, in Western Australia. It is able to set seed and mature under very dry conditions. Once they have become accustomed to the dry diet, sheep eat the inflorescences quite readily, and in so doing distribute the seed. Another eastern Mediterranean clover introduced into Australia is purple clover (*T. purpureum*), an erect annual with purple flowers, which is capable of considerable winter growth and has the ability to persist under close grazing. At high elevations in east tropical Africa, in Ethiopia and Yemen there occurs naturally Kenya white clover (*T. semipilosum*), a low-growing perennial with creeping stems which root freely at the nodes. This growth habit enables it to grow in Kikuyu grass (*Pennisetum clandestinum*) pastures (see p. 134), provided the grass is kept under control. It requires an annual rainfall of over 300 mm and becomes dormant during dry seasons, but considering its distribution, its optimum temperature for growth is surprisingly low at 21 °C.

11.3 *MEDICAGO*

The importance of the Fabaceae as a family containing highly productive forage plants is further reinforced when considering the genus *Medicago*. It contains over 50 annual or perennial species, of which about one half have been tested or developed in agriculture, chief among them is lucerne or alfalfa. Morphologically the members of this genus, or medicks as they are jointly called, differ from clovers by having leaflets with a mucronate tip, a projection of the midrib beyond the edge of the leaf margin. Furthermore, the edge of the stipules tends to be serrated, and the seed pods are generally curved.

Lucerne or alfalfa (*Medicago* spp.)

There are two main perennial species of lucerne, *Medicago sativa*, an upright plant with blue flowers, and *M. falcata*, a more prostrate, yellow-flowered plant. Since these two species hybridise naturally and are frequently crossed by plant breeders, and since they share a good many morphological characteristics, it is more convenient to treat them together.

Strictly speaking, the hybrid nature of very many cultivars should be recognised by using the name *M. media* or *M. varia*, but this is an argument best left to the taxonomists.

Origin and history

Cultivated *Medicago sativa* is an autotetraploid ($2n=32$) but it originated from a diploid, native to an area south of the Black and Caspian Seas and still growing wild in Iran and eastern Anatolia. The early spread of this species is closely linked to the history of the horse and its paramount importance for communication and warfare in the ancient world. It is highly likely that the Kassites who conquered Babylon in the eighteenth century BC used lucerne to feed their horses, although the first written reference to the plant did not occur until over one thousand years later. It was then known by its Iranian name, aspasti, from which the Arabian alfalfa is thought to have originated. During the Persian wars in the fifth century BC the plant was taken to Greece, where it became known as medicai because it came from Media, a part of Iran. The Romans acquired the plant from the Greeks and spread it throughout Europe in the wake of their conquering armies. However, after the collapse of the Roman Empire it survived only in Spain, probably reinforced by further introductions made by the Moslem invaders. In a very real sense lucerne had the same importance in ancient warfare as petrol and aviation fuel have in modern times. From Spain the plant was taken to South America in the sixteenth century and, since it was known by its Arab name, it was as alfalfa that it spread northwards into California and the rest of the American continent. In northern Europe it was referred to as lucerne and widely grown in France, Italy and Germany, but as a result of colonisation it also assumed importance in Australia, New Zealand and South Africa.

Medicago falcata ($2n=16$) had its origin in northern Asia and still occurs naturally in Siberia and other regions with a similar cold-continental climate. Although it was undoubtedly of value to the ancient horsemen of central Asia, it was probably not cultivated since the pods shattered easily and seed was thus difficult to harvest. However, the species spread widely through the migrations of nomadic tribes and invading armies, and thus overlapped in distribution with *M. sativa*. Many natural hybrids will have arisen, despite the difference in chromosome numbers in the two species. The present distribution of these hybrids and of the cultivars produced by the breeder reflects the climatic adaptations of these lucernes. Predominantly *M. sativa* genotypes grow well in Mediterranean to cool-temperate climates and in soils which are not deficient in lime. Genotypes closer to *Medicago falcata* are adapted to cold-continental climates and can tolerate fairly acid soil conditions. A third species, *Medicago glutinosa*, endemic in the Caucasus, has also featured in some hybrid breeding programmes.

The lucerne plant

As in clovers, the first leaf of lucerne to unfold on germination is simple but all subsequent leaves are trifoliate with the central leaflet slightly elevated on a short petiolule. The leaflets are ovoid, the upper part is toothed along the margin, and the midrib projects distinctly (Fig. 11.6). The stipule is deeply serrated and sharply pointed. The whole plant is usually glabrous. A deeply penetrating taproot imparts resistance to drought. In *Medicago sativa* the stems are upright, and in due course these produce in axillary positions roughly conically shaped racemes consisting of numerous flowers, mauve to blue in colour. Pollination by honey bees can be a problem because the flower needs to be tripped so that the stigma becomes receptive. The keel petals fit tightly together, and considerable force is required, especially in cool, damp weather, to open them up. On tripping, the anther column shoots out with explosive violence, causing some discomfort to the insect. Unless forced to do so by massing hives in stands destined for seed, bees do not pollinate lucerne readily. Nectar-collecting bees gain access by biting a hole in the side of the flower. The bumble bee is a more reliable pollinator, although numbers are impossible to control, and in the western United States the leafcutter (*Megachile rotunda*) has been successfully domesticated for this purpose. Following fertilisation *sativa* lucerne produces a spirally coiled dark-brown to black pod, containing usually three to four seeds, except that the very occasional self-pollinated flower bears a short, straight pod. The seeds are about 2–2.5 mm long, yellow-brown in colour, with a distinct projection denoting the position of the radicle. Hard seeds are common. *Medicago falcata* is a more prostrate plant, and it is further distinguished by forming rhizomes below ground. On flowering it produces yellow flowers, and the pods are short and decidedly curved, hence the name sickle medick by which this species is also known. In line with its origin and distribution, *falcata* lucerne is cold-tolerant and resists frosts, but in consequence it is also dormant during the winter and early spring. Hybrids between these two species show a range of flower colours, some of them being distinctly variegated. In their morphological and physiological characteristics they are also intermediate.

Cultivation and uses

The close connection between the history of lucerne and that of the horse emphasises the importance of the plant as forage. It can be grazed,

Fig. 11.6. Lucerne (*Medicago sativa*). (*a*) base of plant showing crown buds and axillary shoots, (*b*) trifoliate leaf and stipules, (*c*) inflorescence, (*d*) flower, (*e*) developing seed pods.

provided management is well controlled, but it is probably in the form of hay that lucerne acquired its best reputation. Effective nodulation and absence of weeds are two key factors in successful establishment. The seed needs to be inoculated with an effective strain of *Rhizobium meliloti* which is specific for lucerne and requires a soil pH of at least 6.0, but preferably above 6.5, for survival and growth. Lime-coated, inoculated seed has been used with success on acid soils. Although the mature plant is drought-resistant, it is advisable to sow lucerne in the spring so that the seedlings are well established before the dry summer. Establishment and longevity of the stand depend closely on preventing competition from weeds. For somewhat similar reasons, lucerne is most easily managed if grown by itself and not in a mixture with grasses. The search for a non-competitive companion grass, attuned to the growth rhythm of lucerne, has so far not been successful.

Lucerne is a highly productive plant and, especially on light soils and in dry climates it out-yields other forage plants. However, successful management depends on a thorough knowledge of the plant, particularly if it is to be grazed. As successive stems of lucerne grow and produce leaves and axillary buds, a complex and ill-defined region, the so-called crown of the plant, is gradually formed near soil level, and this provides the main source of regeneration following cutting or grazing (Fig. 11.6). This region is usually inactive as long as the stems above are growing actively but, after some weeks of shoot growth and usually coinciding with the appearance of flower buds, the crown buds begin to elongate. If the shoots above are now either cut or grazed, almost immediate recovery growth is possible from these buds and others situated in the axils of the stubble leaves. Hence lucerne which is defoliated at intervals determined by its physiological development remains productive and vigorous. By contrast, too-frequent utilisation before the crown buds are ready to move slows down recovery and causes root reserves to be depleted. A weakened stand is soon invaded by weeds and becomes unproductive. Apart from management, tolerance to grazing in *sativa* genotypes is also improved by cross-breeding with *falcata* lucerne, but at the same time this causes greater winter dormancy and a late start to growth in the spring.

Whether grazed or fed after conservation, lucerne is extremely nutritious, and the protein content is especially high. As with other forage plants, digestibility and nutritive value decline with maturity but, while it is young and leafy, lucerne will sustain rapid growth rates in grazing animals. At certain stages of growth, care has to be taken to avoid bloat in cattle. This condition is caused by the formation of gas, mostly carbon dioxide and methane, which becomes trapped in a stabilised foam during rumen fermentation. This causes acute discomfort and may lead to death through failure of circulation and respiration. Pastures high in red or white clover are also prone to cause the same problem. Apart from reducing legume intake, bloat can be prevented by spraying swards with paraffin or drenching cows with an anti-foaming agent. Plants with a high tannin content such as sainfoin, lotus and most tropical legumes do not cause bloat. Some cultivars of lucerne also have oestrogenic effects. Dried lucerne meal is an important constituent of poultry feeds, not only because of its value as a source of protein but also because the presence of carotene and xanthophyll imparts a dark-orange colour to the egg yolk. An unknown growth factor, beneficial to poultry, is also thought to be present.

Plant breeders have endeavoured to improve the productivity of lucerne and to reduce its weaknesses. In particular, they have paid attention to resistance to pests and diseases, notably bacterial wilt, eelworms and various species of aphids which can be very serious in effect. Despite these problems and the exact requirements of management, lucerne remains a plant capable of producing copious quantities of nutritious forage. Especially under dry conditions, it far exceeds other pasture species in production and quality.

Other medicks

The genus *Medicago* includes a number of annual species that are jointly referred to as burr clovers. These are indigenous to the Mediterranean area and occur either naturally or are introduced in regions with this type of climate. Most of their growth occurs during a mild winter but, by the time the dry summer has arrived, flowering has finished and seed has been set. These species thus escape drought and tend to persist by re-seeding themselves. Not only do they provide grazing during the winter, but their dry fruits high in protein constitute a valuable feed reserve for the summer drought period. Several species of burr clover such as *Medicago denticulata*, *M. minima*, and *M. lacinita*, are unsuitable for areas grazed by sheep because the pods have hooked spines which become entangled in the wool. There are, however, several species with straight spines, and some of these have become popular in Australia. One of these is barrel medick (*M. truncatula*), which is represented by several cultivars. This is a semi-prostrate self-pollinating annual whose main distinguishing feature is the barrel-shaped pod consisting of four to six spirals. It is suitable for low to moderate rainfall areas and a variety of soils. Other annual medicks which have attracted attention in Australia are *M. littoralis*, *M. tornata*, and *M. rugosa*, all adapted to mild winters and dry summers. Black medick (*M. lupulina*), a common wild species in Europe, Australia and New Zealand especially on calcareous soils, is often confused with yellow suckling clover (see p. 232) but can readily be distinguished by its mucronate tip.

11.4 *LOTUS*

This genus contains over 100 species, of which only a few are of value in agriculture. The species concerned are leafy perennials with yellow or red flowers. The long, narrow seed pods occur in groups which are said to resemble a bird's foot, a feature which has given rise to several common names. Most members of the genus are adapted to cool climates and fairly acid, damp soils. They are jointly referred to as trefoils.

Birdsfoot trefoil (*Lotus corniculatus*)

This species, a native of temperate Europe and Asia, is grown in parts of North and South America, and in Europe for both grazing and hay production. It is a spreading to erect perennial with a long taproot and numerous secondary rootlets. The plant develops a distinct crown with several stems but, in distinction to other members of the genus, there are no stolons. The leaves are trifoliate, the width of each leaflet being at least half its length, but at the base of the petiole there is a pair of stipules which so closely resemble the leaflets that the appearance is of a five-leaflet compound leaf. Some authors consider minute brown scales at the base of the petiole to be the stipules. Both stems and leaves are usually glabrous although hairy plants occur. Inflorescences which are umbels appear at the end of axillary branches, consisting of four to six flowers which are deep yellow in colour with the standard petal often tinged red. The pods are 3–5 cm long, dark brown and grouped together like the fingers of a hand or a bird's foot. The seeds which are released suddenly as the pod splits are shiny, dark brown in colour and about 1.5–2.0 mm long. Birdsfoot trefoil is relatively drought-resistant, more so than other *Lotus* species though less than lucerne, but it does not readily tolerate poor drainage or excessive irrigation. The soil pH should preferably be about 6.5. It is used as a forage plant in much the same way as lucerne by cutting or grazing. Defoliation should occur at intervals long enough to allow the plant to reach an early flowering stage but, in distinction to lucerne, close cutting or grazing should be avoided, because recovery growth depends on axillary buds on the stems which occur some distance above soil level. Birdsfoot trefoil is most productive in spring and early summer, but tends to die back in the winter when temperatures drop below zero for much of the time.

Greater birdsfoot trefoil (*Lotus pedunculatus*)

The nomenclature of this plant is confusing. Botanically it is synonymous with *Lotus uliginosus* and *L. major*, while its common names include big trefoil, marsh birdsfoot trefoil or simply lotus (Fig. 11.7). It is similar to *L.*

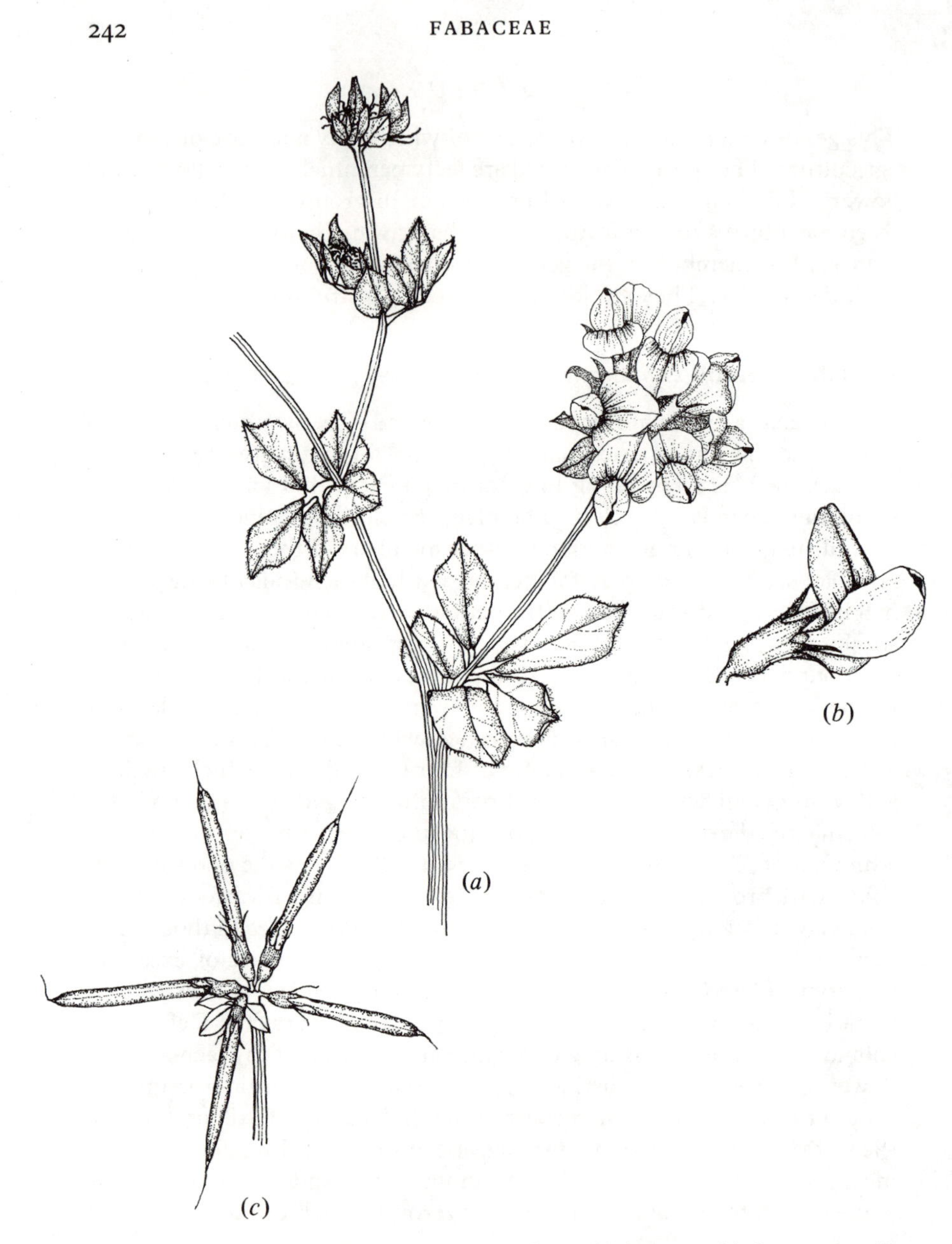

Fig. 11.7. Lotus (*Lotus pendunculatus*). (*a*) plant with leaves and inflorescence, (*b*) flower, (*c*) seed pods.

corniculatus except for its ability to form decumbent shoots in the autumn which spread well over a metre beyond the periphery of the crown, root at intervals like stolons and even penetrate into the soil and become rhizomatous. This ensures that the plant persists and spreads from year to year. The root system is shallow and consists of fine, fibrous roots. Another point of difference concerns the calyx, whose teeth are spreading, whereas those of birdsfoot trefoil are straight. Agronomically *L. pedunculatus* is adapted to wet soils and high rainfall and, because of its ability to tolerate acid conditions and low phosphate levels, it may be considered as a pioneer legume for land development and revegetation of eroded areas. For these purposes in particular the seed needs to be inoculated with the correct rhizobial strain before sowing. Establishment is slow, especially at low temperatures, and the plant is not very competitive at an early stage of growth. In fact, grazing management throughout should be more lenient than that of white clover, although this does not prevent its use as a component of a mixed grass/legume sward. In many situations, particularly at low levels of fertility and in moderate drought, *L. pedunculatus* may out-yield white clover in grazed pastures. One of its attributes of special value in New Zealand, where pasture legumes are of paramount importance, is the presence of tannins in the leaves, which prevents the occurrence of bloat in cattle. Resistance to grass grub and Porina caterpillars is another important characteristic. Plant breeders have improved the vigour of the plant by inducing tetraploidy, and in Mediterranean climates winter activity has been encouraged by crosses with Portuguese lines.

Other trefoils

Slender or narrow-leaf trefoil (*Lotus tenuis*) is remarkable for its ability to tolerate greater soil salinity than almost any other legume. It has also been shown to stand up to very high summer temperatures, and in the winter it continues some growth and does not become dormant. At one time it was considered a subspecies of *L. corniculatus* but is now established as a species in its own right, largely because it is a diploid and not a tetraploid. It can also be distinguished by its narrow leaflets whose width is less than half their length, by the presence of many slender stems, and by distinctly shorter seed pods. Among other members of the genus *Lotus* is hairy birdsfoot trefoil (*L. subbiflorus* or *L. hispidus*), a very hairy plant with thick, trailing stems, adapted to droughty, infertile habitats. Another species of interest is *L. angustissimus*, slender birdsfoot trefoil, which has red, wiry, profusely hairy stems and small inflorescences containing one or two flowers which turn green with age. Both these trefoils may be annual or perennial plants.

11.5 *MELILOTUS*

The 25 or so species in this genus, the sweet clovers, appear to have originated in Asia Minor but are now widely distributed throughout the subtropical to temperate zones, especially in North America. They are annuals or biennials with tall, rather woody stems, narrow trifoliate leaves, lanceolate stipules, and white or yellow flowers in long axillary racemes. An extensive root system imparts resistance to summer drought, but for successful growth, soils require to be neutral or alkaline. Sweet clovers derive their name from the pleasant smell of coumarin, also characteristic of sweet vernal grass (see section 4.11) which is found in all parts of the plant. In poorly cured hay or fermented silage this substance may decompose into compounds toxic to animals because normal blood clotting is prevented.

White sweet clover or Bokhara (*Melilotus alba*)

Also known as honey clover or white melilot, this species is typically biennial, although annual types also occur. It is a tall plant with short stems, reaching 1.5–2.0 m in height and bearing white flowers in which the wing and keel petals are about equal in length. It is grown for pasture, hay and soil improvement, predominantly in North America, but its wide range of adaptability makes it suitable for use throughout the temperate zone. It has become a very important legume in western Canada, where some hardy cultivars grow well at high latitudes. Drought resistance is only moderate, but the plant responds well to irrigation. Cultivars with varying maturity dates have been selected and, depending on genotype and climate, the crop is sown in the spring or in the autumn.

Yellow sweet clover (*Melilotus officinalis*)

The name of this species is more correctly given as biennial yellow sweet clover or as yellow melilot. Typically it is a biennial, even though annual types do occur. Apart from the colour of the flowers, it differs from white sweet clover by the wing petals being longer than the keel. Because it is hardy and starts growth early in the spring, it is valued as a forage plant in cool, temperate regions, especially Canada and the northern United States. Compared with white sweet clover it is superior in establishing in dry conditions but some cultivars at least do not persist well into the second year.

Indian sweet clover (*Melilotus indica*)

This species is also known as yellow annual sweet clover, small-flowered melilot, senji, or sour clover. It is grown under irrigation in India and in the southern USA, but also occurs as a weed of arable land in Argentina, Tasmania and South Australia. It is an annual or sometimes biennial plant with mainly yellow flowers in spike-like racemes. It is grown as a forage plant either alone or mixed with annual grasses, and it requires neutral to slightly alkaline soil conditions for optimum growth. Climatically it is better adapted to hot environments than the other sweet clovers. One of its disadvantages is a fairly high coumarin content which makes it bitter in taste and, wherever it occurs as a weed in wheat fields, it may taint the flour.

11.6 *ONOBRYCHIS*

Of the 80 to 100 species in this genus, sainfoin (*Onobrychis viciifolia*) is agriculturally the most important. Sainfoin originated in central to southern Europe and temperate Asia but is now grown widely on dry, calcareous soils in many parts of the world. It is a long-lived, perennial legume with a stout and deeply penetrating taproot. The erect stems are 30–60 cm high, round and slightly hairy, bearing pinnately compound leaves with six to twelve pairs of leaflets and a single terminal leaflet. The midrib projects beyond the leaf margin, and the stipule is thin and finely pointed, red when young but browning with age. The inflorescence is an axillary raceme, conical in shape, consisting of numerous large flowers. The calyx has five long teeth, and the petals are pink with red veins. The fruit is described as an indehiscent pod, rather flat and semi-circular in shape, marked with a network of vein-like ridges and toothed at one end. It contains a single seed, dark olive-brown in colour, 4–5 mm long, and oval to kidney-shaped. The true seeds are obtained by milling the fruits which are collected when the crop is threshed or combine harvested. Care must be taken to adjust the rate of sowing, depending on whether the fruits or true seeds are being used, especially as the unmilled material could contain a higher proportion of hard seeds.

There are two types of sainfoin, common and giant. Common sainfoin is more widely grown because it persists for a number of years. It flowers sparsely or not at all in the first summer following spring sowing, and gives a hay cut and aftermath grazing each season or it may be grazed in much the same way as lucerne. Giant sainfoin flowers in the year of sowing. It also has a longer flowering season, produces fewer stems but larger leaves, and it normally persists for only two years. It withstands heavy grazing less well than common sainfoin. Both plants are grown in Europe, South and North America, and in South Africa.

11.7 *ORNITHOPUS*

Serradella (*Ornithopus sativus*) is native to coastal Spain, Portugal and Morocco and is adapted to moist, sandy soils and cool Mediterranean-type climates. It is grown as forage or green manure in southern and central Europe. The plant is an annual with weak semi-erect stems bearing pinnately compound leaves with numerous pairs of hairy leaflets. The flowers are pink and, when mature, the pods break up into short, oval pieces, each containing a single seed. A related species, common birdsfoot (*O. perpusillus*) grows wild on sandy soils in southern England. Another species, yellow serradella (*O. compressus*) has been introduced into Australia, where it is used as a winter annual for sheep grazing on moderately deep sandy soils of low fertility.

11.8 *ANTHYLLIS*

Kidney vetch (*Anthyllis vulneraria*) is a perennial legume distributed in temperate areas of parts of Europe, Asia and North Africa. It is drought-resistant and adapted to sandy and calcareous soils, thanks to a deep taproot with numerous branches. During the first year of growth the plant forms a rosette of mostly simple leaves, but later the leaves become pinnate with a disproportionately large terminal leaflet. The stems which are also for the most part produced in the second year are erect, 30–50 cm high and slightly hairy. The flowers are yellow to red and have a distinct inflated calyx. The flat, black pods contain a single green-yellow seed. As a forage plant, kidney vetch is used mainly for grazing by sheep, either as a special-purpose plant or in mixtures.

11.9 *DESMODIUM*

This genus contains over 300 species comprising mainly perennial herbs, subshrubs or shrubs with three, occasionally five leaflets. The flowers are mainly pink or mauve, and the pods break up on maturity into several one-seeded fragments. Most of the species are tropical or subtropical, and a few only are temperate, distributed throughout Middle and South America, Asia and Africa. A few species from South America have been successfully introduced into Africa and Australia where they are assuming some importance as pasture legumes. One of these is *Desmodium uncinatum*, variously known as silverleaf desmodium, Spanish or tick clover, a large perennial herb with trailing stems up to 5 m long and covered with short light-brown hooked hairs. The trifoliate leaves have large leaflets, 5–10 cm long and 2.5–5.5 cm wide, with a silvery spot or strip along the midrib. The flowers are mauve to pink, the sickle-shaped pods break up at maturity, and the segments become attached to animals and clothing. In Queensland the plant grows well in the summer, with a marked peak by early autumn and, despite its origin, it can tolerate moderate frosts. Because of a high tannin content it is not highly palatable but nevertheless it performs a useful function as a grazing plant together with such summer-growing grasses as *Paspalum notatum* or *Setaria sphacelata* (see sections 4.39 and 4.41). Very similar is another species, *Desmodium intortum*, green leaf desmodium or Kuru vine, another introduction from South America cultivated in Australia. This is a finer, less hairy plant with shorter internodes and more leaf. The leaflets have a reddish brown to purple flecking on the upper surface and the narrow pods adhere to animals and to clothing. Agronomically it is used with a wide range of grasses but variable results have been reported concerning its palatability and recovery from grazing at the beginning of growth in the spring. It does not tolerate heavy frost. The plant is commonly grown in northern parts of South America to about 25° South but has spread to many other tropical areas.

11.10 *LESPEDEZA*

Of the 125 or so species in this genus, only three are of agronomic importance, notably in the central and south-eastern United States. Perennial lespedeza (*Lespedeza cuneata*) is an erect perennial resembling lucerne with trifoliate leaves and lavender flowers. It is drought-resistant and persists well on eroded soils of low fertility. Apart from erosion control, its main use is as a hay plant providing two to three cuts each year, as a pasture plant or for soil improvement. Two annual species are also widely grown in the United States, Korean lespedeza (*L. stipulacea*) and Common or Japanese lespedeza (*L. striata*), both for hay, pasture, erosion control and soil improvement. Despite their recent introduction, all three species now cover wide areas of agricultural land in North America, and tests of their performance in other parts of the world are under way.

11.11 *LABLAB*

The lablab bean (*Lablab purpureus*) is probably better known by its older synonym *Dolichos lablab*. This is a tropical annual or short-lived perennial legume with erect or climbing stems and large trifoliate leaves. The flowers are white, blue or purple, depending on cultivar, and the oblong, curved pods have a distinct beak. This plant has an optimum temperature for growth of about 30 °C and a minimum of about 3 °C. Its distribution is thus limited to subtropical to tropical regions, such as east Africa, Sudan, Brazil, the Philippines and parts of Australia. It is a fast growing plant which recovers well from grazing and is palatable to stock. Hay made from it is of good quality but the stems take a long time to dry and this may cause loss of leaf. Disadvantages include its relatively short life and susceptibility to frost. It is also utilised as both a grain legume and for its fleshy pods in Asia and is commonly known as hyacinth bean.

11.12 *LOTONONIS*

Lotononis (*Lotononis bainesii*) is indigenous to South Africa but has been introduced into several east African countries, Brazil, Florida and Australia. It is a perennial plant with creeping stems that produce upright shoots. The leaves are trifoliate and occur in groups of three to five, of which one or two are much larger than the others. For optimum growth it requires high temperatures and, although it ceases to grow below 9–10 °C, it can withstand moderate frosts. It tolerates acid soil conditions and has the further advantage of extracting phosphorus even if soil concentration is low. Rainfall should be in excess of 900 mm. Under good conditions this species is highly productive and it persists well under heavy grazing. Its nutritive value, especially its protein content, has been compared with that of lucerne.

11.13 *MACROPTILIUM*

Siratro (*Macroptilium atropurpureum*) is of Central American origin but has been developed as a forage plant in Australia and other medium rainfall areas of the tropics. It used to be known as *Phaseolus atropurpureus*. Like many other tropical legumes it has trailing, creeping, stoloniferous stems which are supported by tall companion grasses. The leaves are trifoliate, the wing petals are deep purple, the standard is green with a purple tint, and the keel petals are pink. The pod is straight and about 8 cm long and contains about 12 brown to black seeds. Siratro requires warm conditions for active growth but will survive moderate frosts. Drought resistance is achieved by a number of physiological devices. Water potential is at a high level even if the plant is under stress, stomatal control is very quick and sensitive, heliotropic leaf movement reduces incident radiation and, in response to drought, old leaves are shed and replaced by new, thicker leaves. Among its agronomic characteristics are quick establishment, effective combination with grasses, persistance under grazing and high yields of good quality herbage. Although the old plants tend to die after some years, the sward is maintained through establishment of new seedlings and young crowns from rooted stolons. A related species, phasey bean (*M. lathyroides*), an annual or biennial, is an upright plant more suited to wet environments.

11.14 *MACROTYLOMA*

This genus has about 25 species which were formerly classified under the name of *Dolichos*. Two species have been selected for forage production or as pulse crops in tropical or subtropical climates. The first of these is *Macrotyloma uniflorum*, previously known as *Dolichos uniflorus*, which has different common names depending on its distribution and use. In a number of countries in Asia, Africa, in the West Indies and southern USA it is grown as a pulse crop but also used as cover or fodder under the name of horsegram. In Australia it is used as a forage plant and commonly known by its cultivar name, Leichhardt. It is a softly pubescent, twining annual with trifoliate leaves, bearing greenish yellow flowers with a violet spot on the standard and producing slightly curved, downy pods. Agronomically it is a short-season, summer-growing plant, adapted to a wide range of soils which need to be well drained. Selected lines are palatable to stock at all stages of growth, including the pods which may make up a high proportion of total production and may be saved as a feed reserve for the dry season. The seed passes through animals and can thus be spread to native pastures. A related species, *M. axillaris*, known in Australia under its cultivar name, Archer, was also previously classified in the genus *Dolichos*. It is indigenous in tropical Africa, Mauritius, Madagascar and Sri Lanka, but has been introduced into Australia as a forage plant. It requires a frost-free tropical or subtropical climate with high rainfall, even though it can tolerate some drought. Once animals have become used to it, they eat it readily and, since it has a long season of productivity and competes well with other plants, it has found acceptance in parts of tropical Queensland.

11.15 *PUERARIA*

Of the 20 or so tropical species in this genus, two have agricultural importance. Kudzu (*Pueraria lobata*) is a vigorous, climbing plant with large, trifoliate leaves which are smooth above and softly hairy below. The flowers are purple-red with a yellow spot near the base, and the pod is long and thin. It grows best in warm, moist climates, but can withstand periods of drought and some frost. Although sensitive to frequent defoliation, the plant is used widely for cutting or grazing, especially on poor, eroded soils.

Establishment is slow, and full utilisation is best deferred to the second year. It is grown widely in subtropical to warm temperate regions, including the southern United States and Hawaii. Tropical kudzu or puero (*Pueraria phaseoloides*) is also a robust, climbing perennial whose stems are covered with dense, brown hair. The leaves are trifoliate and the flowers are mauve to deep purple, the standard is white with a mauve blotch. The pods are long, slightly curved at the end, thinly covered with stiff hairs and they turn black on maturity. Essentially a plant for the humid tropics and easily killed by frost, puero combines well with tall tropical grasses to form a dense cover crop. It is one of the best tropical legumes for nitrogen fixation, and is thus used for soil improvement in plantations. Although close defoliation is not recommended, puero can be grazed or cut for hay or silage, as it is both productive and palatable.

11.16 *STYLOSANTHES*

This tropical to subtropical genus contains about 25 species of which two or three in particular have achieved considerable importance as herbage plants. Of these, the best known is probably Townsville stylo (*Stylosanthes humilis*), named after a town in northern Queensland and previously referred to as Townsville lucerne. This is an erect to semi-prostrate annual, up to about 70 cm in height, bearing trifoliate leaves with narrow, pointed leaflets up to 2 cm long. The flowers are yellow to orange, and the very characteristic one-seeded pod consists of two segments, of which the upper terminates in a persistent hooked-shaped style. Six or more pods are produced in each inflorescence, and the seeds are kidney-shaped and yellowish brown or darker. The plant occurs naturally in Central America and northern South America, but is now widespread throughout the tropics, and particularly in northern Australia it has become an outstanding pasture legume. It is easy to establish and has been responsible for a remarkable increase in beef production when introduced into native pastures. It may be oversown into existing grassland without cultivation or, more reliably, drilled in mixtures with grasses following cultivation. Adaptability to a wide range of soils is good. Although an annual, it persists for many years by producing plenty of seeds, many of which are hard initially but soften gradually during the dry winter months ready for germination. The seed is also easily spread by passage through grazing animals. Apart from avoiding heavy stocking in the first season to obtain good seed production, Townsville stylo should be grazed intensively. It is not highly palatable when green but sustains very satisfactory animal production in the dry season when the grasses have died down.

The second important species in this genus is fine-stem stylo (*Stylosan-*

thes guyanensis), an erect to semi-prostrate perennial, 60–90 cm in height. The stems are hairy, and the trifoliate leaves have pointed, elliptical leaflets often with hairs along the midrib. The flowers are yellow to orange, and the hairy pods are ovoid, terminate in a very short beak and contain a single seed. This species is native to the northern regions of South America but is now widespread throughout the tropics and cultivated in many countries. It is adaptable to a range of soils, including those low in phosphate, and its climatic range has been extended by selection of different cultivars. It can be oversown into natural pastures or drilled on cultivated land, usually in mixtures with grasses. A high percentage of hard seeds requires mechanical scarification or acid treatment before sowing. Because it is a variable species, cultivars differ considerably in agronomic performance, but most have proved themselves as extremely useful pasture plants. Protein content of the herbage is not necessarily very high, and some lines are not immediately relished by stock. However, most selections have turned out to be vigorous, high-yielding and persistent herbage plants. A third species, *S. fruticosa* should also be mentioned. A perennial, highly branched herb or small shrub, occurring throughout Africa, it is valued for cattle grazing. Another species is *S. hamata*, the so-called Caribbean stylo. It is a highly productive and adaptable perennial which is proving superior to the other members of the genus as a pasture plant in Queensland trials.

Grain legumes

11.17 *ARACHIS*

Although peanuts or ground nuts (*Arachis hypogea*) are often thought of as a tropical crop, extensive areas are grown in China, Japan, the United States and in temperate regions of South America. In continental environments it can be grown from 40° North to 40° South of the equator. There is some argument about the centre of origin of the species but it is now accepted that it arose in South America, probably in upland Brazil. The crop was taken to the Far East by the Spanish following the discovery of the Americas.

The groundnut plant

There are two main cultivated forms of *Arachis hypogea* which differ in their growth habit as a result of their different branching patterns. Upright plants which form sequential branches are described as bunch types, while more prostrate plants with alternate branching are called runner types. Irrespective of growth habit the crop is an annual and is frost-sensitive and can therefore only be grown during the summer months in temperate environments.

Following germination which is neither epigeal or hypogeal, as the cotyledons remain on the surface of the soil, there is rapid elongation of the shoot apex and of two lateral branches, the buds of which are already present on the epicotyl. There is an absence of strong apical dominance as all three branches elongate simultaneously. The radicle grows rapidly and the plant forms a strong tap root and a highly branched lateral root system. A feature of the roots in this species is the absence of root hairs because there is no epidermis.

Branches formed on the stem are of two types: vegetative branches which are monopodial, and much reduced reproductive branches which are formed singly at the nodes of the monopodial branches. Leaves are pinnate with two pairs of entire obovate leaflets per leaf. Leaflets have mucronate tips as in the genus *Medicago* (see section 11.3), and there is a well developed pointed stipule at the base of the petiole. The yellow flowers which are up to 2 cm in length, are borne on reproductive branches singly or in groups. The flowers are self-pollinated and pollination usually occurs early in the morning. Following pollination a process known as pegging occurs. The fertilized ovary becomes geotropic and pushes itself into the soil in a process not unlike the formation of subterranean clover burrs (p. 230). Underground the fertilized ovary elongates to form the nut of commercial trade. The mature pod may contain from one to six seeds of variable size, with a 1000 seed weight from 350 to 1000 g. At maturity the testa is thin and papery and may be white, pink, red, purple or brown in colour.

Cultivation and uses

In some varieties of peanuts there is a short period of dormancy after harvest of the crop. In temperate environments this is of little practical significance as only one crop is grown in each year. Field germination of the crop is usually best from shelled nuts. Besides warm weather peanuts require high rainfall throughout the growing season (500 mm) and the crop has little resistance to drought. The plants require well aerated soils of high calcium status. Peanuts can be grown on soils with a high clay content, but this does not usually occur because of the problems this causes with harvesting the crop at maturity. The crop is harvested by cutting the tap roots and wrenching the whole plant complete with nuts out of the ground. The harvested plant is then left to dry in the field for a few days with the nuts uppermost. The crop is then threshed to separate nuts from plants and, following further drying, it may be hulled to remove nuts from the shells.

Peanuts are the second most important oil seed crop in the world after the soya bean. The seed contains up to 46% oil which is classified as a non-drying oil. The oil therefore contains a high concentration of saturated fatty acids and may solidify at room temperature in cool climates. Following oil extraction a valuable protein concentrate remains which is utilised as stock food. Large nuts are usually eaten whole after cooking and/or blanching. The seed is also utilised in the production of peanut butter which is made by crushing blanched roasted whole nuts. Peanut hay left after threshing may be used to feed ruminant livestock.

11.18 *CAJANUS*

Pigeon pea, Red gram (*Cajanus cajan*)

Although little known in the temperate world the pigeon pea (*Cajanus cajan*), is an extremely versatile and important legume in the tropics. Over 93% of the crop is grown in India where the hulled seed is known as red gram, but in the West Indies the immature pods are picked and processed as a substitute for peas and, because of its drought tolerance, it is an important high protein forage in parts of South America and the Pacific. It is believed that *Cajanus cajan* originated in Africa and it was cultivated in Egypt before 2000 BC. It has been suggested that it was carried to India by traders from Zanzibar and that it was taken from Africa to the Americas by slave traders.

The pigeon pea

The pigeon pea is a short-lived perennial tree which may reach a height of 4 m. Following germination, which is hypogeal, it produces a deep tap root and a lateral root system which is more extensive in bushy than in erect types. Stems are angled and covered in fine hairs. There is considerable variation in the point where branches arise on the main stem and the angle at which branches are carried, and some modern selections are almost unbranched. The leaves are trifoliate with the terminal leaflet on a longer petiolule than the two lateral leaflets. The leaflets which can be up to 10 × 3 cm, are entire, lanceolate and hairy, with the under-surface silvery grey and dotted with yellow resin glands.

Flowers are formed in either terminal or axillary racemes which are up to 12 cm and usually carry several flowers which are up to 2.5 cm long. Flower colour is usually yellow, although in some varieties the standard may be striped or splotched with red or purple. The pods are flattened, up to 10 cm × 1.5 cm, and contain two to eight seeds but more usually four, and they are green, purple or maroon or a combination of green with the other colours. Seeds are smooth, 4–8 mm in diameter, and green when immature but they may be white, grey, yellow, purple, red or black at maturity. Seed colour may be entire or mottled.

In India, pigeon peas are divided into two groups based on growth habit. *Tur* varieties are early maturing, short plants with yellow flowers and green pods which are light coloured at maturity and usually contain only three seeds. These varieties are mainly cultivated on the Indian peninsula. The other large group of pigeon peas is the *Ahar* varieties which are perennial bushy types and are late maturing. The flowers in these lines have purple or red markings on the standard. Pods are hairy and coloured and contain from four to five seeds which are dark coloured or speckled. *Ahar* varieties are more commonly cultivated in the north of India.

Cultivation and uses

With a crop as diverse as pigeon pea it is not surprising that there is considerable variety in the methods by which the crop is grown. It is adapted to a wide range of climates and soils but does best in areas where the annual rainfall is greater than 500 mm and the soils are not calcium deficient. It is not tolerant of waterlogging. There is considerable variation in time to maturity which can be as short as 90 days or more than 250 days. Pigeon peas make very slow initial growth and in India they are frequently intercropped at quite low populations with short duration companion crops such as peanuts, sorghum or millet. By this means the yield of the companion crop is only slightly reduced and the pigeon pea is often grown on residual stored soil moisture. Because of its deep tap root it is very drought- and heat-tolerant.

In pure stands plants are usually grown in rows 1–3 m apart with 30–90 cm between plants within the two. However, experiments in India have given highest yields per unit area at a spacing of 60 cm by 60 cm or about 27 000 plants per hectare. In the West Indies, on the other hand, good yields of green pods were obtained from a determinate dwarf cultivar at 165 000 plants ha^{-1} and plants were shorter and thus easier to harvest mechanically at this population. Common yields of dry seed are 500–1000 kg ha^{-1}, while in India the maximum reported experimental yield is 5000 kg ha^{-1}. One Australian research report gives a seed yield of 7600 kg ha^{-1} from two harvests of a single sowing. The crop is currently being improved at the International Crops Research Institute for the Semi-Arid Tropics (ICRISAT) in Hyderabad, India.

Pigeon pea seed is about 25% protein, 56% carbohydrate and 8% fibre. The seeds appear to be comparatively free of toxins, metabolic inhibitors and flatulence-causing sugars. In Asia it is converted into red gram by milling to remove the testa. Red gram is consumed in a variety of ways usually in conjunction with a cereal. In the West Indies immature seeds are both frozen and canned in much the same way as *Pisum sativum.* In Hawaii, when used as a forage, it has been shown capable of carrying 2.4 to 3.6 head of cattle per hectare and giving liveweight gains of 147–183 kg ha^{-1} $year^{-1}$. It is also used in other parts of the tropics as a temporary shade for crops like cacao or as a green manure crop when branches are lopped and left to decompose and release their nutrients.

11.19 *CICER*

The chick pea (*Cicer arietinum*) is an annual grain legume that is tolerant of cool temperatures. Although thought to have originated in western Asia and familiar to the ancient Egyptians, Hebrews and the Greeks, it was probably carried at an early date into India and parts of southern Europe. It features commonly in the cuisine of the peoples of the Mediterranean basin and the Middle East but the majority of the crop is now grown in India where it is known as Bengal gram.

The chick pea plant

The species may be erect or spreading with numerous branches and it reaches a height of 25–50 cm. The plant is covered completely in clubbed glandular hairs. After germination, which is hypogeal, it develops a tap root and an extensive lateral root system which usually bears numerous large nodules. The first two leaves on the stem are scale-like but further leaves are pinnate with 9–17 leaflets and the midrib terminating with a leaflet. Leaves are about 5 cm in length and individual leaflets may be up to 2 × 1.5 cm. The leaflets are ovate, elliptical or obovate with a serrated margin. The flowers, which are axillary, are usually solitary, borne on peduncles 2–4 cm in length. They are quite small, about 1 cm, and may be white, green, pink or blue. Pods are swollen and oblong up to 3 × 2 cm and contain only one or two seeds, although at times they may be sterile. Seeds vary considerably in both colour and shape from white through yellow, red-brown and black and from smooth to very wrinkled and rough. The 1000 seed weight varies from 170 to 270 g. In India the crop is divided into two types of seed and colour. Coloured seeded varieties which tend to be smaller seeded and have higher yields are known as *Desi*, while the white larger seeded lines which are more commonly grown in the Mediterranean region are known as *Kabuli*.

Cultivation and uses

The crop is tolerant of cool weather and in the Mediterranean region it is therefore grown as a winter annual. In India it is grown in the cool season after the harvest of the major cereal crop of the year, often on residual soil moisture. In the latter situation it may be broadcast into very roughly worked land at a sowing rate of about 35 kg ha^{-1}. It is best suited to light loams but can be grown on clays provided they are well drained. The crop takes from four to six months to mature and, although seed yields of up to 1700 kg ha^{-1} can be obtained, more common yields in India are about 650 kg ha^{-1}.

Chick pea seed is about 20% protein, 5% oil and 60% carbohydrate. Whole or ground seed is used in a variety of ways in Mediterranean cultures. In India it is milled to produce dhal. By milling the testa is removed from the rest of the seed. This may be further milled to produce a flour which is used extensively in the production of confectionery in India in conjunction with ghee and sugar. The exudation from the glandular hairs on the plant is high in malic acid (94%) which is collected in solution with dew by spreading a cloth over the crop at night. The collected exudate has a reputation in Indian folk medicine as being good for digestive upsets and sunstroke. Crop residues are fed to livestock. Although little research attention has been paid to the crop for many years, *Desi* varieties are now under intensive study at the ICRISAT in Hyderabad, India, because of its importance as a cool-season legume in the semi-arid tropics. Larger seeded *Kabuli* varieties are being studied at the International Centre for Research in the Dry Areas (ICARDA) at Allepo, in Syria.

11.20 *GLYCINE*

Only one member of this genus, *Glycine max*, or the soya bean is an important agricultural crop. Another species, *Glycine wightii* originating from Africa, is a minor forage legume in the tropics. The soya bean is the world's major oil seed and is the source of the majority of the protein used to raise the poultry and pigs of North America, Europe and Japan.

Soya bean (*Glycine max*)

The soya bean is considered to have originated in north China or Manchuria. It has also been claimed as the crop with the longest recorded written history. Archaeological evidence from China suggests that it was domesticated during the Chou Dynasty and was carried throughout Asia during the Chang Dynasty. In Asia the soya bean was, and still is, used in the human diet in a variety of forms which include pre-germinated, extracted and precipitated protein curd from ground seed, and fermented with various fungal organisms. Until 1930 China was the major producer of the crop. Although it was known in the United States, its exploitation as a major crop in that country dates from 1928. America now grows more than 26 millions hectares of soya beans and sales of the beans and their products account for about 5% of total United States export earnings. The crop is now a major industrial crop and is grown for its oil, which comprises 18–20% of the seed, and its protein which is of high nutritional quality and is about 36% of the whole seed and about 48% of defatted soya bean meal.

The soya bean plant

Soya beans are generally erect plants varying in height at maturity from 20 to 190 cm. The entire plant is usually covered in short brown or grey hairs. Germination is epigeal, the plant forms a stout tap root from which lateral roots arise and roots may penetrate the soil to 150 cm. Axillary buds on the main stem may develop into branches but this depends on variety and on plant population. Soya bean leaves are alternate and trifoliate with ovate to lanceolate leaflets (Fig. 11.8). Stipules are small and pointed. Inflorescences are borne on axillary racemes which contain from three to fifteen flowers, but determinate types bear terminal racemes which contain more flowers than the laterals. The flowers are small, about 5 mm and are usually white or lilac in colour. The crop is usually self-pollinated, following which small, hairy pods are formed, each usually containing two or three near spherical seeds. Soya bean seed is quite small compared with most other grain legumes, with a 1000 seed weight of 50–400 g. The seed testa is cream, yellow, green-brown or black in colour.

Cultivation and uses

Although soya beans are subtropical they have spread as far north as Canada, and some cultivars have been bred in Sweden. They require warm moist weather during the growing season. The crop is strongly responsive to photoperiod and is a short-day plant. In the United States it is divided

Fig. 11.8. Soya bean (*Glycine max*) showing trifoliate leaves and pods.

into maturity groups ranging from 00, 0 and I in the north to groups IX and X in the south. The plant is so responsive to day length that a change of 160 km in latitude may require the growing of another cultivar.

For maximum yields 500–750 mm of water is required during the growing season. The crop also requires high light intensity for good growth, while optimum growth temperature is 27–32 °C. Yield can be reduced both by low night temperature or by brief periods of high temperature. Most suitable soil types are clays with a pH of 6–6.5. In America soya beans are frequently sown in rows 45–100 cm apart. In other parts of the world narrower rows are used to improve competition with weeds. Average seed yields in the USA have risen from 740 kg ha^{-1} in 1924 to 1600 kg ha^{-1} currently. Soya beans are usually grown in a rotation with maize.

It is in the variety of end products produced from the crop that it has gained its current importance in agriculture. The rapid increase in the area sown during the Second World War coincided with a swing in eating habits in the United States from butter to margarine, and on a world basis the crop now provides 15% of all edible oil consumed and ranks ahead of butter. Once oil is extracted from the seed, the residual soya bean cake is rich in protein and is used extensively in the feeding of all forms of domestic livestock. Although raw soya beans contain trypsin inhibitors, the oil extraction process usually heats the beans sufficiently to render the inhibitors harmless. Whole seed must however be heat-treated before it is safe for consumption by non-ruminants.

Among recent developments has been the production of meat analogues from soya bean protein. The protein from the beans is taken into solution and then precipitated out into a variety of shapes. The precipitated protein can then be flavoured to simulate pork, beef or chicken. The product, which is sold in dry form, has an almost indefinite shelf life and contains no fat, bone or gristle. Milk and cream substitute are also produced from soya beans and most dairy whiteners that are now sold have nothing to do with cows. Other food uses are incorporation into bread, breakfast cereals and meat products.

Soya bean oil is used in industry for the production of paint, oilcloth, printer's ink, linoleum and soap. Phospholipids which are a by-product of the oil industry are used as wetting and stabilising agents in a number of industries. Industrially extracted meal is utilised in the production of synthetic fibres, glue, fabric finishes, water-proofing and as a foam generating compound for fire extinguishers.

Glycine wightii

This is a herbaceous perennial with trailing and climbing stems which frequently root at the nodes and may form crowns below soil level. The leaves are trifoliate and hairy on both surfaces. The axillary inflorescences contain many flowers which are white in colour except for some violet markings on the standard petal. The pods are brown and contain four to five seeds. *Glycine wightii* has some cold tolerance but is best adapted to elevated areas with a tropical climate and a summer rainfall of between 760 and 1500 mm. It is well regarded as a nutritious and reasonably palatable pasture legume which combines well with such grasses as setaria or green panic.

11.21 *LENS*

Like many of the other crops reviewed in this section, the lentil (*Lens culinaris*) is of ancient origin and is thought to be from the Mediterranean basin or west Asia. Although a minor component in western diets it is an important pulse crop in north India and Pakistan where, because of its cold tolerance, it can be grown at elevations up to 3500 m.

The lentil plant

The lentil is an erect, highly branched annual with slender stems which are light green in colour, covered in fine hairs, and reach a height of about 40 cm. The leaves are pinnate with four to seven pairs of leaflets and usually terminate in a tendril. The leaflets which may be opposite or alternate are about 1.3 cm long and may be obovate to lanceolate. The small (about 7 mm) flowers are borne on a slender axillary peduncle in groups of one to four. Flower colour is blue, white or pink. The pods when formed are also small and at about 1.3 cm are not much longer than they are broad, usually containing only two seeds. The seeds are convex on both sides, small and lens shaped, greenish brown to red-brown in colour with a 1000 seed weight of only about 20 g.

As with the chick pea, the species can be divided into two major groups on the basis of seed size and shape. Large seeded cultivars have large flat pods with flattened seeds 6–9 mm in diameter, with yellow to orange cotyledons and large white or rarely blue flowers. This group is normally grown in the Mediterranean region, Africa and Asia Minor. Small seeded cultivars, on the other hand, have seeds which are 3–6 mm in diameter with small convex pods, convex cotyledons and small flowers which are violet blue or pink. These types are found growing mainly in south-west or western Asia.

Cultivation and uses

In India, where the majority of the crop is cultivated, it may be grown either mixed with rice or as a pure stand. In pure stands it is either broadcast or sown in rows 20 cm apart at sowing rates of 35–90 kg ha^{-1}. The crop is tolerant of a wide range of soil types. It is grown in the cool season in the tropics or at high altitudes and as a winter annual in temperate regions. In the United States it is drilled in rows 1 m apart at sowing rates of 13–16 kg ha^{-1}. However, lentil growers in New Zealand sow the crop at 15 cm row spacings at up to 70 kg ha^{-1}. The crop matures in about 14 weeks and the common yield for a dryland crop is about 500 kg ha^{-1}, although under favourable conditions the yield can be as high as 4000 kg ha^{-1}. On soils of good water holding capacity, lentils tend to be unresponsive to irrigation.

Lentils are about 25% protein, 56% carbohydrate and contain very little fibre or oil. The majority of the crop is hulled to produce dhal or split peas, and the seed is a common ingredient in soups. It is also ground into a flour and mixed with cereals to produce cakes, invalid and infant food in Asia.

In Asia, crop residues are an important source of stock feed, and at some times of the year the straw can be almost as valuable as the seed. Besides research work on the crop in the USA and Canada, they are being bred and investigated at the International Centre for Research in the Dry Areas (ICARDA) at Allepo, in Syria.

11.22 *LUPINUS*

There are 1200 to 1500 species in the genus *Lupinus* which has two centres of diversity. From around the Mediterranean basin come most of the annual species of lupins which are now cultivated as crop plants. From the west coast of North America and non-tropical South America come the vast majority of the remainder of lupin species. Only one of these *Lupinus mutabilis* is of major current agricultural interest. Lupins are adapted to sandy, free-draining soils and are often seen growing in disturbed areas. Although some varieties show a degree of frost tolerance they are usually grown as summer annuals in very cold climates.

All wild lupins contain high levels of toxic lupin alkaloids in their foliage and seeds. The discovery of alkaloid-free mutants by German scientists in the late 1920s led to the domestication of a number of species as modern crop plants. Lupins are usually grown either as forage for ruminant animals or for their seed which in all cultivated species has a protein concentration which is as high or higher than that found in soya beans.

General morphology

Cultivated lupins grown for their seed are usually erect annuals which reach a height of 1.0–1.6 m. In the wild they are usually strongly branched. Leaves are palmate with five to nine leaflets which vary from being linear to obovate in shape (Fig. 11.9). Among wild forms the predominant flower colour is blue, and initial flowers are borne on showy terminal racemes on the main stem. If conditions are favourable buds in leaf axils below the terminal raceme elongate and flower. This sequence can repeat itself as long as growing conditions are good, and on any plant there may be a range from newly opened flowers to mature pods. At low plant populations basal

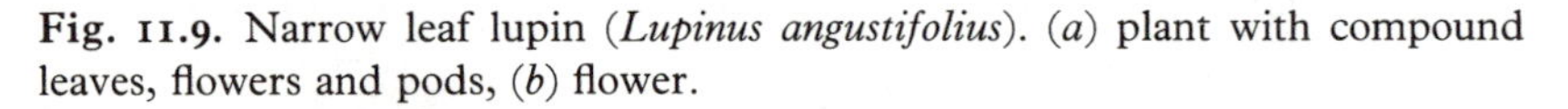

Fig. 11.9. Narrow leaf lupin (*Lupinus angustifolius*). (*a*) plant with compound leaves, flowers and pods, (*b*) flower.

stem branches may also flower. At maturity pods of wild lupins dehisce with explosive force ensuring wide seed dispersal. Many wild lupins are hard-seeded.

Narrow leaf or blue lupin (*Lupinus angustifolius*)

The narrow leaf or blue lupin is of Mediterranean origin. Although it has been cultivated for many years in Germany and in South Africa, it is only since the breeding of non-shattering cultivars of this species in Australia that it has become important as a crop. In Western Australia it is grown extensively for its seed which has a protein concentration of 30–35%. The 1000 seed weight is 150–200 g. Modern alkaloid-free cultivars have been marked with white flowers and buff seed coats to prevent contamination with bitter lines which usually have blue flowers and grey mottled seed. Harvested seed is utilised as a protein source in the production of rations for the pig, poultry and pet food industry. Seed is also used as a protein concentrate to supplement rations for sheep. The crop can be grown as a winter annual in Mediterranean environments or as a summer annual in temperate regions.

White lupin (*Lupinus albus*)

Like *Lupinus angustifolius*, *L. albus*, the white lupin, is from the Mediterranean region where it was known to both the ancient Greeks and the Romans. Before the breeding of alkaloid-free lines, the seed was rendered safe to eat by humans by boiling and soaking in numerous changes of water to elute the alkaloids. Such seeds are still sold by street vendors in Rome today. *Lupinus albus* seeds are considerably larger and flatter than those of *L. angustifolius*, and in some cultivars the mean seed weight is more than 400 mg. The cultural requirements for the crop are similar to those of *L. angustifolius*, although *L. albus* does tend to need more fertile soils than the latter species.

As yet there has been no incorporation of genetic markers to detect outcrossing in this species. Although predominantly selfed, the level of outcrossing may be as high as 9%, and there have been some problems in animal feeding trials using *L. albus* seed caused by the presence of bitter seed in varieties which were supposedly sweet. Since the oil content of the seed is relatively high at 11–18%, and because of its high protein content of up to 45%, there is considerable current interest in Europe in converting *L. albus* into a substitute for the soya bean as the oil is of similar quality.

Lupinus cosentinii

Although this species is of Mediterranean origin, it is only in Australia where it is of any commercial significance. Following its introduction in Western Australia more than 100 years ago, it has become naturalised on the deep sands of the Swan Coastal Plain. The species is used to feed sheep and to build up fertility in a rotation with wheat and other cereals. Because it is hard-seeded, once established there is no need to re-sow the crop. The crop regenerates each season with the onset of autumn rains. Australian plant breeders have produced alkaloid-free lines which are non-shattering and have a range of flowering times. However, as yet no soft-seeded lines have been developed, and so its use as a crop is limited notwithstanding a high seed protein concentration.

Lupinus luteus

Lupinus luteus is a somewhat fleshy yellow-flowered annual lupin species from the Iberian peninsula. It was one of the first lupin species in which alkaloid-free mutants were identified and it has seed with a protein concentration up to 50%. Because of the high seed-protein concentration, there has been a lot of interest both in its agronomy and nutritional quality. Although feeding trials with a wide range of non-ruminant animals have given good results, as yet seed yields are considerably below those obtained from *L. albus* or *L. angustifolius* in the same environment. It is a popular crop in East Germany and Poland.

Lupinus mutabilis

This species was known as the pearl lupin. The common name of the species is now *tarwi* which is its name in the Andean highlands from which it originated. It is an annual species. Like *L. albus* it has a long history of use and was cultivated in the Andean highlands prior to the arrival of the Spanish. Lines with a seed oil concentration of up to 24% combined with a seed protein concentration of 50% have been identified in screening programmes. Research workers in Chile have produced alkaloid-free genotypes of this species which is extremely variable in regard to flower colour, seed colour, seed size and plant height. Fortunately the species is already both soft-seeded and non-shattering.

11.23 *PHASEOLUS*

The taxonomy of the genus *Phaseolus* has recently been revised and as in the pasture legumes, a number of species that were formerly considered to be *Phaseolus* have been assigned to other genera. Among grain legumes there are therefore now only three species that will be considered in detail, all of South or Central American origin and grown for either their dry seed, their immature and fleshy pods or both. The species are *P. coccineus*, the scarlet runner bean, *P. lunatus* the lima bean and *P. vulgaris* the common bean. One other species *P. angularis*, the adzuki bean, is also grown in Japan where its seeds feature in religious ceremonies but it is also used as a food and a forage. The species has similar climatic requirements to the soya bean.

Scarlet runner bean (*P. coccineus*)

The scarlet runner bean is a twining perennial which grows to a height of 4 m. It comes from tropical highland regions and is therefore suited for growing as an annual in cool-temperate climates where it is usually consumed in the form of its tender young green pods. In Central America where it originated, green and dry seed are consumed as are fleshy tubers that form under the soil. The plant forms trifoliate leaves with large ovate leaflets up to 12.5 cm long. Flowers up to 2.5 cm in length are borne on showy peduncles and are either scarlet or white. Pods are also large up to 30 cm and the seeds which are kidney shaped are large with a purple and red coloured testa. In temperate environments the scarlet runner bean is cultivated almost entirely as a garden vegetable.

Lima bean (*P. lunatus*)

Lima beans which are also of Central American origin, vary from small annual bushy types to tall climbing perennials. There is considerable variation within the species in regard to seed colour, shape and growth habit. Twining types which are perennial may reach a height of 4 m, while bush types may be as short as 30 cm. There is also variation in leaf size depending on line, the leaves are trifoliate with acuminate leaflets, and the two lateral leaflets are inserted obliquely into the petiole. The undersides of the leaves are hairy. The inflorescence is a long axillary raceme which bears numerous flowers with usually two to four flowers forming at each node of the raceme. Petal colour is white with a green or violet standard. Pods are up to 12×2.5 cm and contain from two to four seeds, whose colour, size and shape are extremely variable. Testa colours include white, cream, red, purple, brown, and black, and the 1000 seed weight ranges from 450 to 2000 g.

Phaseolus lunatus is quite demanding in its climatic requirements. Seed does not germinate at soil temperatures below 15 °C and seed set is reduced by temperatures above 27 °C. The species is sometimes divided into lima beans, which are perennial with large white seeds, and sieva beans which are annuals with smaller coloured seeds. The latter group are considered to be more tolerant of hot arid conditions than the former. Although fairly resistant to drought these plants require a well drained, well aerated soil with a pH of 6–7. The crop is usually sown at a depth of 2.5–5 cm with a sowing rate of 60–170 kg ha^{-1} depending on seed size. Depending on the variety sown and whether green beans or dry seed are to be harvested, the crop may take from just over three to nine months to mature. Yields of dry seed can be as high as 1.7 t ha^{-1}.

The main constituent of the seed is carbohydrate, except that the seed contains about 20% protein and is therefore an important source of this nutrient in countries where the bean is consumed as a staple. Uncooked lima beans when wet or chewed may release considerable quantities of hydrocyanic acid. The concentration of hydrocyanic acid in the seed increases as seed size decreases and coloured seeds contain higher concentrations than white seed. Cooking the beans destroys the enzyme that is responsible for the liberation of the acid. As a crop plant the species appears to be most important in the United States where it is grown for freezing, canning and the production of dry beans.

Common bean (*Phaseolus vulgaris*)

In western society *Phaseolus vulgaris* is mainly eaten as a green vegetable, immature pods being harvested and cooked. Another common use of the species is as the base material of the ubiquitous canned baked bean which is produced from the dried seed of the crop. In Latin America and in Africa the species is a major dietary staple and it provides an important source of protein for the people of those areas.

Carbon dating of a pod valve of *P. vulgaris* found in the Coxcatlan cave in Mexico indicates that the crop was domesticated by about 5000 BC which was at about the same time as maize. It is thought that multiple domestication may have occurred from a variable and widespread ancestral species in middle America. The species, in common with other new world crops, was carried to Europe during the sixteenth century by the Spanish and the Portuguese, and from there it has been spread to the rest of the world.

The bean plant

The species is highly variable and plants can range from twining cultivars with stems of 20 to 30 nodes up to 3 m in length to determinate bush varieties with only four to eight nodes which are little more than 40 cm in height. Intermediate types between these two extremes also occur. Following germination, which is epigeal, the plant develops a tap root which may penetrate to 1 m. An extensive lateral root system, which is mainly confined to the top 15 cm of soil, is also formed. The leaves are alternate and trifoliate (Fig. 11.10), although the first leaves formed after emergence of the cotyledons are simple. They are borne on long petioles and may be covered in fine hairs. The leaflets are ovate and entire with pointed tips, and the two lateral leaflets are asymmetrical. The individual leaflets can be up to 15 × 10 cm in size.

Fig. 11.10. Common bean (*Phaseolus vulgaris*) showing trifoliate leaves and flowers.

Flowers are borne on short axillary racemes and form near the apex of the peduncle on short pedicels. Flower colour can be white, cream, pink or violet with the standard about 1 cm in diameter. Pods when formed are long and slender up to 20 × 1.5 cm, usually containing four to six seeds but on occasion as many as 12. Pods may range from circular to oval in section. Varieties which have been selected for eating as a green vegetable have extensive parenchyma deposition in pod walls. Pod colour is usually green, but in wax beans it is yellow and in some lines it is purple. Seeds are also extremely variable in colour and many of the descriptive common names for the species are derived from the appearance of the seeds. They may be white, yellow, green, buff, pink, red, purple or black. These colours may be uniform or may be mottled, blotched or striped or with a distinctive coloured eye formed around the hilum, e.g. blackeyed susans. They also vary in shape and in size, with a 1000 seed weight of 200 to 600 g.

Cultivation and uses

Because *P. vulgaris* is a major staple of peasant cultivators in South and Central America and because of the variability in plant type, cultural methods for the crop vary considerably. In the western world indeterminate cultivars, which have to be grown on supports, are now normally only cultivated by home gardeners so as to spread the crop over a period of time. For machine harvesting greater uniformity of maturity is required, and this is achieved by sowing determinate varieties at high plant populations, so that flowering and pod set are compressed into a short period of time. Beans for freezing or canning are then harvested by a once-over machine harvest. For production of dry bean seed the crop is usually cut or pulled, once pods commence to yellow and it is left in wind-rows to further dry in the field before threshing. Threshing bean seed is a skilled job as any soil contamination will cause the seed to be downgraded and sold at a lower price. The testa in *P. vulgaris* is relatively thin, and seeds are easily damaged by threshing and excessive handling. Such damage can cause a considerable reduction in germination of the seed if it is to be used for sowing.

Phaseolus vulgaris is frequently reported in the literature as responding to application of nitrogen fertilizer. It is possible that the facility to fix nitrogen may have been selected against in the production of determinate bush types, as wild indeterminate types in South America nodulate freely and fix large amounts of nitrogen. The crop is frost tender and, in common with most other grain legumes, requires adequate moisture at flowering and pod set. It may therefore respond to irrigation at these times. It is tolerant of a wide range of soil types. Average dry seed yield of beans in the United States is 1.5 $t\,ha^{-1}$ but Belgian farmers obtain nearly double that amount at 2.8 $t\,ha^{-1}$.

The seeds of *P. vulgaris* contain about 22% protein. Very little fibre or oil is present, and the majority of the rest of the seed is carbohydrate. In common with many other legumes the species contains toxic factors, in this case haemagglutinins which destroy red blood cells in animals that eat raw seed. However, as in the soya bean and the lima bean, these toxic factors are destroyed by heat and the nutritional quality of the protein in the seed is then similar to that of other grain legumes.

Besides the immature pods which are commonly called French, snap or wax beans, the dried seed of *P. vulgaris* is sold under such names as Dutch red, red kidney, haricot, navy, flageolet, pinto and yellow eye beans. Although the majority of *P. vulgaris* which is grown in the world is in South and Central America, the species is now grown for freezing and canning throughout the temperate region, and the major exporter of dried bean seed is the United States.

11.24 *PISUM*

The common pea (*Pisum sativum*) is probably one of the best known grain legumes in temperate climates because of its universal popularity as a fresh vegetable. The origin of the pea is not certain but it would seem that it first entered cultivation around the Mediterranean from where it was carried to the rest of the world. Extensive areas of peas are now grown in temperate countries for processing, and it is the leading frozen vegetable in the United States.

The pea plant

Pea seeds are spherical or wrinkled. When planted, germination is hypogeal and following emergence the plant produces a lax hollow stem which cannot usually climb without support. Height attained depends on cultivar, and garden types which are usually supported may be up to 2 m in height while field grown peas may be considerably shorter. Peas have pinnate leaves with one to three pairs of leaflets which are ovate, with the leaf axis terminating in tendrils which aid in climbing (Fig. 11.11). Pea stipules are large and leaf-like. Flowers are borne in their axils and are solitary or in short racemes. The flower colour is white in cultivars grown for consumption as a vegetable, usually in conjunction with a white or green seed coat, or pink to purple with buff seed coats in the hardier field cultivars grown for dry seed production. Pea pods are green and fleshy when the plant is young, and in some cultivars the whole pod is eaten as in French beans. At maturity the pod is papery and contains from two to ten seeds, with a 1000 grain weight of 150–200 g. The whole plant is devoid of hairs and has a bluish waxy appearance.

Cultivation and uses

Cultivation of peas will often vary with the end use. In the home garden where picking is carried out by hand, tall cultivars flower over a long period of time and thus extend the yielding period. For production of processed peas and dry seed uniformity of maturity is important. Cultivars are therefore selected to flower over a short duration and to form two or more pods at each node. For production of frozen or canned peas where tenderness is an important determinant of quality, irrigation is almost essential. The crop is very sensitive to moisture stress at both flowering and pod filling. Although peas do not grow well in very hot environments, they are frost-sensitive and are therefore usually spring sown in temperate countries. The crop requires a well-drained, well-structured soil and is intolerant of waterlogging. In a recent development in England leafless

Fig. 11.11. Pea (*Pisum sativum*). (*a*) plant with compound leaves, tendrils, and pods, (*b*) flower.

peas have been produced or, more strictly speaking, leaflet-less peas, as the leaflets have been replaced by tendrils. These are supposed to be easier to harvest than conventional peas because the intermeshing of the tendrils keeps the crop upright rather than falling on the ground as is the case with normal pea crops. Light interception by the crop may also be improved as a result.

Dry pea seed is about 22% crude protein, which is low compared with grain legumes such as lupins or soya beans but similar to that found in

Phaseolus genotypes and in *Vicia faba*. Although peas are commonly grown for dry seed and are utilised as stock feed, especially now in the EEC, their major utilisation is in human diet. As such they are used fresh, frozen, freeze-dried, or dried. The advent of freezing to which they are eminently suitable has increased the demand for the crop which can now be consumed throughout the whole year. The growing and processing of cultivated peas is a highly mechanised operation calling for considerable skill both by the grower and the processor.

11.25 *PSOPHOCARPUS*

The winged bean (*Psophocarpus tetragonolobus*) is a lowland tropical species which has been attracting considerable scientific attention recently because this one plant has the potential to produce green forage, a green vegetable, a high-protein oil seed and a high-protein tuber. The origin of the crop is not certain but it is currently cultivated in a belt which extends from Sri Lanka across south-east Asia, through Indonesia to Papua New Guinea.

The winged bean plant

The plant is a twining herbaceous perennial which can reach a height of 4 m given suitable support. The root system is shallow with numerous long laterals which run horizontally near the surface of the soil and it is the roots which become enlarged to form tubers. The leaves are trifoliate with large leaflets, up to 15 × 12 cm, which are ovate with acute tips. The inflorescence is an axillary raceme bearing two to ten flowers which can be up to 4 cm across, blue, white or purple in colour. The pods are large up to 30 × 3.5 cm, square in section, and conspicuously winged on their margins from which the species obtains its common name. Embedded in the flesh of the pod are the seeds and each pod may contain from eight to seventeen of them. The seeds are globular in shape, white, yellow, brown or black in colour and have a shiny skin. Their size is quite large with a 1000 seed weight of about 300 g.

Cultivation and uses

The winged bean is well adapted to the humid lowland tropics. It can however be grown in the tropics at elevations to 2000 m and, provided it is irrigated, in more arid tropical environments. It is a short-day plant and does not usually flower if grown at higher latitudes. It is normally sown at the start of the rainy season and must be provided with supports if good seed yields are to be obtained. Experiments have shown that unsupported plants only give half the seed yield but for tuber production supports are apparently not essential. Winged bean may be grown alone or intercropped with sweet potato, sugar cane, taro or bananas. In Indonesia it is sometimes grown on the dry raised areas between rice paddies. Initial growth is slow and weeds must be controlled during the early growth of the crop.

Flowering can commence 50 days from sowing in the lowlands but may take from 90 to 120 days in the highlands. First pods are ready for picking as a green vegetable about 70 days after sowing. Pods are suitable for eating in this way about two weeks after pollination and about a further six weeks are required before a pod is mature. As the plant is a perennial, it will continue to bear pods indefinitely but yield declines over time and most farmers treat it as an annual. The time required to grow a tuber crop is four to eight months from sowing, by which time the tubers are 2–4 cm in diameter and up to 12 cm long. There has been little published on the yield of the crop when cultivated by farmers but experimental yields of seed of 2400 kg ha^{-1} have been obtained. Reports of tuber yields are also not extensive; however, it appears that between 1.8 and 4.0 t ha^{-1} of fresh tubers are produced.

The crop is utilised as immature green pods which are either eaten raw or cooked, as mature seed or as tuber. Green pods are 80–90% water, about 2.5% protein, 3.5% carbohydrate and contain small amounts of fat and fibre. The composition of the mature seed, however, is much closer to that of the soya bean and comprises about 33% protein, 18% oil, 30% carbohydrate and 8% fibre. The mature seed contains both trypsin inhibitor and haemagglutinin. As the seed is usually cooked before eating, these are destroyed by the heat of cooking. The quality of both the protein and the oil are also apparently similar to that of the soya bean. Winged bean tubers contain about 13.5% protein on a wet weight basis and thus yield considerably more protein than any of the other tuberous root crops that are normally grown in the lowland tropics. The tubers are eaten and prepared in a manner similar to that for the potato. After harvest the residual forage from the crop can be fed to livestock.

11.26 *VICIA*

There are two cultivated species in this genus that are utilised agriculturally, *Vicia faba* which has a variety of common names and is known as the broad, tick, horse or faba bean, and *V. sativa* or common vetch. The latter species is of minor importance as a forage crop in temperate climates. It was referred to in the Bible as a crop weed, tares, which is its other common name.

Broad bean (*Vicia faba*)

This species is a cool tolerant annual crop which is the major grain legume of northern Europe. It is thought to have a Mediterranean origin but by the Bronze Age it was widely cultivated in Europe. It has been divided into four types depending on seed size: with a seed weight of less than 300 mg – *paucijuga*; between 310 and 400 mg – *minor*; with a seed weight of 600–1000 mg – *equina* and exceeding 1700 mg – *major*. In spite of this diversity of phenotypes, all forms appear to intercross freely and therefore appear to represent groups of cultivars which have been selected on the basis of seed size. The crop was well known to the Greeks, and Pythagoras enjoined his followers not to eat it. This may have had something to do with the disease favism which is almost entirely confined to males of eastern Mediterranean origin, and may be caused by either eating or inhaling the pollen of *V. faba*. It causes anaemia, haemoglobinuria, jaundice and high fever and can cause death in 24–28 hours especially in young children. It is believed that sufferers have a congenital biochemical defect.

The plant

Vicia faba is an erect annual plant which can grow to a height of 2 m. It appears to be well adapted to soils of high clay content. The stems are square, hollow, and slightly winged, the plant develops a strong tap root down to 1 m with a strongly branched lateral root system and germination is hypogeal. Plants may be sparsely branched and leaves are alternate. The first stem leaves have two ovate leaflets, but higher leaves bear from three to seven pairs of leaflets which are ovate with mucronate tips (Fig. 11.12). Leaves are without tendrils. Inflorescences are formed on short axillary racemes which may contain from one to six flowers. The flowers are large, up to 3.7 cm and, although mainly white, may be flecked with black at about the junction of the wings and the keel. Pod size varies with type, those of broad beans being larger and containing up to nine seeds more than in the field beans. The interior of the pod is filled with a white velvety lining. There is considerable variation in both seed shape and colour. Seeds may be kidney shaped as in broad beans to almost spherical in certain field cultivars. Seed coat colour ranges through white, green, buff, brown, purple and black.

Cultivation and uses

The broad bean is now grown like peas, both as a garden vegetable and as a processed crop, although as the latter it does not enjoy the same popularity.

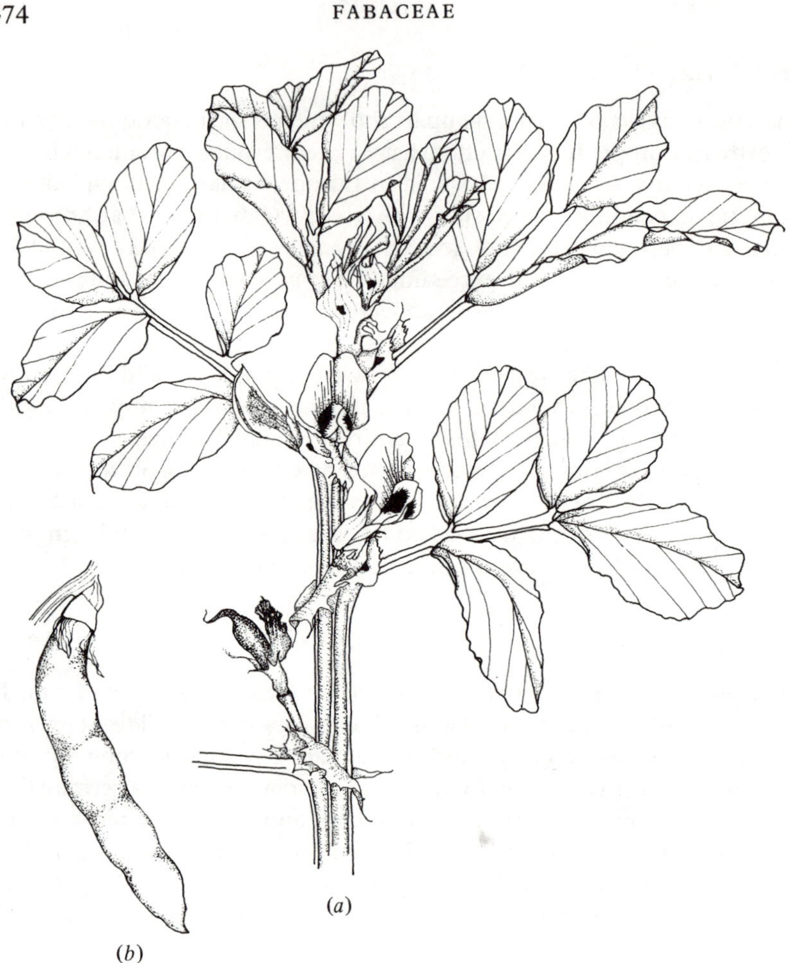

Fig. 11.12. Broad bean (*Vicia faba*). (*a*) plant with compound leaves, flowers and young pods, (*b*) mature pod.

On a broad-acre scale the field bean is cultivated as a source of protein for livestock and human diets in areas where other grain legumes cannot be readily grown. There are two types: winter beans and spring beans. Research work indicates that autumn sown (winter) beans will usually out-yield spring sown crops. In Europe, because of disease problems, the majority of the crop is, however, sown in spring.

To reduce cost of establishment there has been some research aimed at reducing seed size of field-sown cultivars while maintaining yield. A problem with the crop which makes it unpopular with farmers is the wide variation in yield that is obtained from season to season. Average yields in Europe are about 3.5 t ha^{-1}, but experimental yields of more than 7 t have been obtained at high plant populations. There is considerable current interest in the production of determinate cultivars with enhanced pod set at the top of the plant. Also to obtain high yield some degree of outcrossing is required. Because of flower structure, only certain bee species with long tongues usually pollinate the flowers, and it has been suggested that this may limit yield. Plant breeders at ICARDA are therefore currently trying to convert *Vicia faba* into a self-pollinated and self-compatible crop to remove the need for outcrossing.

Apart from its role as a vegetable, *V. faba* is produced as a dry grain crop for use in stock and human feed. The seed has a protein concentration of about 25%, similar to that found in both peas and *Phaseolus vulgaris*. Unlike most grain legumes, the seed appears to be free of major toxic principles. English varieties which have been bred devoid of anthocyanin have been shown to be more digestible to non-ruminant animals than pigmented lines, probably because of a reduced tannin content of the seed. The species is a major staple of diet in the Middle East, particularly in Egypt.

Common vetch (*Vicia sativa*)

Although better known as a common weed of arable land in temperate climates, several cultivars of vetch have been developed as forage plants. This species is an annual with slender, trailing stems bearing pinnate leaves with four to eight narrow leaflets and terminating in a tendril. The stipules are toothed and hairy. The flowers which occur in the axils of leaves are deep purple in colour, and the pods contain eight to ten pale orange to cream seeds. Common vetch is used as a protein-rich forage for cattle and can be sod-seeded into existing grass swards.

11.27 *VIGNA*

The genus *Vigna* now contains two species which are commonly utilised as grain legumes. The cowpea (*Vigna unguiculata*) is an important grain legume of the African lowland tropics, while the mung bean (*Vigna radiata*), also known as black, green or golden gram, is grown and utilised extensively in India and in south-east Asia.

Mung bean: green, golden or black gram (*Vigna radiata*)

Any consumer of Chinese food will be familiar with mung beans (*Vigna radiata*) which provide the source material for sprouted bean shoots that are a common vegetable in many dishes. The crop in its diverse phenotypes is grown extensively in south-east Asia and on the Indian sub-continent where the seeds of black gram are particularly prized by high caste orthodox Hindus.

The mung bean plant

Although formerly divided into two species, *Phaseolus aureus* and *P. mungo*, it is now grouped taxonomically into the single but diverse species, *Vigna radiata*. However two races are recognised, var. *aureus* and var. *mungo*, the former including cultivars which are commonly known as mung beans, golden and green gram, while the latter is commonly called black gram. As could be expected from their former specific ranking, there is considerable morphological variation between the two races even though many features may be common.

Vigna radiata is an erect to sub-erect deep-rooted highly branched annual. The crop grows to a height of 1.5 m and the plants are hairy. The leaves are trifoliate with ovate leaflets which can be as large as 12 × 10 cm and are carried on long petioles. Inflorescences occur in axillary racemes on a long pedicel which may carry from five to 20 flowers depending on race. The predominant flower colour is yellow and flowers are only about 8 mm in size. The pods, which are straight up to 10 × 0.6 cm and can contain up to 15 seeds, are covered in fine hairs and can be buff, grey or dark brown. Seeds are variable in shape, from globular to square in section, and may be green, yellow or black.

The two races can be distinguished because *aureus* lines are longer plants with larger pods which contain more seed but in which the seeds are small, the 1000 seed weight being from 30–80 g. Seed colour in this race is green yellow or blackish. In *mungo* lines, on the other hand, the plants are smaller, the raceme may be branched and bears smaller pods which contain fewer but somewhat larger seeds with a 1000 seed weight of 40–90 g. Seeds are usually black in *mungo* lines. *Aureus* lines are further subdivided, golden gram having yellow seed, a low seed yield and pods that tend to shatter. Green gram has bright green seeds, is more prolific, ripens more evenly and is less prone to shattering. It is green gram seed which is usually utilised in the production of bean sprouts.

Cultivation and uses

Vigna radiata is usually grown at low to medium elevations in the tropics as a rain-fed crop frequently following rice. The plant grows best on good loams with a well distributed rainfall of 750–900 mm per annum. However, it does have some degree of drought tolerance. The species is also able to survive in alkaline and in saline conditions. As with *Cajanus cajan*, (section 11.18), the method of growing the crop varies considerably. It may be sown mixed with rice, intercropped with cereals, sugarcane or cotton, or grown in a pure stand. When sown pure, broadcast crops require 11–17 kg ha^{-1} of seed, while in rows at 25–90 cm apart rates from 5–13 kg ha^{-1} are needed. The crop takes from 80 to 120 days to mature. When the crop flowers under moisture stress a considerable proportion of the seed produced may be hard. Normal yields are only 300–500 kg ha^{-1} but experimental yields of 2700 kg ha^{-1} have been obtained. The species is currently under development at the Asian Vegetable Research and Development Centre (AVRDC) in Taiwan.

Mung bean seed contains about 25% protein, and both the protein and the carbohydrate fractions of the seed are highly digestible and of good nutritional quality. Because of this the crop is highly regarded by Asian consumers and even in the United States more than 11 000 tonnes of mung bean seed are consumed annually. The dried beans, besides being added to soup and mixed with cereals, are ground into a flour and added to noodles, biscuits, bread and snacks. In south India and Sri Lanka ground black gram is fermented with ground rice and utilised in the production of a number of regional dietary staples such as *idli* and *dhosai*. Bean sprouts are produced by soaking whole seed in water overnight and allowing it to germinate in a dark place for a few days. One kilogramme of dry seed is sufficient to produce about 7 kg of sprouts.

Cowpea (*Vigna unguiculata*)

This species is another of the major tropical lowland grain legumes. However, unlike most of the tropical species discussed so far, which have generally been of major importance in Asia, the cowpea is most important in Africa where over 95% of the world's crop is grown - 61% in Nigeria alone. The cowpea is utilised in a variety of ways ranging from the use of young green seedlings as a vegetable through cover crops and forages for livestock to its consumption as a dry bean. The plant is almost certainly of tropical African origin and evidently reached Egypt, Arabia and India at an early date, as there is a written record of it in Sanskrit. It was known to both the ancient Greeks and the Romans, and it was taken to the Americas by the Spanish in the sixteenth century.

The cowpea plant

Like many of the tropical legumes *Vigna unguiculata* is a highly diverse species which at one time was divided into three groups given specific rank based on morphological characters. The three species included *V.*

unguiculata considered to be the most primitive group within the species and, although found in Africa, it is most common in Asia. These are annual plants which are spreading to erect, up to 80 cm in height and have pods up to 12 cm long with seeds up to 6 mm. The next group was *V. sinensis* which has a similar growth habit but larger pods which reach a length of 30 cm with seeds up to 10 mm. These plants are most commonly found growing in Africa. Finally comes the group which probably possesses the longest pods of any of the grain legumes, the former *V. sesquipedalis*. The plants are tall and twining and may reach a height of 4 m and bear pods which are inflated when young and reach a length of 100 cm with seed up to 12 mm long. They are known as snake, asparagus or yard-long beans and are commonly grown in the Far East where they are used as a green vegetable. However, as all these groups can be freely intercrossed they are now regarded as a single species, *V. unguiculata*.

The plants thus range from being erect, sub-erect, prostrate or twining and are all annuals. Following germination which is epigeal, a stout tap root is produced and numerous spreading lateral roots form near the surface of the soil. In twining varieties the stems twine anticlockwise and in all lines the stems may be tinted purple. The leaves are trifoliate and are borne on a stout petiole up to 15 cm long. The leaflets are usually entire, ovate to rhomboidal with acute tips and up to 16 × 11 cm, with the lateral leaflets set obliquely onto the petiole. Flowers are formed in the axils and are borne near the tip of the peduncle which may be 15 cm and usually bears two to four flowers. There is a nectary situated between each pair of flowers on the peduncle. The flowers are up to 2.5 cm across and can be white, yellowish or violet. As indicated above, pod size is extremely variable and can range from 8 × 0.8 cm to 100 × 1 cm, and similarly seed number per pod ranges from eight to 20. Seed shape and colour also are variable and seeds may be kidney-shaped or globular, smooth or wrinkled, and white, green, buff, red, brown or black in various combinations of mottled, speckled, blotched or eyed. The hilum is white and is surrounded by a distinct dark ring. The 1000 seed weight is 100–250 g.

Cultivation and uses

There is some variation in the cultural requirements of the various groups within the species although they are universally a hot weather crop. African lines are more suited to semi-arid conditions than the snake bean types which are more commonly grown in Asia and require a higher rainfall. Although grown on a wide range of soil types in Africa they are not tolerant of waterlogging. There are both short-day and day-neutral lines of cowpea, and the time from sowing to maturity ranges from 60 to 240 days depending on location and variety.

In common with many of the grain legumes cultivated in India the cowpea is frequently grown in combination with other species such as maize, sorghum, millet or cassava. In this situation the seed may be broadcast after the cereal has reached a height of 50 cm at a sowing rate of 22–33 kg ha^{-1}. Following germination the cultivator may adjust the population depending on his perceived assessment of available soil moisture. African experiments have shown that mixed stands may produce more than when the species is grown as a pure stand but this could be partially due to the very high losses of seed by insect predation when grown in pure stands in the tropics. When sown mechanically it is usually established in rows 75–100 cm apart, with seeds 7–10 cm apart within the row at a sowing rate of 17–28 kg ha^{-1}. When sown for forage, seed is broadcast at rates up to 100 kg ha^{-1} often in conjunction with sorghum, sudan grass or maize. Seed yields in Africa are as little as 100–300 kg ha^{-1}, although experimental yields of up to 3000 kg ha^{-1} have been reported, and yields of green snake beans can be as high as 8000 kg ha^{-1}. The crop is currently under improvement at the International Institute of Tropical Agriculture (IITA) at Ibadan in Nigeria.

Dry cowpea seed contains about 25% protein, 57% carbohydrate, 4% fibre and just over 1% fat. The immature pods comprise 86% water and about 3.3% protein and 7.5% carbohydrate. The seed appears to be relatively free of anti-metabolites and flatus producing sugars. The whole plant is utilised in a variety of ways in Africa, seedlings, green leaves, and pods are all consumed as well as the dried seed. In Asia the snake bean is mainly used as a vegetable and in the United States over 18 000 tonnes of the immature green beans are canned or frozen each year.

In Africa the dry seed may be mixed with other vegetables or species to form a starch or gruel which is usually consumed in conjunction with a starchy staple. Seed is also decorticated and ground into a flour and combined with onion and spices to produce fried or steamed bean cakes. An apparent advantage of the cowpea is that the seed does not take as long to cook as dried *Phaseolus* seed, and the species is therefore favoured when firewood for cooking is scarce. Cowpea forage and hay and residues from seed decortication make a high quality source of food for domestic ruminants.

11.28 OTHER LEGUMES

Besides their use in pastures and as grain legumes for food and the production of oil or protein concentrates, there are a few legumes that are used as flavouring agents in food and/or for their medicinal properties.

The first of these is the spice fenugreek (*Trigonella foenum-graecum*). Fenugreek is of southern European and Asian origin. It is a pubescent, highly scented, erect annual and grows to a height of 30 to 60 cm. The leaves which are light grey-green are alternate trifoliate with leaflets 1.5 to 2.5 cm long. Flowers are white and are borne in groups of 1 or 2 in leaf axils on short stems. The plant forms thin pods up to 7.5 cm long which are indehiscent and contain numerous small greenish brown seeds.

Both leaf and seed can be used. The major use of the seed, which may contain up to 9% aromatic oil, is as a component of curry powder and in a number of local medicines. Fenugreek is used extensively in India and in North Africa. The essential oil distilled from fenugreek is used as a flavouring agent in the production of pickles, cheeses, confectionery, and alcoholic beverages.

In the case of liquorice, *Glycyrrhiza glabra*, it is the root which is the commercial product. Liquorice has a long history of use as a medicinal plant and featured in the writings of Hippocrates, Theophrastus and Pliny.

Liquorice is a native of the Mediterranean region and is a perennial, erect branching plant growing to a height of 100 cm. Leaves are pinnate with 9 to 17 ovate leaflets which are 2.5 to 5.0 cm long. The leaves are yellow to green in colour and sticky on their undersides. The flowers which are blue are borne on short spikes. Pods are small (1.2 to 2.5 cm) red to brown in colour and contain three to four seeds.

The commercial product is derived from the root which is brown on the outside and yellow on the inside. Liquorice plants form a tap root which grows down to 120 cm. Stolons arise from the tap root which sends out roots to form a tangled mass. Plants are harvested at the end of the season after tops die off by digging a trench next to the plants and pulling plants into it. Harvested roots are then shade dried prior to marketing.

The major component of the liquorice root is the glycoside glycyrrhizin which is a sweetening agent. However, the root also contains a pharmacologically active terpene.

The best known use of the root is in the production of sweets. However, it is also used in the production of cough medicines and as a flavouring agent in other medicines. It is somewhat surprising therefore that the vast majority of the liquorice used in the United States is as a flavouring agent in tobacco products.

FURTHER READING

Allen, O.N. and Allen, E.K. (1981). *The Leguminosae: a source book of characteristics, uses and nodulation*. Macmillan, London.

Asian Vegetable Research and Development Centre (1988). *Proceedings, 2nd International Mungbean Symposium, Bangkok, November, 1987*.

Aykroyd, W.R. and Doughty, J.C. (1964). *Legumes in human nutrition*. FAO, Rome.

Baker, M.J. and Williams, W.M. (eds) (1987). *White clover*. CAB International, Wallingford.

Bolton, J.L. (1962). *Alfalfa: botany, cultivation, utilization*. Leonard Hill, London.

Bond, D.A., Scarascia-Mugnozza, G.T. and Poulsen, M.H. (eds). (1979). *Some current research on* Vicia faba *in Western Europe*. Commission of the European Communities, Luxembourg.

Caldwell, B.E. (ed.) (1973). *Soybeans, improvement, production, and uses*. American Society of Agronomy, Madison.

Gepts, P. (ed.) (1988). *Genetic resources of* Phaseolus *beans*. Kluwer, Dordrecht.

Gladstones, J.S. (1970). Lupins as a field crop. *Field Crop Abstracts*, **23**, 123–48.

Hanson, C.H. (ed.) (1972). *Alfalfa science and technology*. American Society of Agronomy, Madison.

Harbourne, J.L., Boulter, D. and Turner, B.L. (eds). (1971). *Chemotaxonomy of the Leguminosae*. Academic Press, London.

Hebblethwaite, P.D. (ed.) (1983). *The Faba bean* (Vicia faba *L.*). Butterworth, London.

Hebblethewaite, P.D., Heath, M.C. and Dawkins, T.C.K. (eds). (1985). *The pea crop: a basis for improvement*. Butterworth, London.

Husiman, J., Poel, T.F.B. van der and Liener, I.E. (eds) (1989). *Recent advances of research in antinutritional factors in legume seeds*. Pudoc, Wageningen.

Hutton, E.M. (1970). Tropical legumes. *Advances in Agronomy*, **22**, 2–73.

Jermyn, W.A. and Wratt, G.S. (eds) (1987). Peas: Management for quality. *Agronomy Society of New Zealand Special Publication No.* **6**.

Jones, D.G. and Davies, D.R. (eds) (1983). *Temperate legumes: physiology, genetics and nodulation*. Pitman, Boston.

Knight, W.E. and Hollowell, E.A. (1973). Crimson clover. *Advances in Agronomy*, **25**, 48–76.

Langer, R.H.M. (ed.) (1967). *The lucerne crop*. Reed, Wellington.

Langer, R.H.M. (ed.) (1990). *Pastures: their ecology and management*. Oxford, Auckland.

Maesen, L.J.G. van der and Somaatmadja, S. (eds) (1989) *Plant resources of South-East Asia No 1. Pulses*. Pudoc, Wageningen.

Masefield, G.B. (1973). Psophocarpus tetragonolobus - a crop with a future? *Field Crop Abstracts*, **26**, 157–60.

National Research Council. (1975). *The winged bean: a high protein crop for the tropics*. National Academy Press, Washington.

National Research Council. (1979). *Tropical legumes: resources for the future*. National Academy Press, Washington.

National Research Council. (1984). *Leucaena: promising forage and tree crop for the tropics*. National Academy Press, Washington.

National Research Council. (1989). *Lost crops of the Incas: little known plants of the Andes with promise for worldwide cultivation*. National Academy Press, Washington.

Norman, A.G. (1978). *Soybean, physiology, agronomy and utilization*. Academic Press, New York.

Oram, R.N. (1990). *Register of Australian herbage plant cultivars*. CSIRO, East Melbourne.

Rachie, K.O. and Roberts, L.M. (1974). Grain legumes of the lowland tropics. *Advances in Agronomy*, **26**, 1–132.

Robinson, D.H. (1937). *Leguminous forage plants*. Edward Arnold, London.

Saxena, M.C. and Singh, K.B. (eds) (1987). *The chickpea*. CAB International, Wallingford.

Singh, S.R. and Rachie, K.O. (eds) (1985). *Cowpea research, production and utilization*. John Wiley, Chichester.

Skerman, P.J., Cameron, D.G. and Riveros, F. (1988). *Tropical forage legumes*. FAO, Rome.

Smart, J. (1976). *Tropical pulses*. Longman, London.

Smartt, J. (1990). *Grain legumes: evolution and genetic resources*. Cambridge University Press, Cambridge.

Stace, H.M. and Edye, L.A. (eds) (1984). *The biology and agronomy of* Stylosanthes. Academic Press, Sydney.

Summerfield, R.J. (ed.) (1988). *World crops: cool season food legumes*. Kluwer, Dordrecht.

Summerfield, R.J. and Bunting, A.H. (eds) (1980). *Advances in legume science*. Royal Botanic Gardens, Kew.

Summerfield, R.J., Huxley, P.A. and Steele, W. (1974). Cowpea (*Vigna unguiculata*). *Field Crop Abstracts*, **27**, 301–12.

Summerfield, R.J. and Roberts, E.H. (eds) (1985). *Grain legume crops*. Collins, London.

Sutcliffe, J.T. and Pate, J.S. (ed.) (1977). *The physiology of the garden pea*. Academic Press, London.

Thompson, R. (ed.) (1981). Vicia faba: *Physiology and breeding*. Martinus Nijhoff, The Hague.

Webb, C. & Hawtin, G. (1981). *Lentils*. Commonwealth Agricultural Bureaux, Slough.

Weiss, E.A. (1983). *Oilseed crops*. Longman, London.

Whitcomb, J.R. & Erskine, W. (eds) (1984). *Genetic resources and their exploitation: chickpeas, faba beans and lentils*. Martinus Nijhoff, The Hague.

Whyte, R.O., Nilsson-Leissner, G., Trumble, H.C. (1953). *Legumes in agriculture*. FAO, Rome.

Wilcox, J.R. (1987). *Soybean: improvement, production and uses*. American Society of Agronomy, Madison.

Woodroof, J.G. (1983). *Peanuts: production, processing, products*. Avi Publishing, Westport.

Wynn-Williams, R.B. (ed.) (1982). Lucerne for the 80's. *Agronomy Society of New Zealand Special Publication No. 1*.

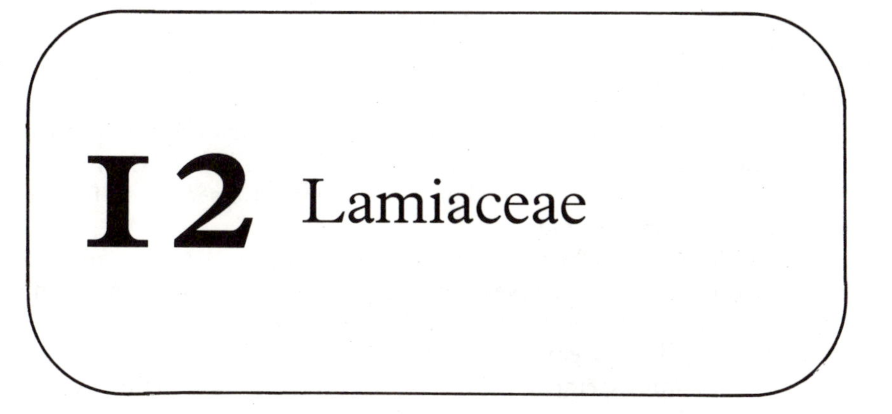

12 Lamiaceae

This family which used to be known as the Labiateae comprises about 180 genera and 3500 species. Although the family has a worldwide distribution, major areas of origin are the Mediterranean region and the Far East. The agricultural importance of the family is related to the fragrant essential oil content of the foliage of many of its members which are used for flavouring of food as herbs, in the pharmaceutical industry, and in the production of perfumes.

Commercial members of the family include basil, lavender, marjoram, the mints, oregano, rosemary, sage, savory and thyme. In general, the commercial product is fresh or dried plant leaf. However, leaves and flowers are also distilled to extract essential oils. There are also a number of temperate weed species which are derived from the family.

Both annual and perennial forms exist. Plant structure varies from herbs to shrubs to trees. Stems are square in section, leaves are opposite or whorled and simple, without stipules. Flowers are axillary or whorled and are usually hermaphroditic. They are zygomorphic with a calyx of five fused sepals and a two-lipped corolla comprising five fused petals. There are two or four stamens. When four stamens are present, the pairs may be of different lengths. The stamens are attached to the petals. The gynoecium is a compound pistil made up of two carpels which appear to comprise four locules. The ovary is superior and the stigma is divided into two at its end. The fruit is usually a nutlet containing a single seed, which generally contains no endosperm. Nutlets are arranged in groups of four.

12.1 *LAVENDULA*

Lavender is the collective common name for a number of species derived from this genus. Most are shrubs and are highly aromatic.

English lavender

In spite of its name, English lavender or *Lavendula angustifolia* is a native of southern Europe and the Mediterranean region. It is extensively cultivated as both an ornamental and as a source of essential oil for the perfume and cosmetic trade.

English lavender is a bushy, branching, perennial shrub which grows to a height of about 1 m. The leaves are opposite, lanceolate, hairy, up to 5 cm in length with a silvery grey appearance. The flowers, which are blue, are carried on terminal racemes up to 20 cm in length.

Lavender requires temperatures of between 7 and 21 °C and thrives best on free-draining calcareous soils in full sunlight. Lavender can be established from seed but is usually established from vegetative material. Initial growth of lavender is slow and it may be some years before it becomes fully productive. Once established, plantings can last up to 30 years. A hectare of lavender can be expected to produce 16 to 22 kg ha^{-1} of lavender oil.

The highest concentration of essential oil is in the lavender flowers. Following harvest by either machine or hand, lavender oil is produced by the steam distillation of lavender flowers harvested while in full bloom. Leaves and flowers are used as a flavouring in salads, desserts, and wine. They are also used as herbal tea and in the flavouring of black tea. Lavender is also used to scent tobacco.

The main use of lavender oil is in the production of perfumes, toilet water and cosmetics. Notwithstanding the Mediterranean origin of the genus, there are at least two old British companies whose long-established reputation depends on their use of the fragrance of lavender in a range of cosmetic and toiletry products. The Yardley family first used lavender in soap during the reign of Charles I prior to the Great Fire of London in 1666, while Potter and Moore, who are associated with Mitcham lavender, first started producing lavender water in 1749.

French lavender

Like English lavender, French lavender (*L. dentata*) is a native of the Mediterranean region and is thought to have originated in Spain. It is somewhat similar in appearance to English lavender but has, as the name implies, leaves with dentate margins. Its agronomic requirements and uses are similar to those of English lavender.

Other lavender species

Other species of lavender which are in commercial production are Spanish lavender (*L. stoechas*), spike lavender (*L. latifolia*) and lavandin (*L. hybrida*), which is a hybrid between English and spike lavender. The latter plant gives a high yield of lower quality lavender oil which is used in cheap perfumes and soaps.

12.2 *MENTHA*

There are probably some 30 species in the genus *Mentha*, the plants of which are commonly known as mints. The mints are all of temperate origin and about half of them are native to North America. Generally they are grown for the aromatic essential oils that are present in their foliage.

Mints are usually erect branching herbs and are generally perennial. They spread by both stolons and rhizomes and, in commercial mint production, plants are usually established from rhizomes. The plants have characteristic square stems which are seldom more than 100 cm in height. Leaves are simple and opposite and may be petiolate or sessile up to 7.5 cm in length. Flowers, which tend to be blue to purple, are borne in clusters in leaf axils or on terminal spikes. (Fig. 12.1).

Mint has been known and used by humans since ancient Greek times and the generic name is associated with Greek mythology. In more recent times, mint sauce is considered to be the ideal accompaniment to roast lamb.

The major commercial species of mint are Japanese mint (*Mentha arvensis*), peppermint (*Mentha* × *piperita*) and spearmint (*M. spicata*). Japanese mint is an erect plant up to 100 cm in height with ovate to elliptical leaves and lilac-coloured flowers. Peppermint is a sterile hybrid between *M. aquatica* and *M. spicata*. There are a number of commercial varieties which vary in their morphology, oil yield and quality. Plants are reproduced by rhizomes and have stems up to 100 cm in length. Peppermint is named after the pungent pepper-like aroma of the oil. Finally, spearmint is an erect plant up to 60 cm in height spreading by stolons.

Mints prefer a climate where the temperature is between 6 and 27 °C with an annual rainfall of more than 300 mm and a soil pH of 4.5 to 8.3. Establishment is from stolons or rhizomes and good weed control is essential. Mint will also probably require both irrigation and nitrogen fertilizer for good production of mint oil. The amount and the quality of oil varies, and oil production reaches a maximum of about 60 kg ha^{-1} in mid- to late summer. Menthol (a major component in quality mint oil) concentration in mint leaf increases with the age of the crop. Mint is

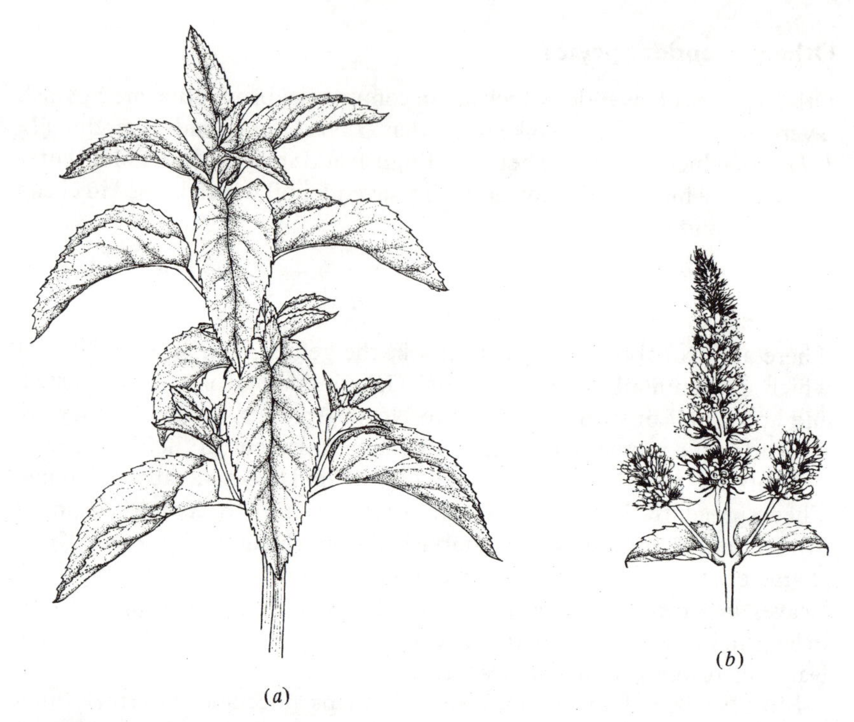

Fig. 12.1. Peppermint (*Mentha × piperita*). (*a*) young plant, (*b*) terminal flowering raceme.

generally cut and left to wilt for a period in the field prior to steam distillation. Properly cared for, a stand of mint can last for many years.

The end use of mint oil depends on the type of plant from which it was extracted. Japanese mint is a major source of menthol which comprises about 60% of the extracted essential oil. The major component of peppermint essential oil is also menthol (approximately 50%). The major component of spearmint oil is carvone at 60%.

Peppermint is an extensively used flavouring agent. Besides its use in chewing gum, sweets, toothpaste, mouthwash, perfumes, cosmetics, and cough remedies, it is used as a flavour mask in pharmaceuticals. Both peppermint and spearmint leaves are put to a variety of culinary uses, being incorporated into herbal teas, jams, salads, sauces and ice cream. Extracted menthol is used in flavouring tobacco products, alcoholic beverages, sweets, perfumes and cough drops.

Other cultivated mints include Bergamont mint, a cultivar of

M. × *piperita* which has a lemon aroma and taste, water mint *M. aquatica*, Scotch mint *M.* × *gentilis* a hybrid between Japanese and spearmint, pineapple mint which is a cultivar of the same hybrid and Corsican mint *M. requienii*.

12.3 *OCIMUM*

Although there are a number of cultivated species from the genus *Ocimum*, only one, *Ocimum basilicum*, or basil, is of economic importance. Like many of the other members of this family basil has a long recorded history. In Italy it has been associated for a long time with romance and, at a more prosaic level, it is the major flavouring component of the sauce pesto which is served with pasta. In India, basil is a sacred herb and is dedicated to the gods Vishnu and Krishna.

Basil, which is of Indian and Asian origin, is a fragrant, highly branched, annual plant which grows to a height of about 60 cm. Leaves are petiolate, ovate, up to 5 cm long, with a purplish tinge. Flowers are white to purple and borne on moderately dense terminal racemes.

Basil can be grown in a temperature range of 7 to 27 °C with a rainfall greater than 600 mm and soil pH in the range 4.3 to 8.2. Basil is frost sensitive and grows best in full sunlight in well-drained soils. It is normally established from seed and, depending on location, is harvested from two to five times in the season. Harvesting commences just prior to the start of flowering which occurs about 10 to 16 weeks after sowing. After harvest, leaves and flowers are dried at low temperature prior to packaging or distillation to obtain essential oil.

There are two types of essential oil derived from basil on the world market. European sweet basil produces the best-quality oil and it is high in methylchavicol and linalool. Oil from Reunion basil, mainly grown in Africa, smells of camphor and, although it contains methylchavicol and some camphor, contains little linalool. Fresh and dried basil leaf is used in sauces, stews, salad dressings, poultry dishes, confectionery and in the French liqueur, Chartreuse. Besides being used to replace basil leaf in a range of food products, the essential oil is used in the production of perfume, soap and shampoo.

Other species of lesser commercial importance include holy basil (*O. sanctum*) from Asia and Australia, *O. canum*, lemon basil (*O. citriodorum*), camphor basil (*O. kilimandscharicum*), *O. gratissimum* a south-east Asian clove scented perennial woody shrub growing to 2 m, and tree basil (*O. suave*) from India and Africa which is a perennial woody shrub growing to a height of 3 m.

12.4 *ORIGANUM*

There are two commercial species in the genus *Origanum*, marjoram (*Origanum majorana*) and oregano (*O. vulgare*). The former was known as 'joy of the mountains' by the ancient Greeks and was associated with wedding ceremonies as it was reputed to be precious to Aphrodite. The latter species was initially noted for its medicinal powers and was recommended by Pliny for the treatment of spider and scorpion bites.

Marjoram

Marjoram (*O. majorana*) is a bushy perennial which grows to a height of 30 cm. Stems are square, hairy and branched. Leaves are opposite and ovate borne on short petioles and are grey-green in colour up to 2.5 cm long. Flowers are white, pink or pale lilac in groups of three to five on short spikes.

Marjoram requires temperatures of 6 to 28 °C, rainfall in excess of 500 mm and a soil pH of 4.9 to 8.7. It is best suited to well-drained fertile loams. Although a perennial, it is usually grown as an annual because of its lack of frost tolerance. Plants which may be harvested two to three times a year are usually harvested at full bloom. Harvested material is dried in the shade to reduce loss of colour and aroma. The main components of the essential oil obtained from marjoram include terpen-4-ol, γ-terpinine and α-terpineol. The fresh and dried marjoram leaf is put to a variety of culinary uses and the essential oil is used in perfumes and cosmetics.

Oregano

Oregano (*O. vulgare*) is an erect, hairy, perennial plant growing to 75 cm. Stems are square. Leaves are petiolate, opposite and ovate up to 4 cm long. Flowers are purple, borne on terminal spikelets up to 3 cm.

Because much oregano is collected from wild stands, it has been estimated that up to 40 different plant species are used in the commercial production of oregano. However, true oregano thrives best in a temperature range of 5 to 28 °C with rainfall in excess of 400 mm and a soil pH of 4.5 to 8.7. Cultivated oregano can be established from either seed or root material on alkaline, light, dry, well-drained soils. Plants are harvested two to six times a year. The essential oil produced from oregano contains predominantly carvacrol and thymol.

Oregano is a frequent component in tomato sauces used in Italian food and is thus familiar to fanciers of both pizza and pasta. The herb is also included in meat products, particularly sausages. The essential oil is used in foods, cosmetics and alcoholic liqueurs.

12.5 *ROSMARINUS*

There is only one commercial species in this genus, *Rosmarinus officinalis*, or rosemary. Rosemary has a long history of use as both a herb and as a medicinal plant. In ancient Greece, students wore garlands of rosemary while studying as it was believed to improve the memory. During the time of the Black Death, the burning of rosemary branches in the home was reputed to keep the plague away.

The plant is an evergreen perennial shrub of Mediterranean origin growing to a height of 180 cm. It has long (3.0 cm), thin, linear, simple leaves which are hairy on their undersides. Flowers which are borne on axillary racemes are pale blue.

Rosemary is not cold tolerant and it grows best in a temperature range of 9 to 28 °C. It has a degree of drought tolerance and, probably reflecting its origin on calcareous soils of the Mediterranean basin, can tolerate soil pH as high as 8.7. Major producers are located around the Mediterranean.

Depending on location, rosemary foliage is harvested once or twice a year. Following harvest, the leaves are dried in the shade. For production of essential oil, rosemary flowers, leaves, and stems are either steam distilled or extracted with organic solvents. In cooking, dried rosemary leaf is used in a variety of, usually meat-based, dishes. The essential oil is used in the production of perfumes, soaps, shampoo and deodorants.

12.6 *SALVIA*

The genus *Salvia* contains over 700 species which are widely distributed in both temperate and warmer regions. Many are grown as ornamentals but a lesser number are grown as herbs or for medicinal purposes.

The plants are annual, biennial or perennial ranging from herbs to sub-shrubs to shrubs with entire leaves having variable leaf margins. The flowers are of various colours and sizes and are borne in groups of two or more on spikes, panicles or racemes or rarely in leaf axils.

Sage (*Salvia officinalis*)

The major commercial species in the genus is sage (*Salvia officinalis*). Sage is of Mediterranean origin and was reputed in ancient times to be associated with immortality and increased mental capacity.

Sage is a hardy perennial subshrub. Stems are square in section and covered in fine down growing to a height of up to 75 cm. Leaves are petiolate, opposite, oblong, greyish green in colour and velvety to touch, up to 5 cm long. Flowers are borne on axillary racemes containing from four to eight flowers which may have white, pink, purple or blue petals. The calyx is purple tinted.

Sage grows best in the temperature range 5 to 26 °C. It is sensitive to both excessively high temperatures and to cold. Plants are killed by frosts of more than −10 °C. It requires an annual rainfall in excess of 300 mm and a soil pH between 4.2 and 8.3. It grows best in warm, dry regions in fertile clay loams in full sun.

Although plants can be established from seed, for commercial production, stands are usually established from vegetative material of selected clones. Plants last from two to six years and the first harvest can be taken in the first year. Usually two to three harvests are taken each year just before flowering. Harvested leaf material is dried in the shade to retain colour and oil content and quality.

Essential oil is obtained by steam distillation. There is a range of components in sage essential oil, which can be up to 2.5% of total dry matter, including α-thujone, camphor, linalool, 1,8-cineole and others.

Sage leaf is used in sausages, minced meat products, poultry stuffing, fish, honey, salads and stews. Sage is also used as a flavouring and antioxidant in cheese, pickles, processed foods and beverages. The oil is used to extend the keeping quality of both meats and fats and in perfumes, cosmetics and as an insect repellent.

Other *Salvia* spp.

For such a large genus, surprisingly few species are of commercial importance. Greek or wild sage (*S. fruticosa*) is imported, in large quantities, into the United States as a substitute for *S. officinalis*. Spanish sage (*S. lavandulifolia*) is a small shrub of minor importance, while *S. multiorrhiza, S. lyrata, S. tomentosa and S. divinorum* have all been used as herbal medicines with various reputed properties.

12.7 *THYMUS*

Thymus is a relatively small genus containing about 50 species of predominantly Mediterranean origin. The plants are small erect or prostrate, shrubs or subshrubs with a strong mint-like odour. Leaves are small and entire. Flowers are borne in small groups in either axillary or loose terminal clusters.

Thyme (*Thymus vulgaris*)

Common thyme was reputed by the ancient Greeks to be a home of the fairies, and beds of it were grown in their gardens as a matter of routine. On a more practical note, they used to burn it in their houses to drive out stinging insects. Thyme was seen to represent style and elegance to the Greeks, chivalry in the Middle Ages and the republican spirit in France. Like rosemary it was reputed to keep off the plague.

Thyme is an aromatic, short, bushy, branching shrub growing to a height of about 25 cm. Leaves are linear to lanceolate, short, sessile and hairy up to 1.25 cm. Flowers, which are pink, are borne on terminal clusters, which may contain few to many flowers.

Thyme grows in a temperature range of 7 to 25 °C, requires a rainfall above 400 mm and a soil pH of 4.5 to 8.0. It is adapted to dry, calcareous soils with good drainage. Thyme is established from either seed or seedlings. In cold environments, plants may need to be mulched to protect them from frost injury. Thyme is harvested whilst in flower. Because of the low stature of the plants it is difficult to machine harvest. Stands of thyme are replanted every two to three years as plants become woody after that time.

There are three principal varieties of thyme: English, French and German, which differ in leaf shape, colour and most importantly in essential oil composition. The essential oils are obtained by steam distillation of dried plant material which contains 2 to 5% oil. The three types of thyme produce thyme oil which is predominantly (42 to 60%) phenols, mainly thymol; origanum oil which is 63 to 74% phenols, mainly carvacrol; and, finally, lemon thyme oil which contains citral.

Thyme is used in a wide range of food products including cheeses, soups, dressings, meat and fish dishes and sauces. Besides being used in the food industry, the essential oils are used in toothpaste, mouthwashes, cough medicines, cosmetics and perfumes.

Other *Thymus* spp.

Other less important economic *Thymus* species include *T. zygis*, *T. serpyllum*, *T.* × *citriodorous*, *T. capitatus* and *T. spicata*.

12.8 OTHER LAMIACEAE

Other minor commercial species from this family include catnip (*Nepeta cataria*) which is a coarse perennial plant that grows to a height of 30 to 100 cm. It has been known to humans for more than 2000 years, but its current major use is related to its strange attraction to cats. It is used in the production of toys for domestic cats and appears to act a a pheromone.

Lemon balm (*Melissa officinalis*) is a lax, loosely branched perennial plant with a distinct lemon aroma. The plant grows to a height of about 60 cm. It has a reputation for being highly attractive to bees. Lemon balm essential oil tends not to be widely used, as perfume manufacturers can use cheaper alternatives. Lemon balm leaves, fresh and dry, are used for a variety of culinary purposes and in the flavouring of alcoholic beverages.

Finally, savory includes two commercial species: *Satureja hortensis* (summer savory) which is a herbaceous annual of southern European origin and *S. montana* (winter savory) which is a woody perennial native of Europe and North Africa. Savory leaf is put to a variety of culinary uses while essential oil from the plants is used in the production of perfume.

There are also a number of minor weedy species among the Lamiaceae, amongst which are henbit (*Lamium amplexicaule*), red dead nettle (*L. purpureum*) and other species from the genus *Lamium*. *Marrubium vulgare* or horehound as it is commonly known, can be a troublesome woody weed. However, it also has properties as a herb and is used in the production of cough remedies, and as substitute for hops (Section 8.1) in the brewing of beer. Other weedy Lamiaceae include selfheal (*Prunella vulgaris*) and various *Stachys* spp. or staggerweeds.

With the exception of horehound most are lax plants growing to a height of about 60 cm and could not be considered to be major weeds of cultivation.

FURTHER READING

Atal, C.K. and Kapur, B.M. (eds) (1982). *Cultivation and utilization of aromatic plants*. Council of Scientific and Industrial Research, Jammu-Tawi.

Festing, S. (1984). *The story of lavender*. London Borough of Sutton, Sutton.

Kowalchik, C. and Hylton, W.H. (1987). *Rodale's illustrated encyclopedia of herbs*. Rodale Press, Emmaus.

Little, B. (1986). *The complete book of herbs and spices*. Reed, Frenchs Forest.

Margaris, N., Koedam, A. and Vokou, D. (eds) (1982). *Aromatic plants: basic and applied aspects*. Martinus Nijhoff, The Hague.

Sanecki, K.N. (1985). *Fragrant and aromatic plants*. Cassell, London.

Simon, J.E., Chadwick, A.F. and Craker, L.E. (1984). *Herbs: an indexed bibliography 1971–1980*. Elsevier, Amsterdam.

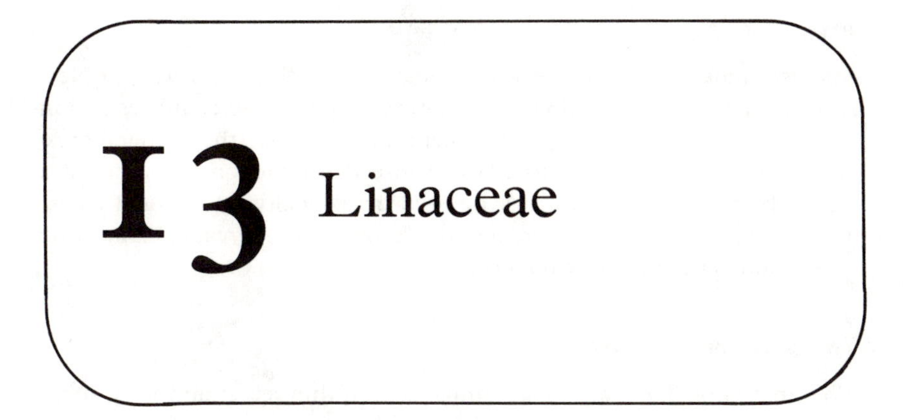

13 Linaceae

The *Linaceae* are a small family comprising only about 150 species, of which only one, *Linum usitatissimum* is of agricultural importance. This crop is grown either for its fibre, linen flax, or for its oily seed, depending on the cultivar used and the cultural and climatic conditions. The taller linen flax is grown in cool, temperate regions, more than one-half of it in the USSR and much of the remainder in Northern Ireland, Belgium and other northern and east European countries. The shorter and more quickly maturing linseed prefers warmer climates and is cultivated in parts of South America and the USA, Canada, India, and the USSR.

13.1 LINSEED AND FLAX (*LINUM USITATISSIMUM*)

The cultivation of linen flax goes back to the dawn of civilisation, as shown by the remains of the prehistoric Swiss lake dwellers. The ancient Egyptians had a high regard for linen, and they used it not only for clothing but also for the wrapping of mummies. They also embalmed bodies with linseed oil. The Greeks and Romans also relied extensively on linen, as shown in the writings of many classical authors, and the crop is also mentioned in the Bible. In Asia, notably in India, linseed oil played a part in religious rituals, and in fact there are strong indications that at least the fibre type of *Linum* originated there and that it spread northward and westward over the centuries. The Phoenicians are thought to have taken the plant to Europe, and it is also possible that it spread from Russia to Finland, and from there to northern Europe. There is evidence that the Romans established a centre for linen manufacture near Winchester, and the continued cultivation of the crop in England is shown by the fact that tithes were levied for linen flax in the twelfth century and that in some areas a defined proportion of land had to be devoted to this crop. The early colonists took it to North America, and later it found its way to New Zealand. It continued to flourish as a fibre crop in these countries and in Britain until the beginning of this century, but since then other fibres, notably cotton have taken its place. Linseed production continues at a steady level in many countries, boosted occasionally by national emergencies and probably maintained in the future by the threat of exhausting other, non-renewable raw materials.

The linseed or flax plant

The linseed or flax plant is an annual with a thin, erect and wiry stem. Alternate on this stem are small, glabrous, lanceolate leaves with an entire margin, which are grey-green in colour. There is little branching except near the top where terminal, cymose inflorescences appear on the branches (Fig. 13.1). The individual flower measures about 2 cm across. It has five small, lanceolate sepals and five bright blue but sometimes pink or white petals. There are five stamens, nectaries are present, and the ovary has five erect styles. Originally it is divided into five chambers, each with two ovules, but later each chamber is almost completely converted into two by a false septum. Despite the presence of nectaries, the flowers are usually self-pollinated, although cross-fertilization is possible and does occur. The fruit is a rounded capsule, light brown in colour and retaining its seeds until threshed (Fig. 13.2). The normal diploid chromosome number is 30, occasionally 32, which allows the plant to be crossed easily with some other members of the genus. Auto-tetraploids of those species with 18 chromosomes could also be considered for cross-breeding purposes. The seeds are oval in outline, flattened, about 4–5 mm long, shiny and chestnut brown or in some cultivars orange-brown in colour. The outer epidermis of the testa consists of large cubical cells which absorb water readily and then produce a sticky mucilage. This can be demonstrated easily by soaking linseed for a short time in water. Other properties of the seed are described below.

Linseed

Linseed has an oil content of about 35–44% and also contains some 20% of protein. It is thus a most valuable commodity which can be used in a number of ways. The seed itself or a cake made from it after oil extraction is used as a concentrate stock food, especially for young animals and those intended for show. It is, however, important to be aware of the presence of the cyanogenic glucoside linamarim which in the presence of the enzyme linase hydrolyses to form poisonous hydrogen cyanide. Use of hot water in

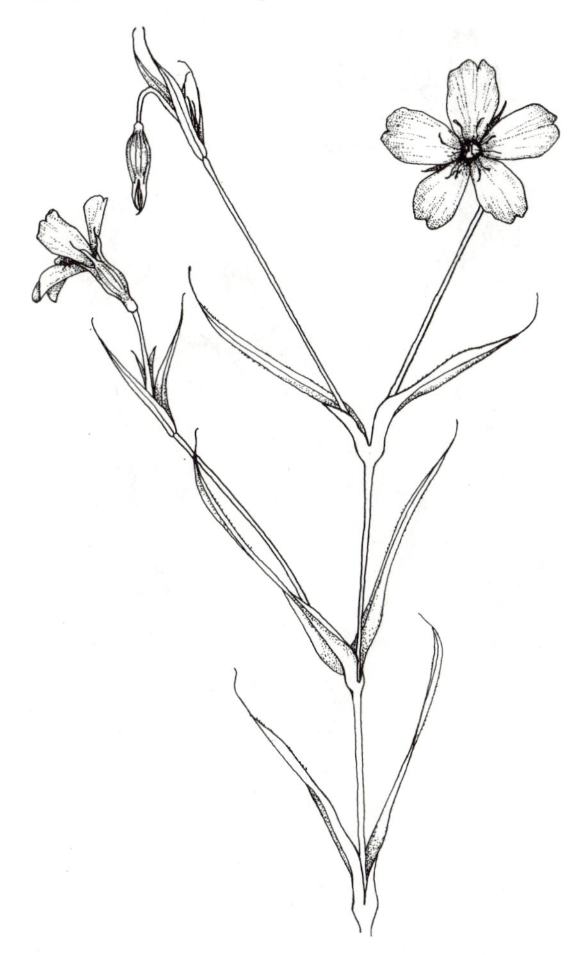

Fig. 13.1. Linseed (*Linum usitatissimum*) showing leaves and flowers.

the preparation of the feed prevents any problems in this respect. More important, however, is the use of linseed as a source of oil which the ancient Egyptians developed long before the plant was utilised as a source of fibre. For extraction, the seed is rolled, crushed, moistened and heated in a large cooking kettle, following which the oil is expressed by rotary mills and hydraulic presses. The crude oil is refined in prolonged storage or by acid or alkali treatment, it is decolourised and waxes are removed. The particular property of linseed is that it produces a drying oil, with a high level of unsaturation as shown by a high iodine number, which quickly forms an impervious layer when used as the basis of an oil paint or varnish.

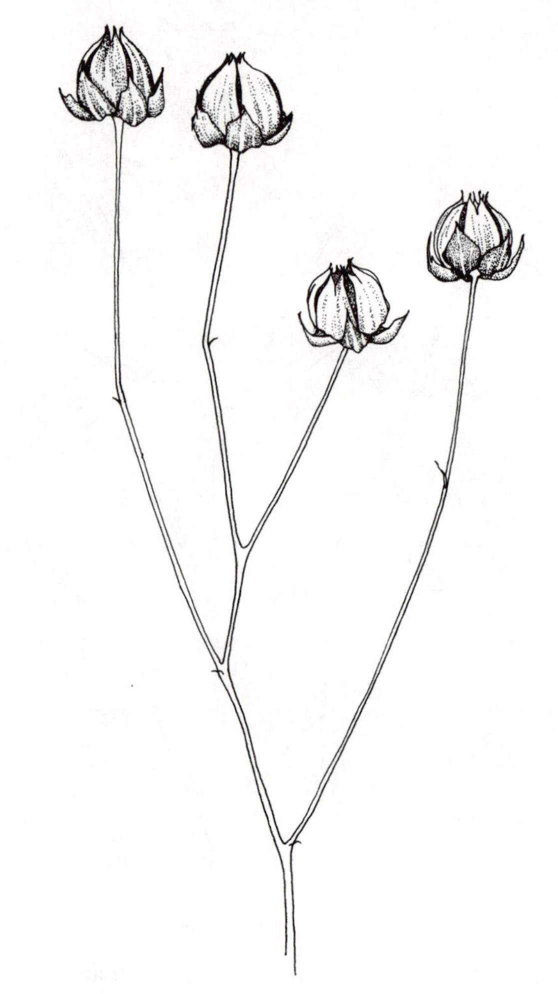

Fig. 13.2. Linseed (*Linum usitatissimum*) capsules.

It is also used as a raw material for oil cloths, floor coverings and printer's ink, although the advent of synthetic products from the chemical industry has depressed the demand for linseed oil. Only increased use of paints throughout the world has so far been able to offset this effect. Paints based on linseed oil have the disadvantage of turning yellow with age. The approximate fatty acid composition of linseed oil is:

stearic and palmitic acids	6–16%
oleic acid	13–36%

linoleic acid	10–25%
linolenic acid	30–60%

The high percentage of linolenic acid is especially noteworthy, for linseed is one of the few plants from which this acid can be obtained in a state of high purity. On the other hand, presence of this acid at these levels prevents linseed from being used in the rapidly expanding market for edible oils, because autoxidation causes substances with bad flavours to be formed. It is therefore encouraging to note that through induced mutations and recombinations plant breeders have recently succeeded in reducing linolenic acid content to acceptably low concentrations. After extraction of the linseed oil, the residue is turned into a meal or pressed into a cake or nuts and used as a protein concentrate for livestock.

For the production of linseed the crop should be sown as early as possible in the spring to obtain rapid establishment. Seed is sown at a depth of 2–3 cm in narrow rows. The seed rate is low, ranging from about 30–50 kg depending on locality, so as to encourage branching of stems near the top and thus greater flower and seed production. Weed control through previous cultivation is most important, because subsequent chemical control requires careful choice of herbicide. The crop attains a height of about 80 cm and is harvested when fully ripe, usually some six weeks after the appearance of the last flowers. Depending on local climate and practices, linseed is combine-harvested or windrowed and then threshed. Yields vary considerably, with national averages ranging from 300 to 650 kg ha^{-1}, although in cool-temperate regions up to 2000 kg ha^{-1} would be expected. Over 2 l of oil are obtained from 10 kg of seed. The straw residue can be used for the manufacture of upholstery, paper and cardboard.

Linen flax

The stems of the linen flax plant have the usual dicotyledonous structure with a central pith surrounded by a ring of vascular bundles. In addition there is outside each group of phloem elements a set of fibre bundles which form a ring in the cortex extending throughout the length of the stem. Each bundle consists of overlapping strands, which individually measure about 20 μm in diameter and 5 to 70 mm in length with an average of about 40 mm. About ten to 40 such strands make up the cross section of a bundle. The middle lamella of these fibres is made up of pectin and some lignin is deposited with age. However, the bulk of the thickening is composed of cellulose, which imparts to linen flax its properties of strength, flexibility and softness. Linen fabrics are well known for their durability and strength, and yet they are soft to the touch, glossy in appearance, and easily washable.

For fibre production linen flax is sown at a higher seed rate, 90–120 kg ha^{-1}, so as to achieve a dense stand of only slightly branched stems with the maximum length of fibre. A typical crop is about 120 cm tall. Efficient weed control is essential, but fertilizer requirements are low. The crop is harvested before the seed is ripe, usually when the petals have been shed, and ideally it should be done by pulling rather than cutting the stems. Machines for this purpose have been designed. The straw is deseeded, even though seed yields are low, and allowed to dry in sheaves. The first stage in the extraction of the fibre is called retting, really a process of controlled rotting. Originally this consisted of immersing the stems in stagnant water for eight to ten days or allowing them to be exposed to the prolonged action of dew. Modern methods consist of retting in tanks of warm water for two to six days during which micro-organisms, mainly *Clostridium* spp., act upon the soft tissues of the stems. In particular the middle lamella of the cell walls breaks down, so that the fibre strands become separated. This is followed by scutching, mechanical beating, and hackling, the combing of the retted material, resulting in the removal of the xylem and other tissues and the isolation of the cellulose fibre. After further processing, the fibre is now ready for spinning into thread and weaving into linen fabric. Yields are very variable but should be in the vicinity of 700 kg ha^{-1} with improved cultivars.

FURTHER READING

Berger, J. (1969). *The world's major fibre crops, their cultivation and manuring.* Centre d'Etude de l'Azote, Zurich.

Godin, V.J. and Spensley, P.C. (1971) *Crop and product digests 1. Oil and oilseeds.* Tropical Products Institute, London.

Hocking, P.J., Randall, P.J. and Pinkerton, A. (1987). Mineral nutrition of linseed and fibre flax. *Advances in Agronomy* **41**, 221–96.

Kirby, R. (1963). *Vegetable fibres: botany, cultivation, utilization.* Leonard Hill, London.

Matheson, E.M. (1976). *Vegetable oil seed crops in Australia.* Holt, Rinehart and Winston, Sydney.

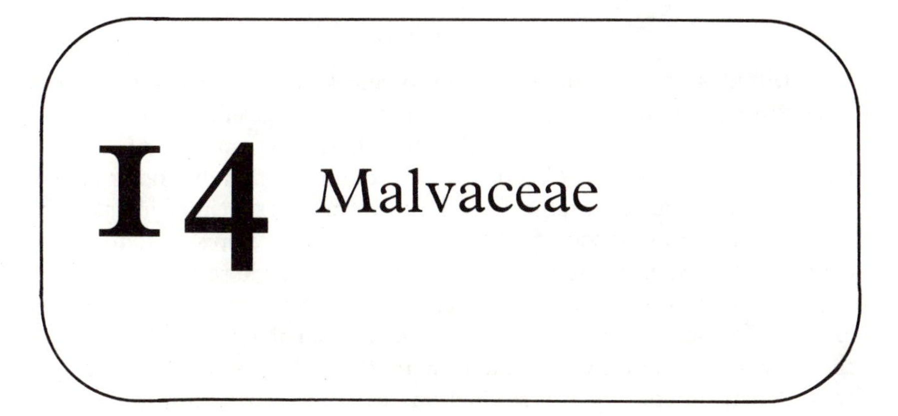

14 Malvaceae

The Malvaceae consist of about 50 genera and 1000 species, herbs, shrubs and trees, distributed throughout the world. Two genera are of major economic importance: *Gossypium* containing several species of cotton, and *Hibiscus* which, apart from its ornamental species, provides a significant fibre crop and a tropical vegetable. Other genera include *Malva* containing several species of weeds, the mallows; *Plagianthus* represented in New Zealand by the ribbonwood tree; *Hoheria*, the lacebark, another tree in the same country. Another genus of interest is *Abelmoschus* which contains the widely grown vegetable okra, and also the musk mallow, which is cultivated in Java, West India and elsewhere to produce aromatic seed for the extraction of musk oil or ambrette for perfume.

14.1 *GOSSYPIUM*

This is a large and variable genus containing about 30 diploid ($2n=26$) and four tetraploid ($2n=52$) species. The latter group includes the two most important species, cultivated as fibre crops and for the oil contained in the seed, *Gossypium hirsutum* which makes up about 95% of world production, and *G. barbadense*.

Origin and history

The different genomes making up the various species in this genus appear to have originated in several parts of the world including Central America, North and South Africa, the Arabian desert, south-east Asia, and Australia. The main cultivated species which are amphidiploid are thought to have developed in northern South and Central America, although one genome seems to be of African origin and was probably derived from seed that drifted across the ocean. Archaeological evidence from Peru and Mexico suggests that domestication of indigenous tetraploids had occurred by 4000–3000 BC. In the Old World, diploid species of cotton, *Gossypium herbaceum* and *G. arboreum*, were cultivated in the Indus Valley and in Africa at much the same time, the oldest known specimen in Pakistan dating back to about 3000 BC. Over the centuries the crop spread to many countries, for example the ancient Nubian kingdom where cotton was first spun and woven in Africa. More recently, the Spanish and Portuguese explorers recognised the superiority of the perennial, tetraploid species of the New World, and it was through their efforts that these plants became widely distributed. These introductions were in their turn displaced by three annual types from America, Upland cotton (*G. hirsutum*), Sea Island cotton (*G. barbadense*) and Egyptian cotton (*G. barbadense*), of which the first has become of outstanding importance. Upland cotton was grown for home use in the south-eastern United States until the end of the eighteenth century when the saw gin (see below) was invented, following which it became the most important and widely cultivated plantation crop in the same area. The demand for cheap labour to grow and harvest this crop led directly to the slave trade from Africa and, in the long run, to the American Civil War. With the eventual abolition of slavery, cotton production declined dramatically in the USA, and other countries such as India, Russia, China, and quite a number of Asian, American, and African states shared the world crop. Sea Island cotton, derived from a distinct species yielding very long and fine fibre, probably arose i[illegible] Carolina in the late eighteenth century and was cultivated in th[illegible] and the West Indies until it proved too vulnerable to one [illegible]ts, the cotton boll weevil. Egyptian cotton, another type b[illegible] same species, can be traced back to a single plant growing i[illegible] but with strong influences coming from Sea Island lines which were introduced from the southern USA. Its main production centres now are Egypt, Sudan and the USSR.

Total world cotton production has exceeded 14 million tonnes during recent years, of which more than half comes from the USA, USSR and the Peoples Republic of China. One important aspect in the last few decades has been efficient control of insect pests, resulting in national average yields of over 1300 kg of lint/ha in some countries.

The cotton plant

The cotton plant forms a small bush growing to about one metre in height, depending on environmental conditions. It has an unusual pattern of branching in that at the base of each leaf two buds may develop. One is a

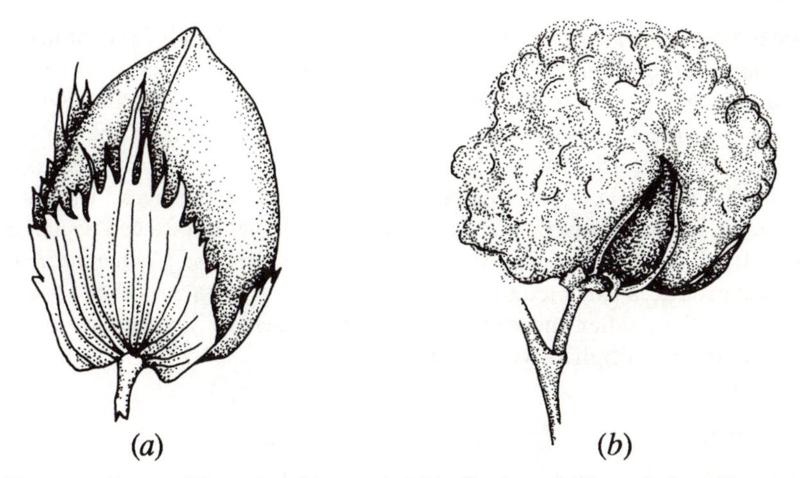

Fig. 14.1. Cotton (*Gossypium hirsutum*). (*a*) boll subtended by epicalyx, (*b*) mature boll with lint fibres exposed.

true axillary bud which normally gives rise to a vegetative branch bearing lobed leaves but no flowers. Higher up on the stem the accessory buds tend to develop into fruiting shoots which have the appearance of a single, straight branch but are in fact sympodial by being composed of a succession of short, lateral branches. Meanwhile the main stem continues to grow producing leaves but no terminal inflorescence. The flowers occur singly and, since they are distributed over several fruiting branches, the time interval between the appearance of the lowest and uppermost flowers may well be protracted. Each flower is subtended by an epicalyx composed of three leafy, highly serrated and toothed bracts which persist beyond maturity. The calyx of five sepals is much reduced and cup-shaped, and the corolla consists of five large petals, tightly folded in the bud but once opened lasting for only about a day. In Upland cotton the petal colour is creamy white and in Sea Island and Egyptian cotton deep yellow with a red or purple basal spot. There are between 100 and 150 stamens inserted on a conspicuous tube which surrounds the style. The superior ovary consists of three to five united carpels, each with eight to twelve ovules. Nectaries are present and serve to attract insects, and humming birds where present, but nevertheless the majority of the seed produced in Upland cotton is the result of self-pollination. The cotton fruit is an ovoid capsule, called the boll, subtended by the epicalyx (Fig. 14.1). It contains 20 to 40 seeds, dark brown in colour and covered with two kinds of hairs of different length. The long hairs which may be up to 4 cm in length, referred to as lint, are loosely attached and yield the fibre for which the crop is grown. The other type of hair, much shorter and strongly attached, is called fuzz. Lint fibres are composed of a central tube or lumen, strengthened by spiral layers of cellulose thickening. On maturity, the boll opens along the sutures where the carpels meet, exposing the lint which fluffs up to give the typical appearance of cotton wool. Upland cotton (*Gossypium hirsutum*) is a relatively short plant bearing large, lobed and hairy leaves. Its main characteristic is the length of the lint hairs which does not exceed 2.5 cm in short-staple and about 3.4 cm in long-staple cultivars. *Gossypium barbadense* produces extra long-staple lint measuring 3.5–4 cm. This species also tends to be taller and to produce glabrous leaves. The boll is darker green and well supplied with oil glands.

Cultivation and uses

Cotton requires an average temperature of at least 21–22 °C during the growing season and makes little progress below 15 °C. A latitude of 40° or less marks the limits of its geographic distribution. Ample sunshine is important, especially during early growth and flowering. A minimum of 500 mm of rainfall is required during the growing season, or alternatively irrigation needs to be provided. Soil type is not highly critical but the land must be thoroughly cultivated to establish a good seedbed. Because of its branching habit, cotton can be grown successfully over a range of plant densities, but it should be sown as early as possible to obtain a long growing season. It is best to remove the fuzz hairs mechanically or by acid treatment, so as to ensure that the seed will run freely from the drill. Good weed control is essential, and so is the prevention of pests and diseases. Cotton requires a long growing season. For example, the interval between fertilization and the opening of the boll is about 50 to 70 days. When the crop is judged mature the bolls are picked by hand or by machine. Hand harvesting still accounts for a large percentage of the total crop, for it produces the best quality and highest recovery. One picker can harvest 50–110 kg of seed cotton per day and between 110 and 200 bolls are required for 1 kg. However, wherever labour costs are high, such as in the United States and Australia, it has become necessary to harvest mechanically. Two types of machinery are available, a spindle picker and a stripper, which under ideal conditions can harvest 140 kg of seed cotton per hour. However, these machines are not selective and harvest a lot of plant debris and weeds as well as cotton bolls. Chemical defoliation prior to mechanical harvesting is often necessary.

After harvesting, the lint fibres have to be removed from the seed, a process referred to as ginning. Improved machinery for this purpose was invented in the United States towards the end of the eighteenth century, especially the use of circular saw blades rather than hooks for the removal of the fibre. The fuzz hairs tend to stay on the seed coat unless special machines or acids are used. The lint is cleaned through the removal of foreign matter, baled and sent away for spinning. Because it is mainly cellulose, cotton fibre imparts strength and yet softness and fineness to fabrics made from it. The remaining seed is extremely valuable, because cotton is a dual-purpose plant which yields not only fibre but also oil. Commercial cotton seed is quoted to contain 18–24% oil and 16–20% protein. Cottonseed oil is used almost entirely in the food industry for the manufacture of salad and cooking oils, cooking fats and margarine. The main constituents are linoleic, palmitic and oleic acids. After oil extraction the seed is further processed into cotton cake for the feeding of livestock. One possible problem with cotton seed is the presence of 1–2% of gossypol, a polyphenol which not only imparts a dark colour and strong scent to the crude oil but is toxic to pigs and poultry, but not to horses, sheep and cattle. The concentration of gossypol is greatly reduced through heat treatment or, better still, plant breeders are selecting plants without pigment glands which produce this substance. One possible use of gossypol, which is under investigation, is as an oral male contraceptive, for it appears to inhibit spermatogenesis without causing any undesirable side effects.

FURTHER READING

Berger, J. (1969). *The world's major fibre crops, their cultivation and manuring.* Centre d'Etude de l'Azote, Zurich.

Godin, V.J. and Spensley, P.C. (1971). *Crop and product digests 1. Oils and oilseeds.* Tropical Products Institute, London.

Kirby, R. (1963). *Vegetable fibres: botany, cultivation, utilization.* Leonard Hill, London.

Matheson, E.M. (1976). *Vegetable oil seed crops in Australia.* Holt, Rinehart and Winston, Sydney.

Munro, J.M. (1987). *Cotton.* Longman Scientific and Technical, London.

Poucher, W.A. (1979). *Perfumes, cosmetics and soaps.* Chapman and Hall, London.

Prentice, A.N. (1972). *Cotton, with special reference to Africa.* Longman, London.

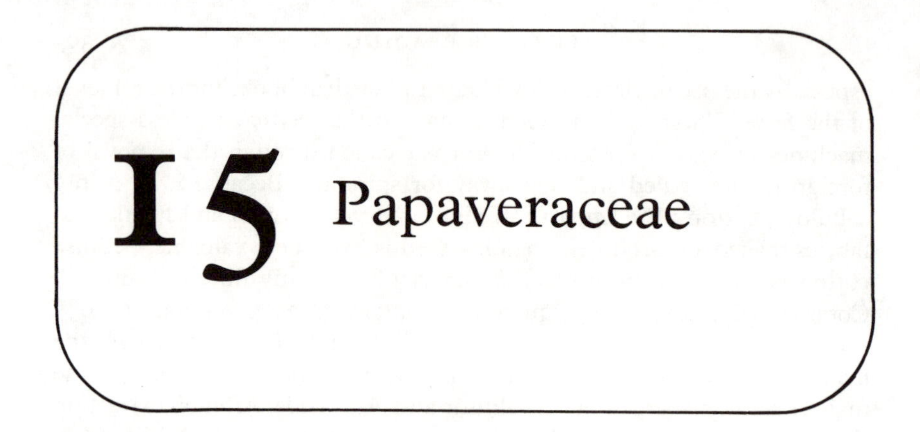

15 Papaveraceae

Although in Europe this family is best known for its red weed species, it contains one crop of economic importance and social significance, the poppy. This plant is grown for its seeds which yield a valuable, edible oil, but its reputation rests more notoriously on opium derived from its alkaloids which include morphine, codeine and papaverine.

15.1 POPPY (*PAPAVER SOMNIFERUM*)

The poppy is thought to have originated under domestication from a wild progenitor in Asia Minor. The powerful properties of opium must have been recognised at a very early stage. The plant has consequently been cultivated for thousands of years and the drug morphine used in medicine and as a narcotic. It was certainly well known to the Persians, Egyptians, Greeks and Romans. Arab traders took crude opium and seed westward to Europe and eastwards into Asia. Further spread was continued by the Spanish and Portuguese, and this together with the notorious opium trade conducted in the Dutch and British colonial empires has ensured that the plant and its effects are universally known throughout the world. As a crop the poppy is now represented by different cultivars, some cultivated for oil, others for opium. The main producing countries are Turkey, USSR, and some Balkan states. Cultivation for opium is usually under strict government control.

The poppy is an annual, with stems about 0.5–1.5 m high, bearing lobed, serrated, glaucous leaves covered with stiff hairs. The flowers are large and solitary. There are two sepals, four petals, white in colour with a purple patch at the base, numerous stamens and a globular ovary of nine to 14 united carpels. The fruit is an oval to round capsule, 5–6 cm in diameter, blue-green in colour and containing 800 to 2000 seeds (Fig. 15.1). In

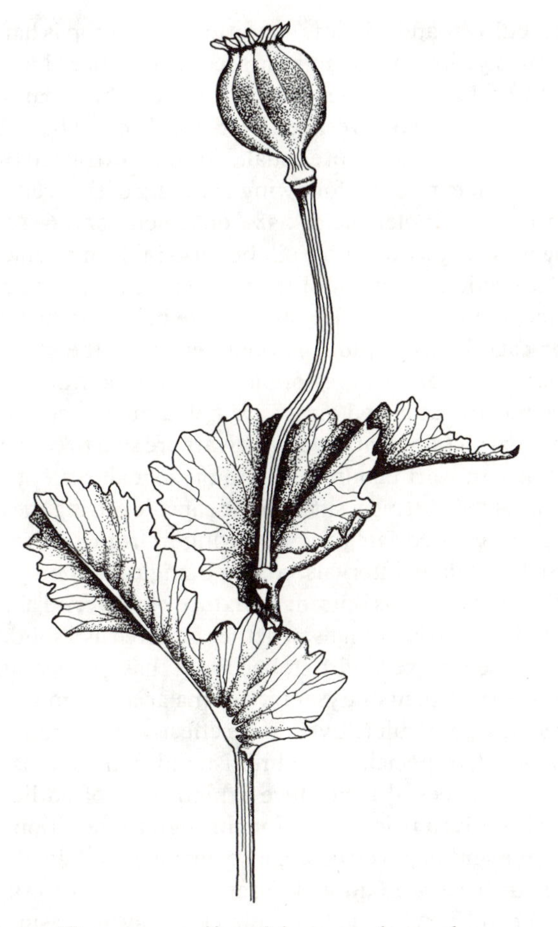

Fig. 15.1. Poppy (*Papaver somniferum*) leaves and capsule.

primitive types of poppy the seeds are scattered through a ring of pores below the stigmatic cap but in bred varieties the capsule remains closed even when the seeds are ripe. The minute seeds measure about 1 mm in diameter and their colour ranges from black to grey-blue and brown.

Poppies are best grown in the warm, temperate zone so as to avoid damage by frost. Good rainfall is required after sowing in the spring but hot and dry weather is preferred after the plant is well established. Because the seed is so small, it must be sown close to the surface and, in order to achieve the recommended low seed rate of 1.5–2.0 kg ha^{-1}, heat-treated seed is often mixed with the fresh seed. The recommended plant density is about 65 m^{-2}. The growing season takes about four to five months depending on cultivar and climatic conditions. The crop is harvested when the capsules turn yellow-green and the seeds can be heard to rattle within. Unless reaped by hand, poppies are cut by a combine harvester or by a forage harvester and then threshed. Seed yields of 3 t ha^{-1} are possible, although 1.2–1.8 kg ha^{-1} are more usual. The oil is extracted by pressure or obtained by solvent extraction following crushing of the seed. Its fatty acid composition is 65% linoleic acid, 25% oleic acid, and 6–10% saturated acids. Its use is mainly as an edible oil, but also in the manufacture of paint and soap. The crude oil content of poppy seeds is between 40 and 55% of the dry matter. Poppy seed is also used by the baking industry and in feed mixtures for cage birds. Opium is obtained from the capsules and the straw. On a large scale and where poppies are grown as a dual purpose crop, the capsules and the attached straw are baled after seed harvesting and dispatched for extraction. Improved cultivars are said to contain just under 0.5% morphine. In warm countries, and on a small scale, the harvesting technique consists of cutting a slit into the capsule during the day and then collecting the coagulated latex the following morning. Several such harvests are possible at daily intervals, so that a single plant may yield 3–6 g of exudate. This latex contains opium: a mixture of powerful alkaloids which have prolonged effects on humans, whether ingested as a solid, dissolved in a drink or smoked mixed with tobacco. It has proved invaluable in medicine for the treatment of dysentery or malaria but, more importantly, as a quickly acting pain killer. Even more effective is morphine, one of the alkaloids involved, especially if administered intravenously. Unfortunately, both substances also produce a high level of addiction, as does heroin which is derived from morphine through the addition of two acetyl groups. Codeine and papaverine are also alkaloids obtained from opium. Morphine yields have been quoted to range from 30 to 100 kg ha^{-1}. High labour costs of hand harvesting the opium gum may increasingly lead to the industrial extraction of morphine from the straw of poppy oil crops.

These drugs are causing severe social problems not only today for in 1839 and 1865 the poppy was at the centre of the so called opium war when the Chinese tried to stop opium from being used by Europeans as an article of commerce in exchange for silk and spices. Long before then poppies were grown for their narcotic properties by the ancient Sumerians, Greeks and Romans. Hypnos was the Greek god of sleep and Morpheus the god of dreams. Less romantically, our present concern is how to control the trading of opium and heroin illegally produced in the Golden Triangle of Burma, Laos, Thailand and the Golden Crescent – Pakistan, Afghanistan, Iran.

FURTHER READING

Atal, C.K. and Kapur, B.M. (ed.) (1982). *Cultivation and utilization of medicinal plants*. Regional Research Laboratory, Council of Scientific and Industrial Research, Jammu-Tawri.

Godin, V.J. and Spensley, P.C. (1971). *Crop and products digests 1. Oils and oilseeds*. Tropical Products Institute, London.

Lööf, B. (1966). Poppy cultivation. *Field Crop Abstracts*, **19**, 1–5.

Veselovskaya, M.A. (1976). *The poppy, its classification and importance as an oleiferous crop*. Amerind Publishing Company, New Delhi.

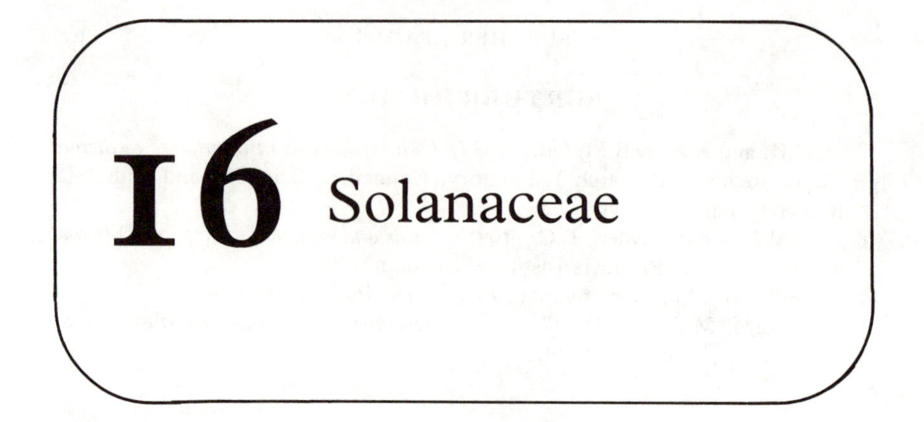

16 Solanaceae

The family Solanaceae is mainly of New World origin and, although without a doubt members of the family were used by the natives of the region before the arrival of the Europeans in America, the recorded history of economic members of the family seems to date from the discovery of the Americas by Columbus in 1492. Although there are about 75 genera and over 2000 species within the family very few are of economic importance. The major food crop is the potato (*Solanum tuberosum*) which is second after the cereals in tonnage of crop harvested in the world. In terms of economic importance next would come tobacco (*Nicotiana tabacum*). The long recognised habit-forming nature of smoking the leaf of this plant has provided governments throughout the world with a major source of revenue since James I of England tried to discourage the habit by setting a high duty on its importation. A further crop of both horticultural and agricultural importance in the family is the tomato (*Lycopersicon esculentum*) which, although extensively cultivated as a salad vegetable, is also grown on extensive areas for the production of soup, juice, sauces, tomato concentrate and canned tomatoes. Finally the chillies (*Capsicum annuum* and *C. frutescens*) provide the heat for the curries of Asia as well as being an important source of vitamin C as the raw vegetable.

In recent years, there has been increased interest in a number of South American members of this family from the Andes region. These are mainly fruits and are in the process of commercial development in other parts of the world. The species involved includes *Cyphomandra betacea* (the tamarillo or tree tomato), *Physalis peruiviana* (the golden berry or cape gooseberry), *Solanum muricatum* (the pepino), and *Solanum quitoense* (the naranjilla).

Members of the Solanaceae often contain high concentrations of narcotics, frequently in the form of alkaloids, and in many parts of the world these are extracted for use in the pharmaceutical industry or other less legal pursuits. It is reputed that the infamous Lucretia Borgia used an extract from *Atropa belladonna* to poison husbands who had ceased to be of interest to her. The same species today is still used in the formulation of drugs to combat cold symptoms, and pure extracts of atropine are used as an antidote to overdoses of organo-phosphorus insecticides and related substances.

In the tropics a number of solanaceous species are serious weeds such as *Solanum torvum* or 'devil's fig', a small spiny tree which grows into thick impenetrable thickets. Fortunately in temperate regions, although there are a number of common solanaceous weeds, these can usually be controlled by herbicides apart from in crops like potatoes and tomatoes.

Members of the family Solanaceae range from herbs through shrubs to small trees. They have alternate leaves which can be simple as in tobacco or compound as in the potato and tomato. The leaves have no stipules and in many species are covered with numerous fine hairs. The flowers are hermaphrodite and usually actinomorphic, with the petals of the corolla fused into five lobes which may appear to be folded, touching or overlapping. The stamens, which are normally attached to the inside of the corolla, are equal in number to the lobes of the corolla and alternating with them. The ovary terminates in a single style which in many species within the family terminates in a stigma surrounded by a cone of stamens through which it must grow to be pollinated. The fruits of the Solanaceae are either berries varying considerably in size or capsules. Large numbers of endosperm-containing seeds are formed.

16.1 *CAPSICUM*

There are four cultivated members of this genus. *Capsicum annuum* and *C. frutescens* are extremely common throughout Asia and Central America for the production of chillies. In temperate agriculture with the possible exception of Japan, it is only the large sweet pepper or capsicum and the powdered dried product of this fruit known as paprika that are commonly utilised. The classification of the genus is confused, and at one time a system involving seven varieties of *Capsicum annuum* was accepted, but as all these freely inter-cross it would seem they are the result of phenotypic rather than genotypic variation. The hot chillies of Asia are probably not grown much in temperate regions because of a lack of demand rather than any climatic unsuitability for the crop.

Two further species are grown in the Andes region of South America. They are a lowland species, *C. baccatum*, which was formerly considered to be a variant of *C. annuum*, and *C. pubescens* which is a highly cold-tolerant perennial growing at elevations up to 2900 m. Both species produce extremely hot fruits.

Capsicum annuum

The fruits produced from this species can vary considerably from the long hot chillies of India to the rather fat bell pepper cultivated as a salad vegetable. In common with other species discussed in this chapter, this species is of South American origin and was carried back to Spain by Columbus on his first voyage. It was spread rapidly from there by both the Spanish and the Portuguese.

Capsicum annuum is an erect annual plant which grows to a height of 150 cm and is strongly branched. The plant has a tap root and an extensively branched root system. Leaves are very variable in shape but remain simple. The flowers are usually solitary and appear to be axillary but, because of growth habit, are in fact terminal. The calyx is bell-shaped with five teeth, and the corolla wheel- to bell-shaped with five to six parts which are white or green in colour. There are five or six stamens inserted near the base of the corolla. The fruits are borne singly at the nodes and are extremely variable in relation to length, width, colour and pungency.

Cultivation and uses

In temperate regions capsicums are usually grown for their large fruits or bell peppers. Like tomatoes and potatoes they are frost-sensitive and thus only annual varieties can be grown. As the crop is usually hand-harvested it is planted out from pre-established seedlings at wide spacings preferably into well drained soils. Flowering commences one to two months after planting and fruit should be ready to harvest from one month later. The crop can therefore only be grown in areas with more than three months of frost-free weather.

Fresh green, red or black capsicums are grown and sold as a salad or cooking vegetable. In common with other cultivated Solanaceae they contain a high concentration of vitamin C (50–280 mg per 100 g). In Spain the crop is used in cheeses and for stuffing of olives. Large quantities of paprika are grown and processed in Hungary by grinding the dried fruits of red cultivars into a powder. Hungarian goulash is famous all over the world. Hot chillies ground to a powder give cayenne pepper, and the pickled pulp is used in the production of tabasco sauce. A wide range of fiery pickles from India depend upon chillies for their heat. Again in common with other members of the family they are used in the preparation of drugs. The extracted compound capsaicin is used as a counter irritant in ointments applied to sprains and bruises.

Capsicum frutescens which is little grown in temperate regions is a short lived perennial similar to *C. annuum* but has two or more fruits at each node. Its fruits tend to be smaller and are conical or cone-shaped and their taste is extremely hot.

16.2 TOMATO (*LYCOPERSICON ESCULENTUM*)

The tomato, like the potato, is thought to have had its centre of origin in western South America. From here it spread as a weed throughout the continent and to Central America. It was probably domesticated in Mexico and it was from there that it was taken to Europe by the Spanish, following their conquest of that country in 1523. From Spain it spread to the rest of the world.

For many years tomatoes were considered to be only a horticultural crop and in northern Europe considerable amounts are still grown each year in glasshouses for the fresh tomato market. However, with the development of the food processing industry large quantities of tomatoes are now used for the production of sauces, chutney, soups, canned tomatoes, tomato concentrate and powder. The demand from the food industry is such that glasshouse grown tomatoes would be too expensive and extensive areas of field grown tomatoes are produced particularly in California, in the United States. The tomato crop can therefore be considered as much a field as a horticultural crop.

Plant structure

Tomatoes are annuals which grow to 70–200 cm in height. Although when young plants are erect, as they mature they are unable to bear their own weight and fall onto the ground thus becoming prostrate. Stems and leaves are covered in numerous short hairs, and when bruised emit a very distinctive odour. Field sown tomatoes develop a strong tap root which branches profusely to give a dense root system. The leaves are large and compound, coarsely toothed, (Fig. 16.1) and consist of small and large leaflets. Inflorescences arise from leaf axils and contain from four to twelve flowers, which are hermaphrodite with a superior ovary. The corolla which has six lobes is usually yellow, and the stamens (usually six) form a distinct cone which in many varieties covers the tip of the stigma, ensuring a high degree of self-pollination. In cultivated varieties the pistil is made up of five to nine fused carpels which have a fleshy central placenta. Following

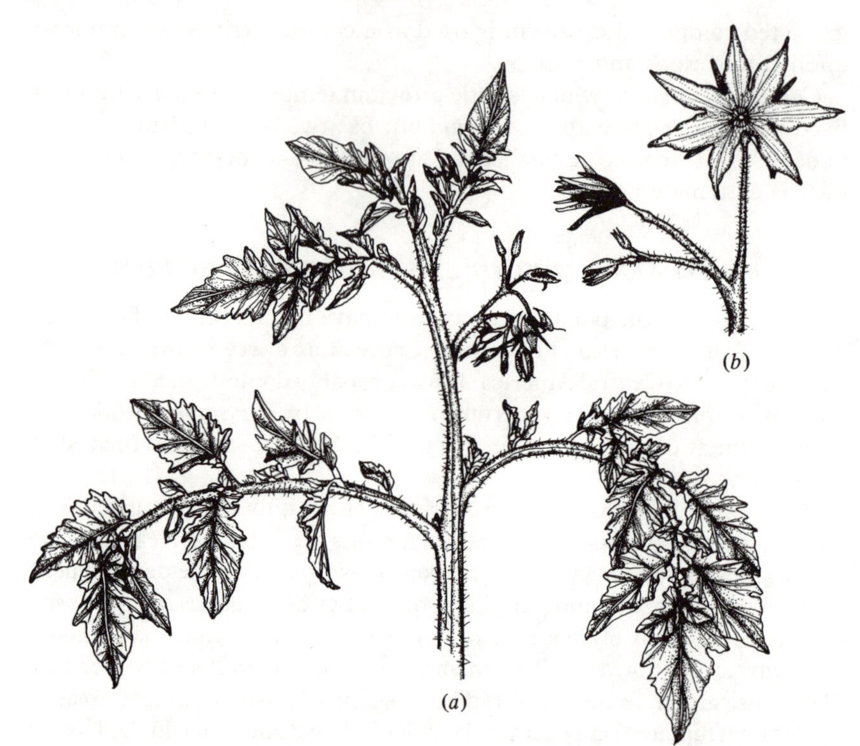

Fig. 16.1. Tomato (*Lycopersicon esculentum*). (*a*) leaves, (*b*) flowers.

pollination the tomato fruit forms, botanically a berry which contains a large number of seeds. At maturity fruit is usually red in colour due to the presence of lycopene and carotene. However, yellow-fleshed forms of tomato also exist.

Cultivation and uses

Tomatoes are frost-sensitive and for outdoor cultivation require at least three months without frost. They are best suited to sunny environments with a well distributed rainfall where night temperatures are above 10 °C. Warm moist cloudy environments improve vegetative growth but increase the risk of plant diseases. For horticultural production, whether grown under glass or outdoors, tomatoes are usually established from seedlings. In recent years there has been the development of grafted plants particularly for the home garden where large numbers of fruits per plant are required. Transplanted seedlings are usually given support to which they are tied and, as the plants elongate, axillary vegetative shoots are removed by hand to leave only the leaf at each node. Usually the shoot apex is also removed when the plant reaches a height of 150–200 cm, and in this way the number of flowering trusses left on each plant can be controlled. Maturity of the crop and thus harvesting, which has to be done by hand, is also spread.

For field sowing very different methods are employed, with hand labour inputs being reduced to a minimum. Plants are usually established by sowing pelleted seed through a precision planter directly into the field at the required population. Whereas with transplanted plants, because of hand harvesting, rows are widely spaced, in field crops the rows are sown close together to suppress lateral branching and to reduce the number of flowering trusses that form. By using this method and perhaps spraying the maturing plants with the hormone ethrel a high degree of uniformity in crop maturity can be obtained. Crops are then harvested by machines in a once-over process. Machines for harvesting of tomatoes are extremely sophisticated and expensive and even have electronic devices to sort the fruits on the basis of colour. To cope with the demands of machine harvesting American plant breeders have now bred plants with thicker skins, higher dry matter content and a pedicel which fractures easily on harvest.

In Israel, where water is in short supply and crops are irrigated, sheets of pre-punched black plastic complete with a trickle irrigation pipe and whiskers are laid by tractor over the top of beds of pre-established seedlings. The holes in the sheet determine crop population while the sheet controls weeds and reduces water loss from the soil surface.

The tomato berry has a high water content (90–94%). In terms of its importance in human diet the main benefit is probably derived from its high content of vitamin C at about 25 mg per 100 g. There is also no doubt that it contributes considerable aesthetic value to our food by its bright colour and that it has a very pleasant taste.

16.3 *NICOTIANA*

There are two species of the genus *Nicotiana* that are of agricultural importance. *Nicotiana tabacum* is used mainly in the production of tobacco products which are smoked, chewed or sniffed, while *N. rustica* is used for the extraction of nicotinic and citric acids for the pharmaceutical industry. Both species are of South American origin, and prior to the arrival of Columbus they were grown and used in the West Indies, Central America and the northern part of the South American continent. Within 100 years of the arrival of the Spanish both species had spread throughout their possessions in South and Central America and to the rest of the world. By the start of the seventeenth century it was being grown in Europe, Africa, the Middle and the Far East. Initially *N. rustica* was cultivated for smoking but it has since been replaced almost entirely by *N. tabacum.*

Tobacco (*Nicotiana tabacum*)

Plant structure

The plant is upright with a stout stem which under suitable conditions grows to a height of 3 m. The tobacco plant develops a tap root from which arises an extensive branched root system. The stem is usually unbranched but laterals may elongate from buds in the leaf axils following the removal of the shoot apex during cultivation. The leaves which are large and simple are arranged spirally around the stem, and in most varieties 20 to 30 leaves are formed, which are sessile and acute to lanceolate in shape and range from 5 to 75 cm in length (Fig. 16.2). They are covered in fine hairs, some of which are glandular and exude gum which gives the leaves a sticky feeling. Lower leaves tend to be larger than those set higher up the stem and the crop matures from the base of the stem towards the top.

Tobacco flowers are borne in a lax terminal raceme which contain up to 150 flowers. The calyx is five toothed and from 1–2.5 cm in length. The corolla is considerably longer than the calyx and also has five lobes, usually pink, but may be white or red. There are five stamens attached near the base of the corolla and arranged in two pairs of different lengths, with the fifth single stamen of different length again. The ovary is superior and has two cells, each of which contains numerous ovules. After pollination the ovary develops into a capsule which at maturity contains large numbers of extremely small seeds. Each plant produces up to a million seeds, so that the 1000 seed weight is only 80 mg.

Cultivation of tobacco

Although tobacco is of tropical origin it has been carried to many parts of the world and is now cultivated from northern Europe to the South Island of New Zealand. It is frost-tender and requires at least 90 to 120 frost-free days after transplanting. Because the main harvested crop consists of leaf, high inputs of solar radiation are required with a well-distributed rainfall to prevent checks in leaf growth while the crop is maturing. Because of the risk of damage, tobacco crops are not usually grown in areas that are subject to hail storms or high winds. High-quality tobacco has large leaves with a high proportion of lamina to vein and a low nitrogen content at maturity.

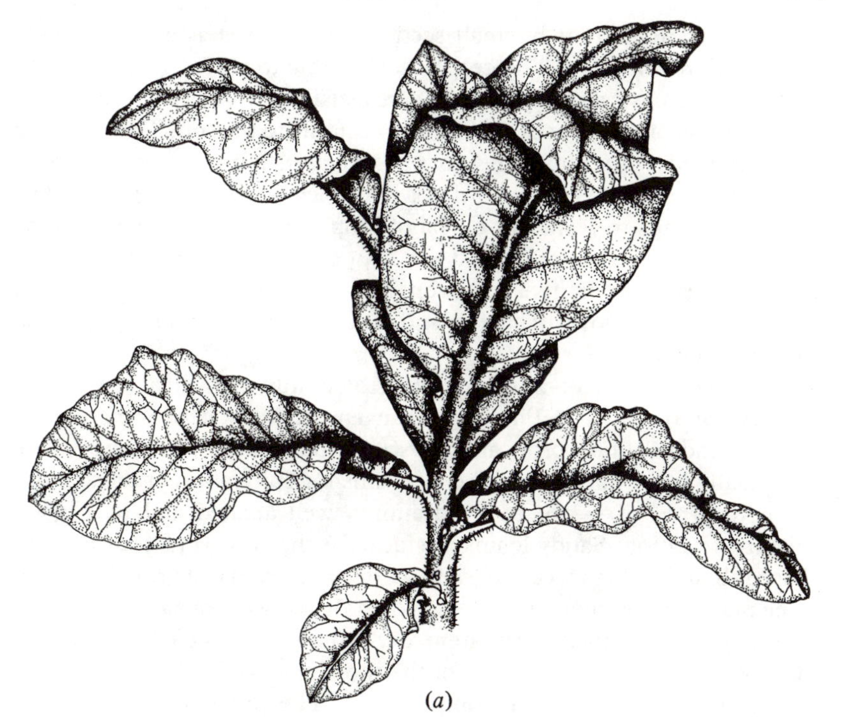

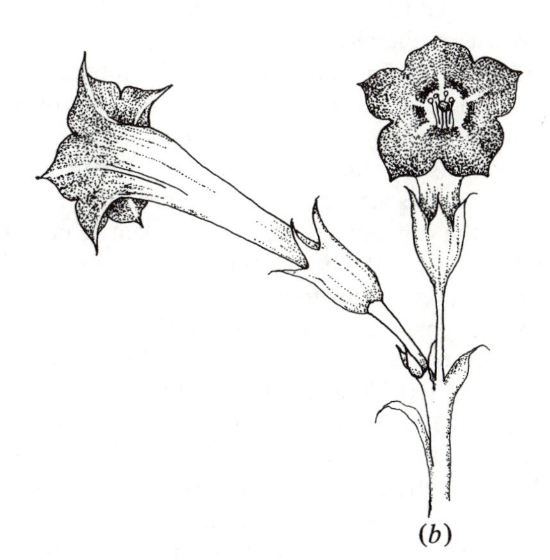

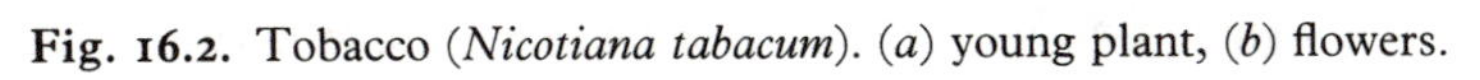

Fig. 16.2. Tobacco (*Nicotiana tabacum*). (*a*) young plant, (*b*) flowers.

Because of its extremely small seed size tobacco is as a rule initially established in a nursery. The crop is susceptible to a wide range of plant pathogens and pests and nursery beds are usually sterilised prior to sowing by application of substances like methyl bromide to the soil. Tobacco seed is also treated with fungicides to prevent seed-borne transmission of pathogens. The seed is so small that a single level teaspoon will establish enough seedlings to plant out 0.4 ha of crop. The seed is sown onto the surface of the soil which is shaded and kept damp. Germination, which is epigeal, takes place in five to seven days. Seedlings are hardened off gradually by the removal of covers, and are ready for transplanting 40 to 60 days after sowing.

Strong seedlings of 15–20 cm are planted out into the field at spacings of 60–90 cm by 90–120 cm. Planting arrangements vary depending on the projected end use of the crop, and rows are wider in crops where leaves are individually hand-harvested. The best soils for cultivation of tobacco are acid with a pH of 5–5.6, freely draining, well aerated and have good moisture retention. Sandy loams are ideal for the crop. The transplanted tobacco should be kept weed free and, to promote even leaf growth, should not be placed under moisture stress. When the grower considers sufficient leaves have been formed on the stem, depending on projected end use, the plants are topped by the removal of the shoot apex which by then usually includes the young terminal inflorescence. The process of topping by removing apical dominance promotes growth of lateral buds. To prevent nutrients being diverted lateral shoots in turn are either removed by hand or supressed by use of chemicals.

For the production of flue-cured or Virginia tobacco, which is used mainly in the production of cigarettes, leaves are individually harvested from the bottom of the plant upwards as they mature, by a method known as priming. Harvesting by this method usually commences 80–100 days from planting 14–21 days after topping. For production of air- and fire-cured tobacco on the other hand, the whole plant is cut off at close to ground level when most of the better quality leaves are ripe. This usually occurs from 40 to 55 days after topping.

Depending on the position of the leaves on the plant the name they are given in the tobacco trade and their quality varies. Leaves closest to the ground are usually the largest but are frequently contaminated with soil or have been damaged. They are known as lugs. From the lower middle of the stem come the cutters, the best quality leaf, large and with a low proportion of leaf vein. Above this the laminae may still be of good quality, but there is an increased proportion of vein and mid-rib, and the size is smaller. This fraction is known as leaf. Finally the leaves from the top of the stem are usually quite small and, because they are often immature at harvest, may contain quite high levels of nitrogen and can be dark in colour. They are of lower quality and are called small leaf or tips. Provided that the crop has been grown under good conditions with adequate soil moisture and high solar radiation it should be possible to produce harvested leaf of good quality. However, before the crop can be utilised it must be cured.

Curing

At maturity tobacco leaf contains 85–95% moisture. During curing, as the colour and aroma of the leaf are developed, the moisture content drops by 10–25%. If the leaf has been well grown at maturity the majority of the dry matter is carbohydrate. This fraction comprising mainly starch can be as high as 50% in tobacco which is to be flue cured but only 25% in cigar leaf. Nitrogen compounds in the tobacco leaf consist of protein, which should be low in flue-cured tobacco, and the alkaloid nicotine ($C_{10}H_{14}N_2$). Nicotine is unique to the genus *Nicotiana* and can comprise 1.5–4% of leaf dry matter in *N. tabacum* and up to 10% in *N. rustica*. This alkaloid is both narcotic and addictive. The narcotic properties are responsible for the feeling of well-being that habituated smokers obtain from smoking, while the addictive properties are well demonstrated by the considerable difficulty that smokers have in giving up their habit, and the withdrawal pangs that they suffer. The characteristic aroma of tobacco is derived from etherial oils and resins that are secreted by the glands of the leaf hairs. Tobacco leaf may also have a high ash content which can affect both combustion and flavour qualities. Leaf with a high chlorine content, for example, is not acceptable to smokers.

There are a number of methods of curing the harvested leaf and the choice depends upon the desired end product. Generally, cultivars selected for growth are also selected on the basis of their final use. Flue-cured tobacco produces Virginia tobacco which is mainly used in the production of cigarettes. For pipe smokers the leaf may be fire-cured, a process in which the leaf is hung in barns where it is cured by the smoke and heat produced from open fires at the bottom of the barn. Air-curing is also conducted in barns, but no heat is applied and the curing process is controlled by the degree of ventilation in the barn. Sun curing, which is used in the production of Turkish or Middle East tobacco, is a combination of air-curing, fermentation of the partially cured leaf in piles, and final drying in the sun. Cigar tobacco, which is rather specialised, comprises three types of leaf: 'filler' which provides the bulk of the cigar, 'binder' which ties the filler together and 'wrapper' which envelopes the whole cigar.

In each different curing method the final changes that occur in the leaf are similar. The process of flue curing will be described as it is the most commonly used. After the leaves are individually harvested they are placed in pairs over a stick, and the sticks covered in leaves are placed into the barn for curing. There is no direct application of fire to the leaf, and hot air is circulated through the barn by large pipes. In the first stage of curing, the leaf is not killed but, because of the high temperatures in the barn, the respiration rate of the leaf is increased. Carbohydrates such as starch are broken down to di- and monosaccharides and oxidised to produce energy. In air-curing there is no application of heat and final sugar content of the leaf is very low. At this time the chlorophyll pigment in the leaf is destroyed, and the pigments carotene and xanthophyll give the leaf the yellow colour which is characteristic of cured tobacco. This part of the process, which is known as yellowing of the leaf, takes from 36 to 48 hours. During this period the temperature in the barn is raised from 35 to 45 °C. The next stage of the curing process fixes the colour. The temperature in the barn is raised to about 50 °C. As a result the enzymes in the leaf cells are destroyed and the protoplasm killed. The leaf lamina begins to dry out and the yellow colour is fixed. This takes from 12 to 20 hours. In other tobaccos the final colour is quite different. In air-cured and fire-cured tobacco polyphenols in the leaf are oxidised to produce a red to brown coloured leaf at the end of curing. Turkish tobacco has a yellow to light red colour and a distinctive aroma. The final stage of flue curing is the drying of the leaf. The process takes about 50 hours and the temperature in the barn is further raised until it is at 60–70 °C.

Following curing tobacco is stacked in piles or bulks of about 2 m in height for a month where further fermentation takes place. Finally, as far as the grower is concerned, leaf is graded depending on its position on the stem, leaf size, freedom from blemish, colour and desired end use. Tobacco is usually sold at auction on sample. Following sale, it is baled and the bales are usually left to mature for one to two years before being used in the manufacturing process.

Tobacco products

The majority of tobacco is consumed as factory manufactured cigarettes. Pipe smoking, roll-your-own cigarettes and cigars account for far less of the crop. The habits of taking snuff (fine powdered tobacco which was inhaled into the nose) and chewing tobacco are now relatively uncommon in the western world. Notwithstanding the long history of the crop, the production of cigarettes is relatively recent. They were invented late in the nineteenth century but it was not until 1910 that the first machine-made cigarettes were produced. Irrespective of end use, a considerable mystique has been erected around the production of individual brands of cigarettes or blends of pipe tobacco. Manufacturers purchase leaf from all over the world and, after cutting and removal of leaf stalks, blend different types of tobacco together to produce their own distinctive brand. Far more than leaf goes into tobacco products and among additives are glycerine, sugar, liquorice, saltpetre, apple juice, rum, whiskey, port wine, and menthol. It is somewhat ironical that, because of the collection of excise duty on tobacco products, many countries know more about adulteration of tobacco than of food-stuffs.

Tobacco smoking and health

An overwhelming mass of evidence now points to a strong association between the smoking of tobacco products, particularly cigarettes, and cancer of the lung and other organs. The habit is also considered to be responsible for numerous other respiratory and cardiac complaints. Until a few years ago there was a lower incidence of lung cancer in women than men. However, recent reports from the United States show that following the upsurge of smoking among women after the 1939–45 war lung cancer has now become the major form of this disease in women as well. Because of this type of evidence over the last few years, there has been a publicity campaign to make people aware of the risks that are associated with smoking.

In an increasing number of western countries, governments are now legislating to reduce the habit of smoking. Smoking is now banned in many countries on long-distance flights and even on some international flights. Bans on television advertising have been followed by bans in the print media and on the sponsorship of sporting events by tobacco companies. Increasingly, with the recognition of the risks of passive smoking by non-smokers, employers are being required to provide smoke-free workplaces and restaurants to provide smoke-free eating areas. In spite of this, there seems to be little evidence that the smoking habit is declining, and in some groups it appears to be on the increase again.

The tobacco industry is very large and considerable vested interests are involved. Quite apart from the capital invested in the industry, governments throughout the world collect very large amounts of revenue from taxing manufacturers and consumers of tobacco. No government yet seems to have accepted the premise that savings in the provision of medical services and invalid pensions might be greater than the loss of current revenue if smoking was banned. However, the addictive nature of the habit and the experience of the United States with prohibition of alcohol after the 1914–18 war suggests any attempt at total suppression would be futile. In eastern countries such as India and China the industry seems to be at the stage it had reached in Europe at the time of mechanisation of cigarette production which led to the large upsurge in consumption. It is a sobering thought that tobacco continues to enjoy the dubious reputation of being the most valuable non-food crop grown in the world today.

16.4 *SOLANUM*

Potato (*Solanum tuberosum*)

The potato (*Solanum tuberosum*) is a major dietary staple in nearly all temperate countries. World production is of the same order as the major cereals.

Origin and history

The potato is thought to have originated at high altitudes in the Andes mountains, close to the junction of the present countries of Chile, Peru and Bolivia. From here it was carried throughout South and Central America. After the arrival of the Spanish it was introduced into Europe in about 1570. A separate introduction was made into England by Sir Walter Raleigh in 1586. Its introduction into Ireland, which now lends the crop its common name, is considered to have come from the latter source.

Following its arrival in Europe, members of the nobility tried to encourage their landholders to grow and eat the new crop. It was not popular for a long time, and it was not until the eighteenth to early nineteenth century that it became widely grown. In Ireland it did gain acceptance as a crop, perhaps because it was easy to hide from marauding war-like bands, or because it was more suited to the cool damp climate of Ireland than most cereal crops. The crop rapidly became the main staple of the Irish peasantry, and records indicate that Irish farm workers in the last century consumed more than 3 kg of potatoes per day. Late blight of potatoes (*Phytophthora infestans*) struck the Irish potato crop in 1845. Because of the genetic uniformity of the plant associated with its vegetative reproduction, the potato was highly susceptible to the disease. A major famine followed and within three years one million Irish had died, a further million had emigrated and three million were unemployed.

Plant structure

The potato is propagated from tubers which are called 'seed' potatoes although they are in fact vegetative stem structures. From the eyes of the

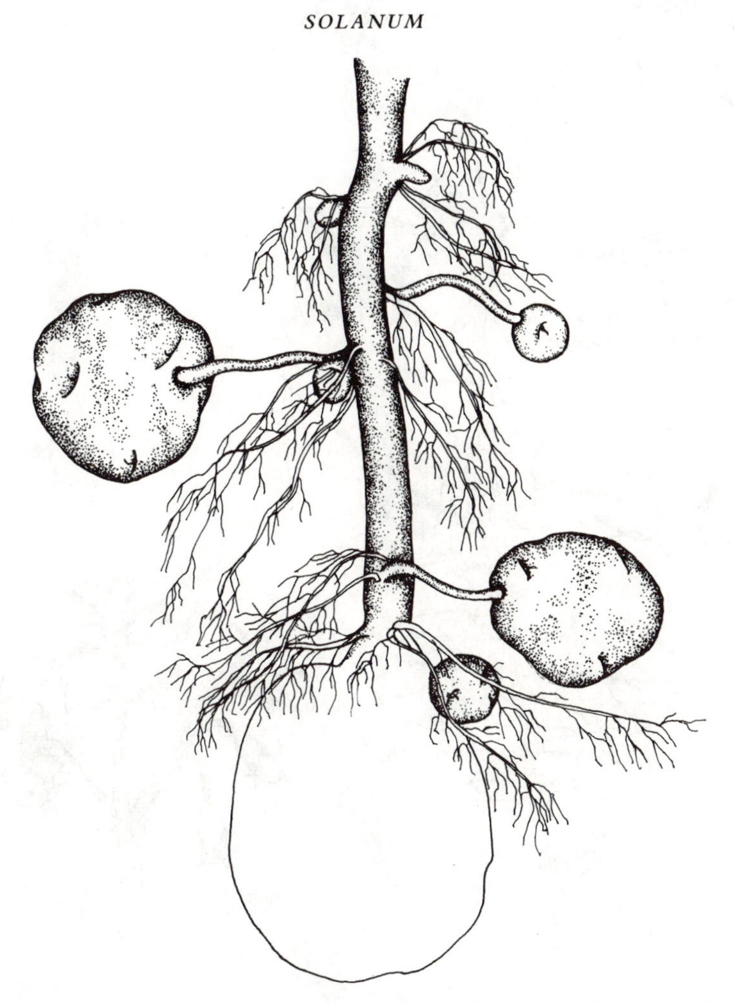

Fig. 16.3. Potato (*Solanum tuberosum*). Stem arising from 'seed' tuber with roots, rhizomes and developing tubers.

'seed' (Fig. 16.3) stems grow which bear adventitious roots and also produce rhizomes that grow horizontally through the soil. When the elongating stem reaches the soil surface it continues to grow and forms a series of pinnate compound leaves (Fig. 16.4). The stems that arise from the 'seed' potato are individual plants and, once the mother tuber dies, there is no physical connection between stems that have arisen from the same tuber. The stems branch and, when flowers are formed, these occur in the leaf axils and may be white, yellow or purple in colour, conforming to the typical flower structure of the family. Fruit, if it forms, is a small green

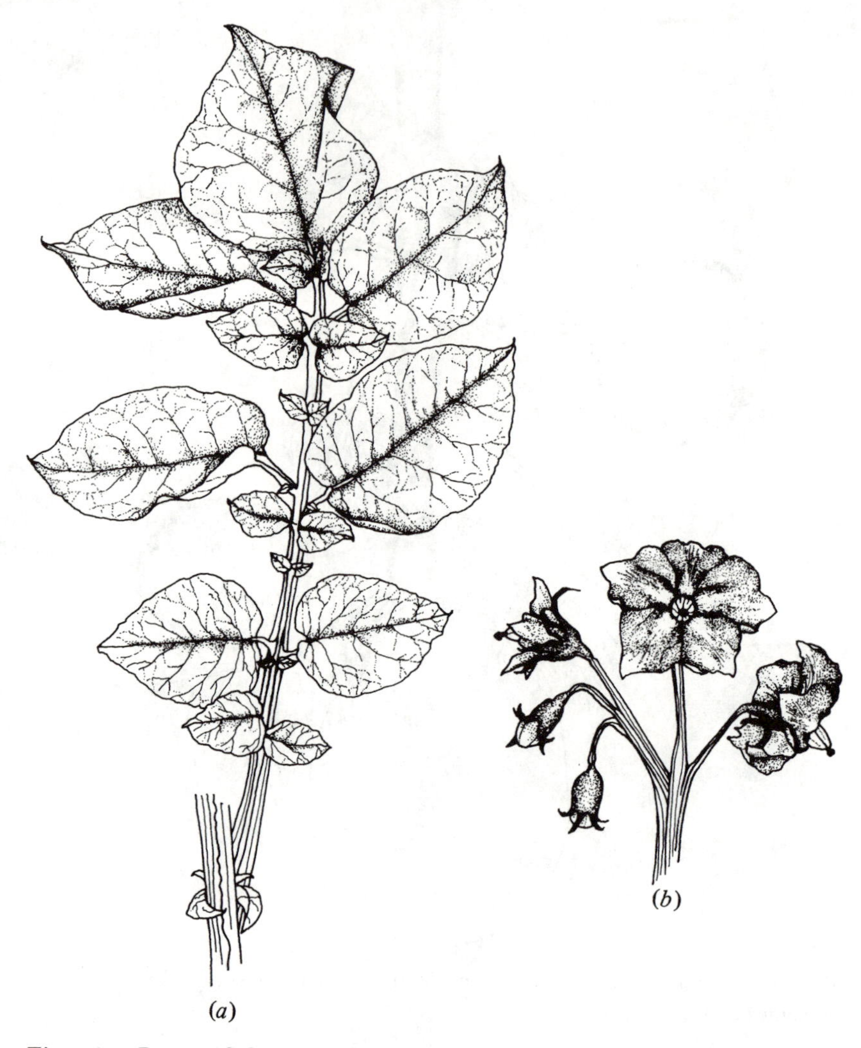

Fig. 16.4. Potato (*Solanum tuberosum*). (*a*) compound leaf, (*b*) flowers.

berry, not unlike an unripe tomato in appearance. The true seed of the potato is generally of interest only to plant breeders in the production of new cultivars. However, it is possible to produce a crop of potatoes from true seed. In temperate regions crop yields from true potato seed have been low and have seldom exceeded more than 30 t ha^{-1}. In the tropics, because of the savings of freight and the generally poorer yields of potatoes, production from true potato seed appears to have more promise. The final height of the plant is between 30 and 100 cm.

In a process which is not yet fully understood the horizontal underground rhizomes begin to swell at their tip with the onset of tuberisation. Towards the end of the growing season the foliage of the potato either dies off or is killed by frost and the new tubers which comprise mainly starch-filled parenchyma cells are then ready for harvest.

Cultivation and use

Probably the most important aspect determining yield in the potato crop is the conditions under which the 'seed' potato is stored prior to planting. Manipulation of storage temperature can lead to considerable differences in the number of shoots, and the number of nodes that are produced on the shoots when seed is planted. Generally low temperature storage is best. A process of pre-sprouting potatoes in light known as chitting is sometimes used in Europe to produce lots of short sprouts with numerous nodes to increase the earliness and yield of the crop.

Assuming that 'seed' has been well stored and that there are adequate soil nutrients and moisture, the factor that determines the final yield of the crop would appear to be leaf area for it is the leaves which produce the sugars that are stored in the tubers. Therefore, any factor that reduces leaf area also reduces final yield. However, it is possible to have too much leaf and high inputs of fertilizer nitrogen on potato crops can sometimes lead to a delay in the onset of tuberisation, and yield may be reduced. Given high leaf area, maximum potato yields are obtained from areas that have low night temperatures combined with warm bright sunny days and an adequate supply of soil moisture. Highest potato yields, are therefore obtained in temperate countries.

Manipulation of plant population in the field by increased planting can alter the size of potatoes produced. In the production of 'seed' potatoes which preferably are small to reduce the chance of disease, cut tubers are used for planting or, to obtain large numbers of small potatoes for canning as 'new potatoes', high populations of seed potato are sown.

The potato tuber contains 70–80% water and the main dry matter component of the crop is starch. The tubers also contain small amounts of fat, protein and minerals. Because of their high water content and their bulk, potatoes are expensive to transport and therefore do not enter world trade to the same extent as other major food crops. Besides being a major source of food energy in many countries, particularly those that experience long winters, and where citrus fruits are not readily available, the potato can be an important source of vitamin C in the human diet. In some countries in Europe, during the 1939–45 war, there were outbreaks of the Vitamin C deficiency disease scurvy following reduction in intake of potatoes.

Traditionally the potato has been a major dietary staple in European countries where it has been prepared for home consumption. In the last few years there have been considerable changes in the method of marketing potatoes. Often now potatoes are washed and bagged before they appear in the retail trade, but an even bigger change has been the degree of processing of the crop that now occurs before it is passed to the consumer. In the United States more than half the potato crop is processed. Major dietary products produced are potato crisps, deep frozen French fries, dehydrated potatoes, and canned new potatoes. As a result of this move of the potato towards being a convenience food, consumption, which had been falling in most countries, has started to rise again as living standards have risen. Besides their domestic use, potatoes are used in the production of industrial starch, alcoholic beverages and industrial alcohol. Undersize potatoes which are not suitable for use as 'seed' are often fed to pigs.

One result of the swing to processing of potatoes has been increased research into storage of the crop without loss of dry matter while it is awaiting processing. Depending on the end use to which they intend to put the potatoes, processors require growers to comply with strict specifications as to variety, fertilizer use and water applications to the crop. Potatoes required for production of chips for example, must not have a high content of free sugars, as they brown badly when immersed in hot oil. Similarly, tubers which are to be used in the production of dehydrated potato are required to have a high dry matter concentration. Irrespective of end use, processors require the crop to be of a uniform size and shape, so as to minimise wastage during the peeling process and to avoid the need to remove eyes from processed potatoes.

Two recent developments in plant breeding have increased potato yield and reduced production costs. Using tissue culture, it has been possible to produce pathogen-free seed tubers for commercial planting. Yield increases of the order of 30% have been obtained. Further, using genetic engineering, genes for herbicide resistance have been incorporated into potato plants which has allowed improved weed control in the crop.

Potato tubers that are allowed to grow too close to the surface of the soil or are stored in the light may develop chlorophyll under the skin and become green. Such potatoes contain high concentrations of the toxic alkaloid solanine and must not be eaten. Because of the presence of the same alkaloid green potato tops should not be fed to livestock.

Aubergine (*Solanum melongena*)

This species, commonly known as aubergine, egg plant or brinjal, is an erect bushy shrub that is grown for its fruits which are utilised as a vegetable. Although in tropical countries it can show weak perenniality, in temperate regions it is grown as an annual because of its susceptibility to frost. Plants are usually established from seeds which are small. Stems are erect and branching and covered with fine grey hairs. Final height is 50–150 cm and some plants develop spines. Leaves are large, simple and densely hairy. Flowers may be either solitary or borne in cymes of from two to five flowers. The flowers are up to 5 cm in diameter, have five to six lobes and are purple to violet in colour. Following pollination, the fruit which is a berry is formed. In most temperate varieties the fruits are large (up to 500 g) purple to black, and egg shaped. In tropical varieties fruits are often longer and thinner and can be yellow or white in colour. The cultural requirements of *Solanum melongena* are very similar to those of the tomato. In temperate regions they are nearly all sold for utilisation as a fresh vegetable and are therefore usually hand-harvested. In the tropics, besides their use as a vegetable, they are used to produce chutneys and pickles.

Solanum aviculare and *S. laciniatum*

These two species are natives of New Zealand where they are known by the Maori name of *poro-poro*. They are currently under both agronomic and pharmaceutical research in eastern Europe, as a source of the alkaloid solasidine. This chemical is a precursor in the production of corticosteroids used in the manufacture of cortisone and oral contraceptives. Although all parts of the plant contain solasidine, the highest concentration is found in the berries. The crop is currently under commercial cultivation in Hungary. Another plant capable of producing physiologically active steroids is *Dioscorea*, the wild yam growing in Mexico and South Africa.

16.5 OTHER SOLANACEAE

A number of species of the family Solanaceae are minor weeds of temperate agriculture. Those that share the common name of nightshade are generally all derived from the genus *Solanum*, except for *Atropa belladona* which has the appropriate common name of deadly nightshade. Perhaps the most common is *Solanum nigrum* or black nightshade which is a small plant with simple leaves which grows to a height of about 30–50 cm and has small purple to black berries about the size of a pea. Henbane (*Hyoscyamus niger*), besides being a source of the drug hyoscamine is also a weed of temperate agriculture. Its other claim to fame is that it was used extensively by plant physiologists working on the effects of photoperiod and vernalisation on flowering. In ancient times mandrake (*Mandragora officinarum*) was invested with magical properties because of the fancied resemblance of its divided root to human form.

Apart perhaps from tobacco, the member of the family enjoying the most current notoriety is the thorn apple or *Datura stramonium*. Extracts of this plant, which is a relatively common agricultural weed in some parts of the world, were for a time popular with the hippy culture in America because of their hallucinogenic properties. In fact many members of this family are of interest to the pharmaceutical industry.

FURTHER READING

Akehurst, B.C. (1981). *Tobacco*. Longmans, London.

Andrews, J. (1984). *Peppers: the domesticated capsicums*. University of Texas Press, Austin.

Atherton, J.G. and Rudich, J. (eds) (1986). *The tomato crop: a scientific basis for improvement*. Chapman and Hall, London.

Burton, W.G. (1989). *The potato*. Longman, London.

Gould, W.A. (1983). *Tomato production, processing and quality evaluation*. Avi Publishing, Westport.

Halasz, Z.C. (1963). *Hungarian paprika through the ages*. Corvina Press, Budapest.

Harris, P.M. (ed.) (1978). *The potato crop*. Chapman and Hall, London.

Hawkes, J.G. (1990). *The potato: evolution, biodiversity and genetic resources*. Belhaven Press, London.

Hawkes, J.G., Lester, R.N. and Skelding, A.D. (eds) (1979). *The biology and taxonomy of the Solanaceae*. Linnean Society of London. Academic Press, London.

Heiser, C.B. (1969). *Nightshades, the paradoxical plants*. Freeman, San Francisco.

Hill, G.D. and Wratt, G.S. (eds) (1985). Potato growing: a changing scene. *Agronomy Society of New Zealand Special Publication No.* 3.

Jones, C. (1981). *Growing tomatoes: with a note on cucumbers, peppers and aubergines*. Penguin, Harmondsworth.

Kingham, H.G. (ed.) (1973). *The UK tomato manual*. Growers Books, London.

Kowalchik, C. and Hylton, W.H. (1987). *Rodale's illustrated encyclopedia of herbs*. Rodale Press, Emmaus.

International Potato Centre. (1984). *Potatoes for a developing world: a collaborative experience*. International Potato Centre, Lima.

Lisinska, G. and Leszczynski, W. (1989). *Potato Science and Technology*. Elsevier, London.

National Research Council. (1989). *Lost crops of the Incas: little known plants of the Andes with promise for worldwide cultivation*. National Academy Press, Washington.

Peirce, L.C. (1987). *Vegetables; characteristics, production and marketing*. John Wiley, New York.

Smith, D. (1979). *Peppers and aubergines*. Grower Books, London.

Smith, O. (1977). *Potatoes: production, storing, processing*. Avi Publishing, Westport.

Somos, A. (1984). *The paprika*. Akademiai Kiado, Budapest.

Talbot, W.F. and Smith, O. (eds) (1987). *Potato processing*. Van Norstrad, New York.

Tso, T.-C. (1972). *Physiology and biochemistry of tobacco plants*. Dowden, Hutchinson and Ross, Stroudsburg.

17 Physiological basis of yield

In the preceding chapters of this book we have become acquainted with the major crop plants of the temperate and subtropical regions of the world, we have learnt something of their structure and life cycle and we have begun to appreciate their place in agriculture. This knowledge, useful though it must be accepted to be, is but the first step along a much longer path which ultimately leads to the production of field crops and the utilisation of their end products. We have, in a sense, merely acquired the necessary vocabulary of a language but we still have to learn its grammar before we can begin to speak it. The next move is to fit the plants into an agricultural system - a complex amalgam, determined by climate, soils, management practices, economic constraints and consumer demand. This is undoubtedly quite a different task, requiring expert knowledge of a great variety of fields covering a very broad spectrum. Not only that, but the factors to be considered will vary widely from region to region. Crop production must ultimately be viewed in a local, geographically defined context, even though there will undoubtedly be re-occurring themes which are common to most situations. However, before we reach this level of organisation, we should do well to consider some universally applicable principles underlying the growth and productivity of plants wherever they may be cultivated and irrespective of the purpose for which they are being grown. These universally applicable principles arise from a study of the physiological basis of yield, bringing together knowledge of plant structure and development with an understanding of basic physiological processes. How does the crop plant fit into its environment, how do external factors operate in determining yield, and how can the system be manipulated to achieve maximum production? These are some of the questions which must engage our attention in this final section of the book. Only a generalised survey will be possible in this context, and any more detailed

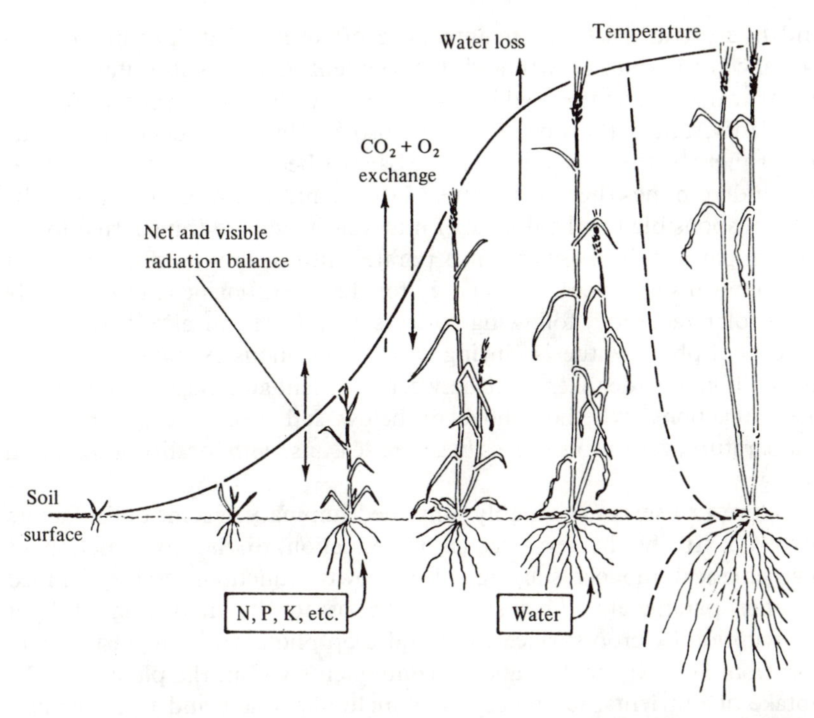

Fig. 17.1. Diagrammatic representation of the growth of wheat from germination to maturity, showing environmental inputs and exchange processes above and below soil level. (From Milthorpe, F.L. and Moorby, J., 1979.)

information should be obtained from the specialised books that are quoted at the end.

17.1 PLANT AND ENVIRONMENT

The complex interactions between plant and environment are depicted at their simplest in the diagrammatic representation of Fig. 17.1. It shows a typical crop plant, such as wheat, following its life cycle from germination to maturity, as it passes through its various phases of leaf production, tillering, floral initiation, stem elongation, ear emergence and grain development. These changes occur against a never-constant background of external influences, and indeed are made possible and directed by the inputs and exchange processes of the environment, both above and below ground. Of these, water is in a special category, for 70–90% of the plant consists of water and, although it functions as a substrate in photosynthesis and hydrolytic reactions, maintenance of turgor is its paramount importance to normal growth and development. Changes in water supply, whether natural or provided by irrigation, thus have a profound effect on the plant. At the same time as water is absorbed by the roots, it is lost to the atmosphere by the leaves, and it is the balance between these two processes which determines the water status of the plant. The roots are also in the main responsible for the uptake of minerals which are transported to the various parts of the plant and incorporated into an almost infinite range of biochemical substances. The energy for these reactions comes ultimately from solar radiation, following conversion into chemical energy by the process of photosynthesis, during which continuous exchange of oxygen and carbon dioxide takes place between the plant and the atmosphere. All these reactions, whether above or below soil level, are governed by temperature, varying as it does between seasons, with locality and diurnal changes.

Figure 17.1 serves as a highly simplified summary of some of the factors involved. It should serve as an introduction to the construction of mathematical models starting with growth functions and quantified weather parameters. This should take us to the modelling of light interception by crop surfaces, leaf and crop photosynthesis, respiration, partitioning of dry matter and its components within the plant, and the uptake of nutrients. Crop responses including water and fertilizer use, temperature interactions and other environmental effects are also best understood through mathematical treatment. All this should serve as the baseline for the construction of empirical and mechanistic models of crop growth and yield, and place us in the position of being able to explain and predict with some degree of assurance.

These interactions with the environment are complex enough, if we assume that the plant is healthy and not affected by other organisms. In nature this would be an unusual occurrence, even if the most stringent measures are taken, and it would thus be wrong to think of the crop plant entirely in isolation. Pests and diseases are almost universal, but in addition there are many beneficial associations between higher plants and other organisms, such as mycorrhizal or symbiotic relationships, which are of fundamental significance. If these are not specifically mentioned here, it is not to detract from their importance but to simplify an already multifactorial system.

17.2 YIELD AND YIELD COMPONENTS

Although the plants described in this book are all grown by man for a specific purpose, they represent many different families with contrasting morphological and physiological characteristics. Greater diversity could hardly be imagined. Crop plants may be annual, biennial or perennial. They may be herbs, shrubs or trees. Many are cultivated for the sake of their leaves, either to be eaten by man himself or, like the herbage grasses and clovers, to be fed to animals that produce milk, meat, wool, leather and other commodities. Many others are grown for their storage organs, be this part of their vegetative or reproductive structure. A large number of plants fall into this category ranging from root and tuber crops to cereals. Even if the end product is the same, for example a grain, there can be great variety, because some grains are almost entirely of starch while others are largely protein or oil or a combination of both. Protein and oil differ greatly in composition from species to species, and often among cultivars. Many plants do not figure prominently in our diet, and furthermore not by any means do all crops serve as food for man or his animals. Chief among these are the fibre crops like cotton, linen flax or hemp, but many others serve as raw materials for industrial processes, resulting in the manufacture of very many different products: plant spices, flavouring, drugs, stimulants, scent, dyes, rubber, insecticides and a whole host of other items useful to man.

The word used to describe the quantity of material produced by the plant is yield, measured in some appropriate unit of weight. But in view of the great diversity of plant products, it will be clear that yield assumes different meanings for different types of crop. By the same token it follows that different physiological processes are involved in producing the products concerned. Not only that, but even in the same type of plant the final yield is the result of very many processes, each making its own specific contribution. From the mere practical point of view all that matters is probably the total amount of useful product measured at the farm gate or the factory, but scientific enquiry must not be confined to the final harvest alone. This was recognised many years ago when it was stressed that yield should not be considered as a single figure but analysed in terms of the components that contribute towards it. For example, in cereals, yield depends on the number of grains produced and the weight of the individual grain. More specifically and going into greater detail, we can enumerate the important yield components for different crops, as follows.

Wheat: yield = number of ears per unit area × number of spikelets per ear × number of grains per spikelet × mean weight per grain

Potato: yield = number of plants per unit area × number of tubers per plant × mean weight per tuber.

Sugar beet: yield = number of plants per unit area × mean weight per root after topping × percentage sucrose content of root.

It would, of course, be easy to subdivide each component further or to recognise the many partial processes involved in determining the magnitude of each of them. But even at this simple and admittedly superficial level it is now possible to think of yield, not as a statistical measure of plant productivity but as the end result of a number of contributing and inter-related components, each determined by its own set of controlling factors.

17.3 PLANT DEVELOPMENT

If we accept the argument that final yield is made up of a number of components, and that each of these is determined individually during the growth of the plant, it follows that we must investigate when the crucial events occur that lead to their formation. Not only must we know the particular yield structures involved but we must enquire into the time when they are first laid down and how quickly they develop. This means that we have to study in considerable detail the development or ontogeny of the plant, often going right back to the early stages of growth when some yield components may be initiated. The objective is first to understand the sequence of important developmental events and then to attempt to manipulate them at the most opportune time, in an endeavour to improve individual components and hence ultimately to raise final yield. However, yield components are strongly inter-related and frequently compensate. Improvements in yield are most commonly the result of positively correlated increases in major components.

This approach is illustrated in Fig. 17.2 which shows the time sequence of some developmental processes and the yield components which each finally controls. The example is based on winter wheat growing in New Zealand but the same principles would apply to any other crop of wheat, even if the time scale would differ depending on the environment and time of sowing. It will be seen that, from germination onwards, the plant produces leaves and tillers and that tillering is spread over a period of several weeks. The length of this time will be governed by climate, notably photoperiod and temperatures, and the number of tillers produced will vary with water supply and minerals. Primary and maybe some secondary tillers will be formed. With the advent of the longer days of spring the plant changes from the vegetative to the reproductive condition, an event indicated by the appearance of double ridges on the apical meristem and the subsequent formation of spikelet primordia (see section 4.1). At least three things should be noted in relation to this phase. In the first place it is comparatively brief, lasting for a much shorter period of time than the tillering period. Secondly, the number of spikelets is now determined, well ahead of ear emergence and anthesis. The third consequence is that internode elongation now begins, and not only is this correlated with the end of tiller appearance but also with the death of some of the tillers already

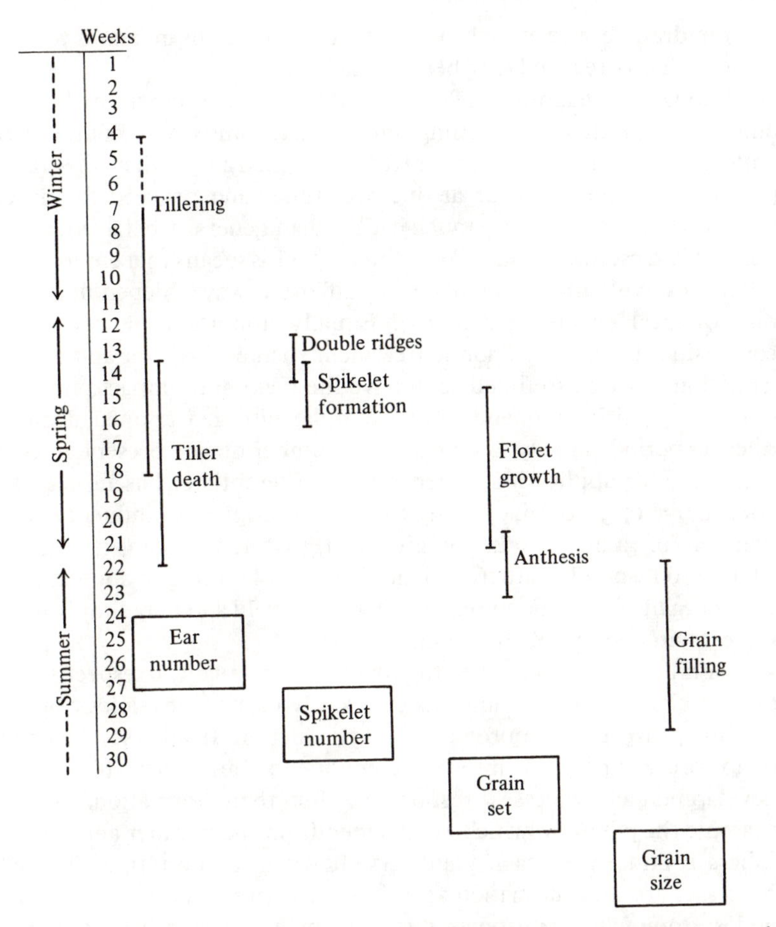

Fig. 17.2. Approximate time and duration of physiological events and corresponding yield components in winter wheat. (After Scott, W.R., 1977. PhD thesis, Lincoln University, New Zealand.)

formed. Secondary tillers and those formed most recently are the most vulnerable and tend to die over the next few weeks, leading up to ear emergence and beyond. These two processes, tiller formation and tiller death, determine the number of ears per plant. Meanwhile the spikelet primordia develop further, and florets are initiated and grow. By the time ears emerge and anthesis occurs, a proportion of florets in each spikelet is ready for pollination and some or all of these form a grain. This determines the grain set and, in conjunction with spikelets per ear, the total number of grains produced. During the final period the grains swell in size as carbohydrates arrive at each site, so that grain weight increases to its final value which is reached just before maturity.

Although the magnitude of these yield components is extremely variable and although there are strong inter-relationships and compensations among them, the important picture to emerge is that the relevant physiological events occur at different times and proceed for different durations. Most take place much earlier than generally believed, and well before yield can be measured or estimated. This means that environmental influences will affect the plant in different ways, depending on the physiological events in progress. Equally important, however, is the conclusion that any agronomic treatment such as fertilizer application or irrigation must be so timed as to have the maximum impact on the most sensitive yield components. For example, nitrogen applied during the tillering period is likely to increase the number of ears present at harvest, whereas later application, when florets differentiate, has no significant effect (Fig. 17.3). Similarly, any treatment designed to influence spikelet numbers or grain set must be given early while the plant is still plastic enough to respond. Late treatment, for example nitrogen applied close to ear emergence could not be expected to raise yield and may simply increase the protein content of the grain.

A similar approach applies to any other crop plant, irrespective of the type of yield produced, and indeed a study of crop physiology opens up exciting prospects of improving plant production. In all cases the first step is to break up yield into its components, and then to study the developmental processes responsible for their formation. Take for example the potato in which yield depends on the number and size of the tubers produced by each plant. As shown in section 16.4, the critical processes are tuber initiation at the end of rhizomes and the formation of nodes from which rhizomes may originate. Once tuber numbers are determined, it is then a question of having a large enough leaf area to ensure ample supplies of carbohydrate for the growth of the developing potatoes at all available sites. In principle, the same considerations apply to all other cases in which the end product is a storage organ, be this a grain, root or modified stem. While the plant is young, our main concern should be to provide conditions suitable for the laying down of yield structures but, once this has been achieved, the emphasis changes to the provision of carbohydrates and other reserve materials to fill the sites that have been established. The analogy with capital investment and growth should be obvious. Only when it comes to herbage grasses and forage plants in general do we encounter a major group of plants with a different yield structure. These plants are intended to withstand a series of successive harvests, or almost continuous utilisation as in grazing, during which leaves and stems are removed. The operative yield components in these

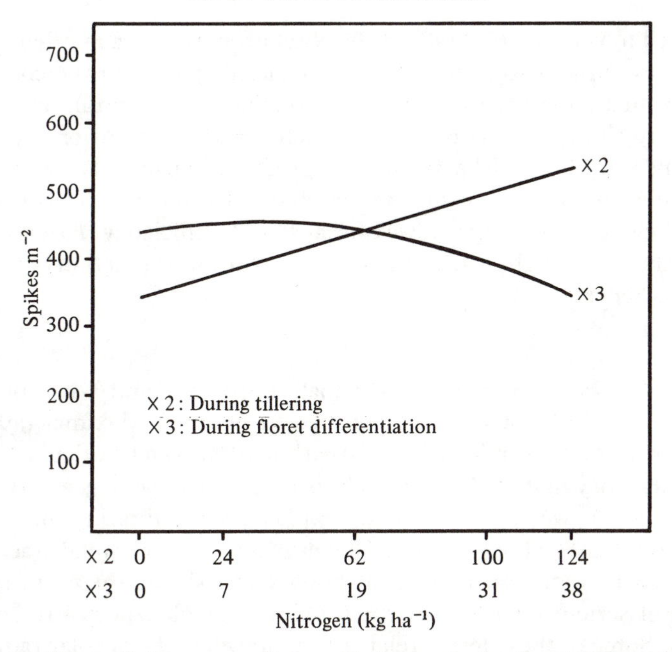

Fig. 17.3. Effect of nitrogen applied at tillering or during floret differentiation on the number of spikes per unit area in a crop of winter wheat (from Langer, R.H.M., 1979. *New Zealand Wheat Review*, 14, 32–40).

cases are the buds from which recovery growth takes place, and subsequently the size attained by the new crop of leaves. In the grasses, regrowth depends on basal axillary buds from which new tillers arise, together with continued leaf production from existing shoots whose apical meristem is below the level of defoliation (section 4.1). In legumes resumption of growth usually comes from axillary or crown buds alone, and both their number and physiological condition will affect the speed and degree of recovery, as for example in lucerne (section 11.3). However, the number of new shoots and leaves, their rate of growth and eventual size ultimately depend on the availability of carbohydrates. Thus, irrespective of yield structure, we have to look to photosynthesis as the basic process governing plant production, given an adequate supply of water and minerals.

17.4 CROP PHOTOSYNTHESIS

Many thousands of articles and books have been written on the subject of photosynthesis, covering every conceivable approach and facet, much of it going well beyond the scope of this book. Although we must of necessity be interested in very many details of this vital process, we can perhaps argue that the essential feature to receive special recognition in the context of crop production is the question of converting the maximum amount of solar energy into plant dry matter, of which as high a proportion as possible should be useable as the harvested end-product. In particular, we must be concerned with any opportunities that may exist to make this conversion more efficient. At its simplest, we can say that the efficiency (E) with which the plant uses the light energy intercepted by it for production of new dry matter is represented by

$$E = \nabla W / \mathcal{J} \qquad [17.1]$$

where ∇W is the gross amount of dry matter produced and $\mathcal{J}$ the amount of light energy intercepted in a given period of time. Maximal light use efficiency in the conversion of CO_2 to carbohydrate can be calculated to be around 8 μg dry matter $\mathcal{J}^{-1}$ although its more usual value is about 2.5 μg $\mathcal{J}^{-1}$. As long as water and nutrient supplies are not limiting, light is the major factor controlling growth. However, only a very small fraction of total incident radiation is fixed in photosynthesis, as there are several sources of serious energy loss which reduce the efficiency of the overall process. Some of these losses relate to the wavelengths of solar radiation, because less than half of the total is photosynthetically active. Other major losses are caused by the energy requirements of plant metabolic processes. After all losses have been taken into account, only about 2% of total radiation has been estimated to be converted by the plant into dry matter used for growth or storage. It is upon this minute fraction of total radiation that our life depends, and hence we must use all possible devices to make the best possible use of it.

One of the most effective ways of achieving this objective is to ensure that the maximum amount of incident radiation is intercepted by leaves and other green tissue and that it does not strike bare ground. In other words, the leaf area index (L), the ratio of leaf area present to ground area covered by the plants, needs to be high enough to assure the fullest possible utilisation of light. In the early stages of crop growth this is not likely to be the case, but as more leaves are formed the leaf area index rises to an optimum value at which interception exceeds 95% (Fig. 17.4). As still more leaves appear, L continues to rise, but canopy photosynthesis does not increase any further because the basal leaves no longer receive adequate light and soon senesce. A balance is achieved between formation of new leaves and leaf death, with photosynthesis not changing very much or, as in some grass swards, declining. At one time it was thought that net photosynthesis should actually decline with increases in leaf area index beyond the optimum, since respiration rate would continue to rise and thus

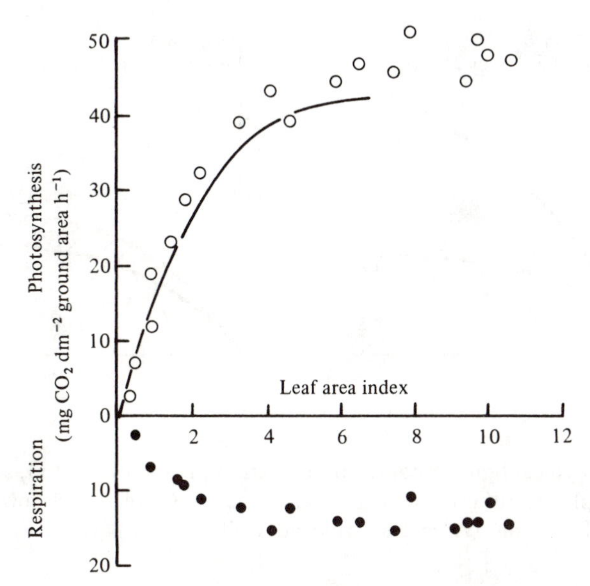

Fig. 17.4. Relation between leaf area index, photosynthesis and respiration of wheat crops. (From Evans, L.T., Wardlaw, I.F. and Fischer, R.A. in Evans, 1975.)

more than offset photosynthetic gain but this has not been confirmed by experiments. The optimum leaf area index varies with canopy architecture and total incident radiation. In species with horizontally inclined leaves, such as clovers, L for maximum crop photosynthesis is lower than in plants with more upright leaves, largely because self-shading prevents a greater proportion of radiation from being intercepted. At low sun angles a horizontal leaf surface may be more effective. The vertical distribution of leaves is also important. For example, as the stems of a cereal crop elongate and the leaves become separated in space, light penetrates more readily, canopy photosynthesis increases, and carbohydrate concentration in the tissues rises. Green stems and, later on, the ears also contribute to total photosynthesis. Optimum leaf area index also varies seasonally, reaching its peak in mid-summer and declining with total radiation towards autumn.

Considerable species and even cultivar differences have been found in the rate of photosynthesis per unit leaf area, and at one time there were hopes that this could become a potent selection criterion for plant breeders to produce superior genotypes. For example, in perennial ryegrass it was found that plants with small mesophyll cells had higher rates of photosynthesis. However, many other factors appear to be involved before such differences can be translated into increased crop productivity, and it is

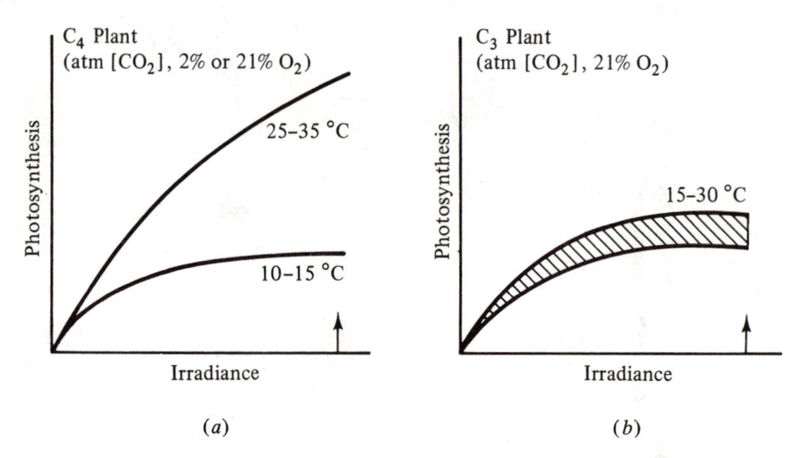

Fig. 17.5. Typical light response curve at atmospheric levels of O_2 and CO_2 for single leaf photosynthesis of (*a*) C_4 plants and (*b*) C_3 plants. Arrows on *x* axis indicate full sunlight. (After Edwards and Walker, 1983.)

probably quite significant that during the course of evolution modern wheats have declined in leaf photosynthetic rate compared with their primitive ancestors, whereas leaf and grain size has increased. On the other hand, there exists among species a major difference in the carbon pathway during photosynthesis, which has a major bearing on productivity. Plants of the temperate region but also crops like rice follow the Calvin (C_3) cycle during which carbon dioxide is initially incorporated into 3-carbon compounds. In tropical (C_4) grasses this step is preceded by CO_2 fixation in the 4-carbon dicarboxylic acids. These differences are associated with other characteristics relating to enzyme configuration and leaf structure. Thus in C_4 plants stomatal resistance is greater than in C_3 species, efficiency of dry-matter production per unit water transpired is greater, and the stomata continue to keep open up to very high light intensities. For instance, C_3 grasses reach light saturation at one-third to one-half sunlight, while at high temperature C_4 grasses are not saturated until full light is reached (Fig. 17.5). Maximum photosynthetic rates per unit leaf area in C_4 plants can be at least twice as high as in C_3 species, photorespiration is absent, and mesophyll resistance to CO_2 uptake is low. However, the strong effects of light on photosynthesis are greatly influenced by other factors, notably temperature. As Figure 17.5 (*a*) shows, high temperatures are required to enable C_4 plants to take advantage of high levels of irradiance, and in fact C_3 plants may have superior photosynthetic rates in cooler conditions and when light is limiting (Fig. 17.5 (*b*)). Carbon dioxide concentrations also influence the responses of these two types of plants, and this may become increasingly important as atmospheric CO_2 levels continue to rise at a rate estimated to have reached 10% over a recent period of 25 years. Because of their physiological characteristics, C_3 plants will respond by greater CO_2 fixation and dry matter production, and increases of 10–40% in primary productivity have been predicted, if current levels of carbon dioxide concentrations in the atmosphere were to be doubled. The accompanying increase in temperatures, the so-called greenhouse effect, will also affect plant production which will differ depending on the type of carbon pathway during photosynthesis. No doubt the geographical distribution of C_3 and C_4 plants will also change.

Irrespective of these future predictions, let us remember that C_4 plants are capable of very high photosynthesis and growth rates, but in cool conditions and with moderate light levels C_3 species perform at their best. It appears that these two carbon pathways should be considered as adaptations to different climates rather than an opportunity for raising crop productivity in a wide range of conditions. Important as these physiological differences are at present, we should not forget the question whether the potential for enhanced photosynthesis and crop growth rates in C_4 plants will become increasingly important, if the world's climate changes as predicted. Rising temperatures should have this effect, but increasing carbon dioxide concentrations in the atmosphere will tend to favour C_3 plants.

We are now in a position to return to the statement that, as long as water and nutrient supply are not limiting, the rates of carbon assimilation and dry matter accumulation are proportional to the intercepted radiation. Fig. 17.6 shows this to be borne out by total plant production in such diverse crops as barley, potatoes, sugar beet and apples. The close similarity in the behaviour of quite different species emphasises the importance of the relationship shown in equation 17.1 which shows that ultimately dry-matter production depends on the light intercepted by the foliage and the efficiency with which it is used. In accepting this conclusion we must, however, be aware that water stress or shortage of nutrients will undoubtedly depress photosynthetic efficiency.

17.5 HARVEST INDEX

Total dry-matter accumulation is of course not synonymous with agronomic yield, unless we are dealing with crops like forages in which the whole plant is harvested. In most cases only a fraction of the total is economically useful, and this is the part that is measured in agricultural practice and quoted in experimental results and statistical returns. It can

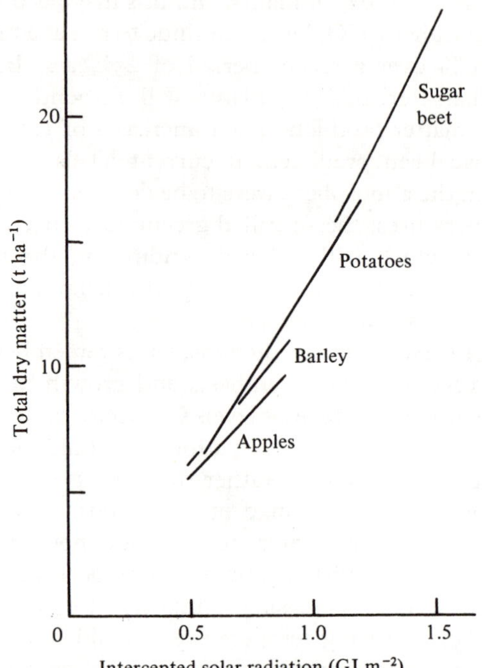

Fig. 17.6. Relation between total dry matter produced by various crops and radiation intercepted by foliage throughout the growing season. (From Monteith, J.L., in Cook, G.W., Pirie, N.W. and Bell, G.D.H., 1977.)

not, however, be considered separately from total yield, whether above or below ground, and in fact the two are linked by what is known as the harvest index (HI). For example, in grain crops we can say that grain yield (Y_{gr}) is the product of total or biological yield (Y_{biol}) and the harvest index.

$$Y_{biol} \times HI = Y_{gr} \qquad [17.2]$$

The same concept applies to other crop plants as well, depending on the nature of the end product. It could refer to the yield of potatoes, carrots, or any other root crop, it could be extended to include products extracted from total dry matter, for example sugar from sugar cane or beet, and it need not by any means be restricted to food plants. It would certainly be quite meaningful to discuss the HI of a crop of linseed, cotton, or sunflowers. In order to accommodate the full range of crops it would probably be better to generalise equation 17.2 by referring to economic yield (Y_{econ}) rather than grain yield alone.

$$Y_{\text{biol}} \times \text{HI} = Y_{\text{econ}} \qquad [17.3]$$

As we have seen, total plant production is ultimately limited by the amount of radiation intercepted by the foliage, and it must be the objective of agriculture to get as close as possible to the attainable maximum. But within this total limit there are opportunities for raising yield by improving the HI. This can be done in several ways, although we should be aware that most treatments affecting HI will also influence Y_{biol}, and not necessarily to the same extent or in the same direction. For instance, in cereals it is possible to raise biological yield by applying nitrogen at high population density and in the presence of adequate water. The expected result would be heavy vegetative growth, but also reduced light transmission into the canopy, poor grain set and development and thus a low harvest index. On the other hand, it is possible to demonstrate that the superiority of short-strawed cereals is directly linked with a greater harvest index than in tall cultivars. Short, erect cultivars of rice yielding 4–5 t ha^{-1} have been shown to have a harvest index of about 0.53 to 0.56, compared with 0.39 to 0.42 in tall, leafy cultivars with a grain yield of about 2.4 t ha^{-1}. The same trend occurs in wheat and thus underlines the importance of dwarf and semi-dwarf cultivars emanating from the Mexican plant breeding programme. On the other hand, extremely short plants may also produce little grain. It follows that biological yield and harvest index should be assessed in plant breeding programmes, because they could well be the key to superior performance. Agronomists should also pay attention to these concepts. For example, they might well consider to what extent growth-retarding chemicals which reduce stem length in cereals and grasses grown for seed affect yield through harvest index or some other way.

The physiological basis of harvest index relates to the partitioning of assimilates, the distribution of the carbohydrates produced by the plant. It is customary to refer to the leaves and other green tissue active in photosynthesis as the 'source' and to those parts in which storage occurs as the 'sink'. Clearly there are many diverse sources and sinks within the same plant system, and it is one of the most challenging and only partially solved problems of plant physiology to discover the mechanisms through which this allocation is made. Actively growing storage organs, like developing grains or tubers, attract carbohydrates, but what controls the distribution to competing centres and to what extent demand affects supply are only partially understood.

17.6 TOWARDS HIGHER YIELD

Crop production is the result of the conversion of light energy to chemical energy. Given that we have no control over incoming radiation, our main concern in agriculture must be to ensure that as much light as possible is intercepted by the crop plant, that the photosynthetic processes operate as efficiently as possible, and that the partitioning of total dry matter produced favours economic yield. How are we going to achieve these objectives?

Effective interception of light depends on the rapid development of the leaf system and the maintenance of leaf area index (L) at an optimum level and leaf angle to achieve maximum utilisation.

17.7 MANIPULATING THE LEAF AREA INDEX

In the first place we must realise that different crops have different patterns of leaf development. The biggest difference exists between perennials which have at least some leaf cover throughout the year and annual crops which function only during their particular growing season. But there are further considerable differences within annuals, depending on seed size, rate of germination, crop architecture, time of flowering and a whole host of other factors. A few examples will clarify these concepts.

In annual crops leaf area is often slow to develop and hence, even if growing conditions are favourable, light interception may be poor. Any treatment which could reduce this initial period of inefficiency may thus be expected to increase leaf cover and ultimately yield. Several quite practical strategies are available to achieve these objectives. One of these is the time of sowing, provided of course that soil and climatic conditions allow it to be altered. For example, sugar beet and fodder beet are now planted much earlier in the year than used to be the custom, and the direct relationship between earliness of sowing and final yield can be attributed to the earlier development of the leaf system. Under experimental conditions it has been shown that still better results could be obtained by planting out beet seedlings in the spring rather than sowing seeds at the same time. Even though this may not be a highly practical solution, there are crops in which the same principle is applied on a routine basis. One of these is the potato in which pre-sprouted or chitted tubers are planted to achieve a rapid leaf cover and thus an early crop (see section 16.4). The size of the 'seed' potatoes to be planted also makes a considerable difference to leaf area and, if the increase occurs at a time of high radiation receipts and persists for long enough, improvement in yield would be expected (Fig. 17.7).

Another practical device which can be employed to vary leaf area is to change the density of the plant population or the planting arrangement. A dense planting would be expected to develop a canopy of leaves more quickly and, quite apart from possible beneficial effects in terms of competing with weeds, make better use of light in the early stages of

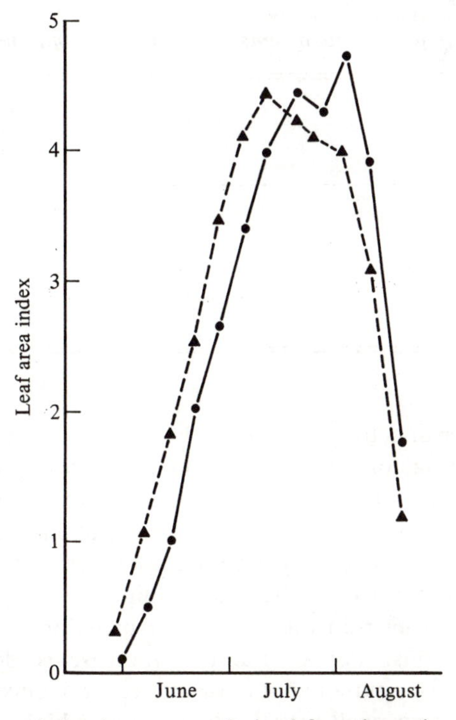

Fig. 17.7. Leaf area index in potatoes derived from large (▲-----▲) and small (●——●) tubers. (From Bremner, P.N. and Taha, M.A., 1966. *Journal of Agricultural Science Camb.* **66**, 241–52.)

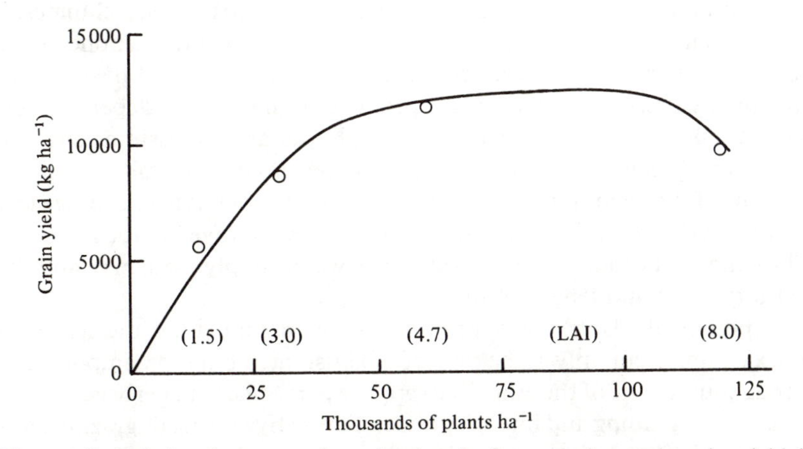

Fig. 17.8. Relation between plant density, leaf area index and grain yield in irrigated crops of maize. (From Duncan, W.G., in Evans, L.T., 1975.)

Table 17.1. *Optimum counts of emerged plants for maximum productivity.*

Crop	Number of plants ha^{-1} ('000)
Wheat	3–5000
Barley, oats	3–4000
Potatoes	48–50
Sugar beet	80–100
Rape	5–800

growth. Whether and to what extent the benefits continue to last, will depend on many factors. Very dense populations would induce a high degree of inter-plant competition, possibly to the detriment of yield. However, if water and nutrient supply are not limiting, many crops respond to increasing density by developing a larger leaf surface, followed by an improvement in yield. The relationship between plant population, and grain yield is shown in Fig. 17.8 for irrigated maize growing under favourable conditions. As plant numbers increase, so does the leaf area index. Yield also rises initially but ceases to respond, once enough leaf has accumulated to intercept all available light. At very high populations there appears to be so much competition for light that some plants fail to produce grain, and consequently grain yield begins to decline. For most crops an optimum plant density has been worked out (see Table 17.1), but this needs to be re-evaluated for different cultivars, districts, and climates.

Although not developed in any detail, it will be obvious that one of the best ways of influencing leaf area is to encourage rapid and healthy leaf growth. Leaf size and rate of leaf production are greatly dependent on mineral supply, notably nitrogen, phosphorus and potassium, and the longevity of leaves is equally strongly under environmental control. In fact, one of the major influences of fertilizer application is to encourage leaf growth, and it is through this effect that yield is improved in the long run. The same can be said with equal validity of water supply, both in relation to leaf formation and length of life.

In perennial plants the opportunities for manipulating the leaf area index do not occur only at the time of establishment but more importantly throughout the life of the crop. This applies particularly to herbage species because both timing and intensity of defoliation by cutting or grazing have a profound effect on leaf area and subsequent growth. From what we have learnt already we can postulate that to maintain a sward at beyond its

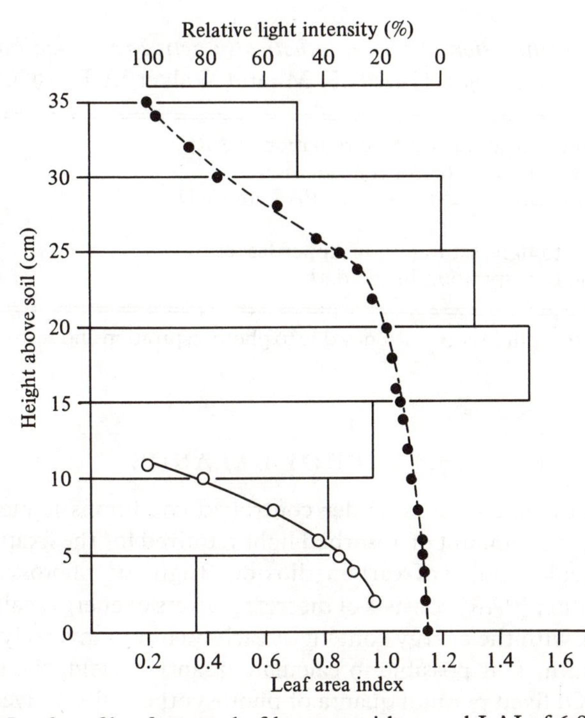

Fig. 17.9. Leaf profile of a sward of lucerne with a total LAI of 6.8 and relative light transmission before defoliation (●——●) and relative light transmission after defoliation to about 12 cm (○——○). (After Keoghan, J.M., 1970. PhD thesis, Lincoln University, New Zealand.)

optimum leaf area would not be the best use of incident radiation, and that this will vary at different times of the year. By the same token defoliation should not be so severe that inadequate foliage remains to allow rapid recovery. Some of these relationships are illustrated in Fig. 17.9 which shows the leaf profile of a sward of lucerne with a total L of about 6.8. Before defoliation the canopy was so dense that only the top 10 cm of the sward received more than about 35% of total incident radiation and leaves died rapidly at lower levels. Light transmission improved dramatically when the sward was cut to about 12 cm, leaving a L of more than 1 receiving light of relatively high intensity. Whether the surviving bottom leaves can suddenly adapt to a greatly improved light condition is another matter, but there is no doubt that recovery was more rapid than in similar swards cut at a lower level. In general we can say with a fair degree of accuracy that the art of grazing management has its scientific basis in the concept of leaf area index.

Table 17.2. *Conversion of photosynthetically active radiation into carbohydrate (C_3 crop)* (Hay, R.K.M. and Walker, A.J., 1989).

Incident photosynthetically active radiation (PAR)	100
15% loss due to reflection and transmission	85
Maximum efficiency of conversion of PAR into CH_2O, 17.7%*	15
30% loss due to light saturation of upper leaves	10.5
40% loss due to respiration in the dark	6.3

*Takes into account loss of efficiency due to photorespiration and respiration in the light.

17.8 ENERGY BALANCE

Photosynthetic efficiency under controlled conditions is measured by determining the amount of absorbed light required for the fixation of each gram molecule (mole) of carbon dioxide. Light or photosynthetically active radiation (PAR) consists of discrete packets of energy, called quanta or photons, with the energy content of each quantum inversely related to its wavelength. It is possible to calculate quantum yield, the number of moles of CO_2 fixed per mol quanta of photosynthetically active radiation, and for the formation of carbohydrate by single leaves a maximum efficiency of about 17% can theoretically be shown. In practice the conversion of incident light to carbohydrate will be considerably less, and some of the losses involved in converting radiation are shown in Table 17.2. The overall efficiency of the process under optimum conditions is not likely to exceed 6%, but this is in excess of what is generally being achieved in practice. The total dry matter accumulation by different crops shown in Figure 17.6 occurred on average at the rate of 1.4 g per megajoule (MJ) of total solar radiation, and this represents a conversion efficiency of 4.9%. The crops concerned were grown under good experimental conditions and suffered no shortages. More realistically, the photosynthetic efficiency of high yielding crops is likely to be between 2 and 4%, but even this estimate ignores the effects of limiting factors such as water stress, nutrient deficiencies or unfavourable temperatures.

Yield improvement through enhanced conversion of solar energy is most likely to be attained by reducing the effects of these limitations, and this is in fact the aim of good crop husbandry. At the same time we can look forward to continuing advances in science and in our understanding of the processes involved. For example, it is possible to select crop plants with improved water use efficiency, and it is of great interest to note that a range of C_4 crops was found to need only about half the amount of water for every gram of dry matter produced than a group of C_3 plants which were also tested. Even more encouraging are the differences in photosynthetic efficiency which appear to occur between these two types of plants. Under the same conditions of solar radiation, maize (C_4) has been shown to attain an efficiency of 4.5% compared with rice (C_3) at 3.1%, and over a whole season mean crop growth rates of $22.0 \pm 3.6\,\text{gm}^{-1}$ have been calculated for C_4 and $13.0 \pm 1.6\,\text{gm}^{-2}$ for C_3 crops. Differences of such magnitude provide opportunities for plant selection and for the eventual production of superior cultivars. Progress will not necessarily be dramatic and not all approaches lead to success. For instance, there have been hopes to select crop plants with low rates of photorespiration, but, so far, superior genotypes have not been clearly identified. Similarly, we must be careful that selections based on differences between C_4 and C_3 plants are fully adapted to the environmental conditions under which a particular crop has to be grown.

Irrespective of any progress that may be achieved, the primary objective is to raise economic yield even though total biological yield will almost certainly also be affected. In other words, harvest index (HI) has to be maintained, at least at present levels. There is, however, good evidence to show that harvest index itself is also capable of being improved, with the result that increases in economic yield can be achieved without any corresponding improvement in total biological productivity. A good example of this approach is provided by an examination of British winter wheat varieties introduced since the beginning of the century.

As Figure 17.10 shows, grain yield of modern wheats tends to be at least 50% higher than in cultivars available some decades ago. This very substantial increase does not appear to have been brought about by greater total dry matter production. Although total biological yield, including roots, was not measured, total above-ground yield in these experiments showed no distinct trend towards any improvement. It follows that improved economic yield of recently released wheats was the result of increased harvest index, which rose from about 0.35 in older to about 0.5 in modern varieties. It is remarkable that improved HI was not in itself a selection criterion, although plant breeders favoured lines with short, stiff straw which were capable of growing at high levels of soil fertility without lodging. Nevertheless the superiority of new varieties in terms of yield and harvest index was expressed irrespective of level of soil fertility.

Further increases in HI appear to be possible, and an increase from the present 0.5 to perhaps 0.62 has been predicted. Clearly there will be limits to the reduction in straw length which can be tolerated without sacrificing photosynthetic capacity, but there may still be further opportunities for

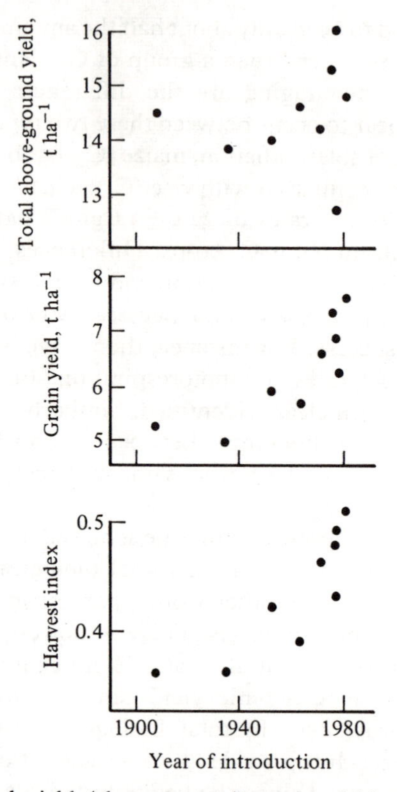

Fig. 17.10. Biological yield (above ground), grain yield and harvest index of British winter wheat cultivars plotted against the year of their introduction. (From Hay, R.K.M. and Walker, A.J., 1989 based on Gifford, R.M. *et al.* (1984), *Science*, **225**, 301–8.)

improving grain number and grain size which are determined at different times (Fig. 17.2) and have a large effect on yield. Similar considerations apply to other cereals, and equally other crop plants are also capable of improvement by raising the harvest index. For example, the yield potential of old peanut cultivars has been shown to be less than that of modern releases, not because of any shortfall in total productivity but owing to an inferior harvest index.

The achievements of plant breeders, viewed in these terms, are remarkable and we can feel justified to look forward to further success. Similarly, resistance to pest and disease has been greatly improved and many yield-limiting attributes have been eliminated. Modern genetical techniques should bring about even more improvements which, together with a high level of crop husbandry, should ensure the continuing supply of food, fibre and other plant products.

FURTHER READING

Cook, G.W., Pirie, N.W. and Bell, G.D.H. (eds) (1977). *Agricultural efficiency*. Royal Society, London.

Cooper, J.P. (ed.) (1975). *Photosynthesis and productivity in different climates*. International Biological Programme 3. Cambridge University Press.

Donald, C.M. and Hamblin, J. (1976). The biological yield and harvest index of cereals as agronomic and plant breeding criteria. *Advances in Agronomy* **28**, 361–405.

Edwards, G. and Walker, D. (1983). *C_3, C_4: mechanisms, and cellular and environmental regulation, of photosynthesis*. Blackwell Scientific Publications, Oxford.

Evans, L.T. (ed.) (1975). *Crop physiology, some case histories*. Cambridge University Press.

Evans, L.T. and Peacock, W.J. (eds) (1981). *Wheat science –today and tomorrow*. Cambridge University Press.

Evans, L.T. and Wardlaw, I.F. (1976). Aspects of the comparative physiology of grain yield in cereals. *Advances in Agronomy* **28**, 301–59.

Fitter, A.H. and Hay, R.K.M. (1987). *Environmental physiology of plants*. (2nd edn). Academic Press, London.

France, J. and Thornley, J.H.M. (1984). *Mathematical models in agriculture*. Butterworths, London.

Green, M.B. (1978). *Eating oil. Energy use in food production*. Westview Press, Boulder.

Hay, R.K.M. and Walker, A.J. (1989). *An introduction to the physiology of crop yield*. Longman Scientific Technical, London.

Landsberg, J.J. and Cutting, C.V. (eds) (1977). *Environmental effects on crop physiology*. Academic Press, London.

Leach, G.L. (1976). *Energy and food production*. International Institute for Environment and Development, London.

Loomis, R.S., Williams, W.A. and Hall, E.A. (1971). Agricultural productivity. *Annual Review of Plant Physiology* **22**, 431–68.

Milthorpe, F.L. and Moorby, J. (1979). *An introduction to crop physiology*. Cambridge University Press.

Pearcy, R.W. and Ehleringer, J. (1984). Comparative ecophysiology of C_3 and C_4 plants. *Plant, Cell and Environment*, 7, 1–13.

Petr, J., Černy, L., Hruška, L. *et al.* (1988). *Yield formation in the main field crops*. Elsevier, Amsterdam.

Tesar, M.B. (ed.) (1984). *Physiological basis of crop growth and development*. American Society of Agronomy/Crop Science Society of America, Madison.

Wareing, P.F. and Cooper, J.P. (eds) (1971). *Potential crop production, a case study*. Heinemann, London.

Index of specific names

Numbers in italic type indicate page on which illustrations appear

Subject index